AF773565

Piotr Mikusiński
Michael D. Taylor

An Introduction to Multivariable Analysis from Vector to Manifold

Birkhäuser
Boston • Basel • Berlin

Piotr Mikusiński
Department of Mathematics
University of Central Florida
Orlando, FL 32816-1364
U.S.A.

Michael D. Taylor
Department of Mathematics
University of Central Florida
Orlando, FL 32816-1364
U.S.A.

Library of Congress Cataloging-in-Publication Data

A CIP catalogue record for this book is available from the Library of Congress, Washington D.C., USA.

AMS Subject Classifications: 58-01, 28-01, 26-01

Printed on acid-free paper.
© 2002 Birkhäuser Boston

Birkhäuser

All rights reserved. This work may not be translated or copied in whole or in part without the written permission of the publisher (Birkhäuser Boston, c/o Springer-Verlag New York, Inc., 175 Fifth Avenue, New York, NY 10010, USA), except for brief excerpts in connection with reviews or scholarly analysis. Use in connection with any form of information storage and retrieval, electronic adaptation, computer software, or by similar or dissimilar methodology now known or hereafter developed is forbidden.
The use of general descriptive names, trade names, trademarks, etc., in this publication, even if the former are not especially identified, is not to be taken as a sign that such names, as understood by the Trade Marks and Merchandise Marks Act, may accordingly be used freely by anyone.

ISBN 0-8176-4234-X SPIN 10832360
ISBN 3-7643-4234-X

Reformatted from authors' files by TEXniques, Inc., Cambridge, MA.
Printed and bound by Hamilton Printing, Rensselaer, NY.
Printed in the United States of America.

9 8 7 6 5 4 3 2 1

CONTENTS

Preface

Multivariable analysis is an important subject for mathematicians, both pure and applied. Apart from mathematicians, we expect that physicists, mechanical engineers, electrical engineers, systems engineers, mathematical biologists, mathematical economists, and statisticians engaged in multivariate analysis will find this book extremely useful. The material presented in this work is fundamental for studies in differential geometry and for analysis in N dimensions and on manifolds. It is also of interest to anyone working in the areas of general relativity, dynamical systems, fluid mechanics, electromagnetic phenomena, plasma dynamics, control theory, and optimization, to name only several.

An earlier work entitled *An Introduction to Analysis: from Number to Integral* by Jan and Piotr Mikusiński was devoted to analyzing functions of a single variable. As indicated by the title, this present book concentrates on multivariable analysis and is completely self-contained. Our motivation and approach to this useful subject are discussed below.

A careful study of analysis is difficult enough for the average student; that of multivariable analysis is an even greater challenge. Somehow the intuitions that served so well in dimension 1 grow weak, even useless, as one moves into the alien territory of dimension N. Worse yet, the very useful machinery of differential forms on manifolds presents particular difficulties; as one reviewer noted, it seems as though the more precisely one presents this machinery, the harder it is to understand. One of the main thrusts of this book is to get around some of these obstacles by first introducing the wedge product in an especially concrete and geometric way — one that is a straightforward extension of the analytic geometric treatment of vectors — and then using this definition of the wedge product to define differential forms again in a particularly concrete and geometric fashion. Nevertheless, despite the simpler, more concrete character of our presentation, it is fully capable of sustaining a rigorous derivation of important properties of differential forms, such as the the generalized Stokes' theorem.

A further goal of our text is to present Lebesgue integration in the simplest way possible. One of the great difficulties for many students studying Lebesgue integration is that they must first master the machinery of Lebesgue measure. Rather than deal with measure theory, we follow the approach in *An Introduction to Analysis* and introduce Lebesgue integration in a more concrete manner. This involves certain series "expansions" of functions and in effect reduces the theory of integration to that of absolutely convergent series. The introduction of the Lebesgue integral pays a dividend in later chapters by simplifying the definition of integration of real-valued functions on manifolds.

The presentation of material in this book falls into roughly three parts: There is first a brief introduction to linear algebra and the elements of metric space theory; this provides our foundation for the study of multivariable analysis. The second section runs through Chapters Three, Four and part of Five, and covers standard multivariable fare in $\mathbb{R}^N$: differentials as linear transformations, the inverse and implicit function theorems, Taylor's theorem, the change of variables theorem for multiple integrals, etc. The third section, starting in Chapter Five and going through Chapters Six and Seven, moves out of $\mathbb{R}^N$ to manifolds and analysis on manifolds, covering the wedge product, differential forms, and the generalized Stokes' theorem.

The material is supported by numerous examples and exercises ranging from the computational to the theoretical, all aimed at bringing the important ideas more fully to life.

We now discuss the presentation in somewhat more detail.

Chapter One briefly develops linear algebra in $\mathbb{R}^N$, at least to the extent needed in this book. This is included, in part, because linear algebra is crucial to a real understanding of multivariable analysis and the background of students may be uncertain. But even if students have a good foundation in linear algebra, there are certain ideas in this chapter that should be reviewed: the Binet–Cauchy formula, the interpretation of the determinant in terms of oriented volume, the formula for the K-dimensional volume of a parallelepiped in $\mathbb{R}^N$, and the quantity $\mathcal{D}(f)$ associated with the linear transformation f. (This last quantity is a simple generalization of the Jacobian and can be thought of as the "volume-distortion" factor associated with f.) These concepts play key roles in later chapters.

In Chapter Two we establish the basic ideas of limits, continuity, topology, and compactness in the general but accessible setting of metric spaces. Our goal is to have these ideas available in successive chapters for use in $\mathbb{R}^N$.

Chapter Three shows that for transformations in higher dimensional Euclidean spaces we can connect differentiation with linear transformations. The chain rule is shown to be expressible in terms of matrix multiplication, and we establish the inverse and implicit function theorems and the higher dimensional version of Taylor's theorem.

There may be those who want to get to the theory of manifolds and differential forms as quickly possible. For them, Section 4.1, *A Bird's-Eye View of the Lebesgue*

Integral, provides an overview of the Lebesgue integral and everything they need to know about integrals for the last three chapters of the book.

Chapter Five contains a proof of the change of variables theorem for multiple integrals, the most complex proof in the book. The notion of manifolds is introduced. All manifolds encountered here are assumed, for the sake of simplicity, to be embedded in some $\mathbb{R}^N$, and essential advantage is taken of this fact in describing integration over manifolds and—later on—in constructing the machinery of differential forms.

Chapter Six is devoted to the introduction of the wedge or exterior product. No appeal is made to tensor theory nor is any distinction raised between covariant and contravariant quantities. Instead the wedge product is built up from elementary machinery that ought to be familiar to anyone who has gone through a matrix theory course. The idea is both to use simple ideas and to build up a strong geometric intuition for the wedge product. (Indeed it is proved that simple K-vectors, that is, those of the form $\alpha_1 \wedge \alpha_2 \wedge \cdots \wedge \alpha_K$, correspond to equivalence classes of oriented parallelepipeds.)

In Chapters Five and Seven, tangent vectors are introduced and handled in the most obvious possible way: If $\mathcal{M}$ is a K-dimensional manifold and the map $f : U \to \mathcal{M}$ defines a local coordinate system, then a tangent vector w to $\mathcal{M}$ at the point $q = f(p)$ is one having the form $w = f'(p)v$ where v is a vector in $\mathbb{R}^K$ and $f'(p)$ is the linear transformation corresponding to the Jacobian matrix of f at p. This makes it possible to talk about tangent K-vectors to $\mathcal{M}$ at q, and these may quite justifiably be pictured as something like K-dimensional postage stamps glued to $\mathcal{M}$ at one corner at the point q in such a way as to be geometrically tangent to the manifold. An orientation of a K-dimensional manifold may then be defined quite simply as a continuous unit, tangent K-vector field on the manifold.

Differential forms make their appearance in Chapter Seven. They do not come as sections in the cotangent bundle of the manifold but as K-vector fields (analogous to common garden-variety vector fields) defined over the manifold. Integrals of differential forms are then obtained by taking the dot product of a differential form with the orientation of the manifold and integrating this real-valued function over the manifold. The standard notation and results then follow ($d^2 w = 0$, properties of pullbacks, and the generalized Stokes' theorem).

Most of the symbolism used in the text is standard, but it may be useful to indicate to the reader some symbols which are nonstandard:

$f^{\diamond}$ (page 34) The dual of a linear transformation f.

$\mathcal{D}(f)$ (page 38) $\mathcal{D}(f) = \sqrt{\det(f^{\diamond} \circ f)}$ where f is a linear transformation from a Euclidean space into one of the same or higher dimension or, equivalently, $\mathcal{D}(A) = \sqrt{\det(A^T A)}$ where A is the matrix of f. This is interpretable as a "volume distortion" factor.

$f'(x)$ (page 81) By $f'(x)$ we mean the linear transformation whose matrix with respect to the standard bases is the Jacobian matrix of f evaluated at x.

$\rightarrowtail$ (page 81) $f : \mathbb{R}^M \rightarrowtail \mathbb{R}^N$ means that the range of f is a subset of $\mathbb{R}^N$ and the domain of f is an *open* subset of $\mathbb{R}^M$.

$\simeq$ (page 122) This is used to indicate that a function f is expandable in a certain way in terms of other functions, as in $f \simeq \sum_{k=1}^{\infty} f_k$.

The text may be used in several ways. We recommend using it as a supplement to a course on single variable analysis. One could start with a brief tour of the high points of Chapters One and Two, followed by a more leisurely trip through Chapter Three with particular emphasis on higher dimensional "derivatives" as linear transformations and on the chain rule. A second use is for a semester-long course introducing students to manifolds and differential forms; several results may be plucked from Chapters One and Three, Section 4.1, *A Bird's-Eye View of the Lebesgue Integral*, and then time and energy may be devoted to the last three chapters.

It is only simple courtesy to acknowledge the effort, time spent, and encouragement of our friends and colleagues, Alexander Katsevich, Heath Martin, Frank Salzmann, and John Synowiec, who critiqued portions of the manuscript. And it is but bare justice to make public notice of such students as Herve Andre, Holly Carley, Melissa Camp, Keith Carlson, Chinyen Chuo, Eric Curtis, Cory Edwards, Abby Elliott, John Hunter, Benjamin Landon, Robert Lange, Timothy Long, Pablo Matos, Mary McDowell, Daniel Moraseski, Kevin O'Hara, John Ortiz, Brad Pyle, Javier Rivera, Rachid Semmoum, Sidra Van De Car, Mark Varvak, and Alexander Zamyatin who (almost) uncomplainingly endured the role of guinea pigs for this text and brought many mistakes and shortcomings to our attention. A vital role in the preparation was played by June Wingler who patiently, carefully, and efficiently turned it into LaTeX for us. We also appreciate the timely assistance of Chinyen Chuo in typing a revised portion of the manuscript. Last, but far from least, we want to thank the people at Birkhäuser, Amy Ross, Tom Grasso and Elizabeth Loew for all their efforts and timely responses to our requests and to express our gratitude to our editor, Ann Kostant, for her encouragement, support, and patience.

P. Mikusiński
M. Taylor

1

VECTORS AND VOLUMES

1.1 Vector Spaces

$\mathbb{R}^3$, the set of ordered triples (x_1, x_2, x_3) of real numbers, is a natural and useful model for physical space. Similarly, $\mathbb{R}^4$ is an obvious model for space-time. More generally, problems in the sciences or engineering that involve N variables are often investigated in the setting of $\mathbb{R}^N$. Such problems often require the standard ideas of analysis: continuous change, instantaneous rates of change, integration, and so forth. To adapt these concepts from a one dimensional to an N-dimensional setting, it is first helpful to introduce some algebraic structure on $\mathbb{R}^N$, the structure of a vector space, and then to consider transformations of Euclidean N-dimensional spaces, particularly the simple and very useful ones known as linear transformations.

It may well be that the reader's first encounter with vectors was to hear them described as quantities that have both magnitude and direction. A real number x has a magnitude, namely $|x|$, and it has a direction, but there are only two choices for this direction. One is the positive direction (to the right in a conventional drawing of the number line) and the other is the negative direction (to the left). Vectors are usually thought of as having a much larger choice of directions. Vectors in the plane have the full range of directions available to a compass needle. Vectors in 3-dimensional space can assume all the directions of a well mounted telescope.

The most important properties of vectors from the mathematician's point of view have been carefully thought out. First, one must be able to carry out two operations: multiplication of vectors by scalars and addition of vectors. In this text by a scalar we shall simply mean a real number. In other settings scalars might be complex numbers or other objects. Second, vectors and scalars must satisfy certain axioms:

Definition 1.1.1 By a *vector space* over the field of reals we mean a nonempty set V, called the set of vectors, which is equipped with a binary operation of addition, $(v, w) \mapsto v + w$, and a scalar multiplication, $(\alpha, v) \mapsto \alpha v$, which satisfy the following:

(a) If $v, w \in V$ and $\alpha \in \mathbb{R}$, then $v + w \in V$ and $\alpha v \in V$.

(b) For all $u, v, w \in V$, we have $v + w = w + v$ and $u + (v + w) = (u + v) + w$.

(c) For all $\alpha, \beta \in \mathbb{R}$ and for all $v, w \in V$, we have $\alpha(\beta v) = (\alpha\beta)v$, $(\alpha + \beta)v = \alpha v + \beta v$, and $\alpha(v + w) = \alpha v + \alpha w$.

(d) For all $v, w \in V$ there exists $u \in V$ such that $u + v = w$.

(e) For all $v \in V$ we have $1v = v$.

It is straightforward to see that a vector space over the reals must have the following properties:

(1) There is a unique vector, which for the sake of convenience we denote by 0, having the property that $v + 0 = v$ for all $v \in V$.

(2) For any vector v we have $0v = 0$, where the zero on the left is a scalar and the one on the right is a vector.

(3) For every vector v there is a unique vector w which satisfies $v + w = 0$. We denote w by the symbol $-v$.

(4) For every vector v we have $(-1)v = -v$.

Just as we did with the real numbers, we may introduce an operation of subtraction of vectors by setting $v - w = v + (-w)$.

The main vector space in which we shall be interested is $\mathbb{R}^N$. It is the preferred setting of scientists and engineers for physical phenomena and the natural setting for partial differentiation and integration.

Theorem 1.1.1 *If we define addition and scalar multiplication by*

$$(x_1, x_2, \ldots, x_N) + (y_1, y_2, \ldots, y_N) = (x_1 + y_1, x_2 + y_2, \ldots, x_N + y_N)$$

and

$$\alpha(x_1, x_2, \ldots, x_N) = (\alpha x_1, \alpha x_2, \ldots, \alpha x_N),$$

then $\mathbb{R}^N$ with these operations is a vector space.

Exercises

1. Prove properties (1)–(4) of vector spaces.

2. Prove that $\mathbb{R}^N$ is a vector space with the given operations of addition and scalar multiplication.

3. Let S be a nonempty set and define V to be the set of all functions $f : S \to \mathbb{R}$. For $f, g \in V$ and $\alpha \in \mathbb{R}$ we define $f + g$ and αf by $(f+g)(x) = f(x)+g(x)$ and $(\alpha f)(x) = \alpha f(x)$. Show that with these operations V is a vector space.

1.2 Some Geometric Machinery for $\mathbb{R}^N$

In making use of the vector space $\mathbb{R}^N$ it is very helpful to have some extra concepts, some extra "machinery," and some intuitive geometric pictures.

An element x of $\mathbb{R}^N$ may be thought of either as a point or as a directed magnitude. In the second case it is perhaps more useful to form a mental picture of an arrow running from the origin 0 of $\mathbb{R}^N$ to x. More generally we may talk about the vector in $\mathbb{R}^N$ from y to x. We may picture this as an arrow or directed line segment running from y to x, and when we wish to be precise, we shall mean the vector $x - y$. At this point we suddenly encounter one of the peculiarities of pictorial thinking. Notice that $x = (x + z) - z$ so that the directed line segment running from 0 to x and the directed line segment running from z to $x + z$ both represent the vector x. This way of thinking leads to a very nice picture of addition of vectors. Notice that $(x + y) - 0 = [(x + y) - y] + (y - 0)$. This means that if we draw a directed line segment running from 0 to y and then a directed line segment which starts at y but represents x (namely, the directed line segment from y to $x + y$), then drawing these two directed line segments one after the other is, in some sense, equivalent to drawing a single directed line segment from 0 to $x + y$. (See Figure 1.2.1.)

By the *magnitude* of $x = (x_1, x_2, \ldots, x_N)$ we mean

$$|x| = \sqrt{x_1^2 + x_2^2 + \cdots + x_N^2}.$$

This generalizes naturally our idea of distance from 0 to x with which we are familiar in dimensions 2 and 3, and in the case $N = 1$, it reduces to the absolute value of x. We also call this the *Euclidean norm* of x.

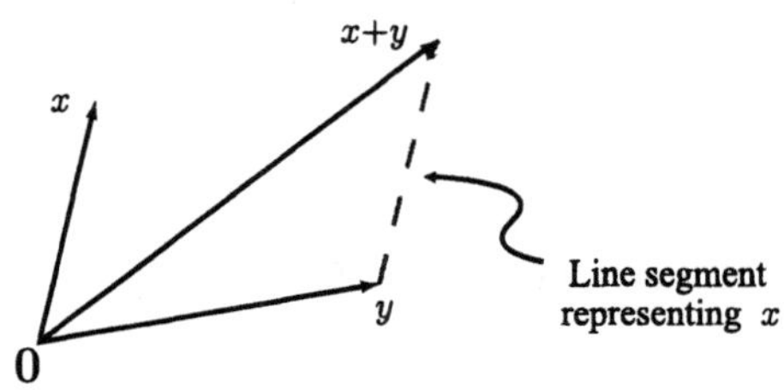

FIGURE 1.2.1.

We next generalize to $\mathbb{R}^N$ the *dot product* which should be familiar to us from dimensions 2 and 3. For $x = (x_1, x_2, \ldots, x_N)$ and $y = (y_1, y_2, \ldots, y_N)$, elements of $\mathbb{R}^N$, we define

$$x \cdot y = x_1 y_1 + x_2 y_2 + \cdots + x_N y_N.$$

Proof of the following properties of the dot product is left as an exercise.

Theorem 1.2.1 *For $x, y, z \in \mathbb{R}^N$ and $\lambda \in \mathbb{R}$ we have*

(a) $x \cdot y = y \cdot x$

(b) $x \cdot (y + z) = x \cdot y + x \cdot z$

(c) $(\lambda x) \cdot y = \lambda(x \cdot y)$

(d) $x \cdot x = |x|^2$.

The geometric content of the dot product lies in the formula $x \cdot y = |x|\,|y| \cos(\theta)$ where θ is interpreted as the angle between x and y. We first give an informal argument to justify this formula.

Let us draw directed line segments from 0 which represent x and y and let θ be the angle between these directed line segments. Next we draw the directed line segment which runs from y to x, and now we have a triangle with one vertex at 0. Note that the third line segment represents $x - y$. By the law of cosines we have

$$|x - y|^2 = |x|^2 + |y|^2 - 2|x|\,|y|\,\cos(\theta).$$

Recall that $|x - y|^2 = (x - y) \cdot (x - y)$. If we multiply the left side of the law of cosines equation and collect terms, we obtain $x \cdot y = |x||y| \cos(\theta)$.

The deficiency in this procedure is that we have no formal definition of the angle between two vectors. The way out of this difficulty is to define θ to be the unique number satisfying $0 \le \theta \le \pi$ and

$$\cos(\theta) = \frac{x \cdot y}{|x||y|}$$

whenever $x, y \neq 0$. Then we do not need to call on the law of cosines to prove a connection between the dot product and θ. But there is another problem with this procedure. We know that $-1 \le \cos(\theta) \le 1$. Can we be sure that $(x \cdot y)/(|x||y|)$ lies in the same interval? Assurance on this point and the final link in our chain of justification is provided by the Schwarz inequality $(x \cdot y)^2 \le |x|^2 |y|^2$.

Theorem 1.2.2 (Schwarz inequality) *For any real numbers $x_1, \ldots, x_N, y_1, \ldots, y_N$ we have*

$$\left| \sum_{n=1}^{N} x_n y_n \right| \le \sqrt{\sum_{n=1}^{N} x_n^2} \sqrt{\sum_{n=1}^{N} y_n^2}.$$

Proof. Since the inequality is trivial if $y_1 = y_2 = \cdots = y_N = 0$, we can assume that $\sum_{n=1}^{N} y_n^2 \neq 0$. First note that for every $t \in \mathbb{R}$ we have

$$0 \le \sum_{n=1}^{N} (x_n + t y_n)^2 = \sum_{n=1}^{N} x_n^2 + 2t \sum_{n=1}^{N} x_n y_n + t^2 \sum_{n=1}^{N} y_n^2. \tag{1.1}$$

The last expression is a quadratic in t and it is easily seen to assume its minimum value at

$$t = -\frac{\sum_{n=1}^{N} x_n y_n}{\sum_{n=1}^{N} y_n^2}. \tag{1.2}$$

Substituting (1.1) in (1.2) and simplifying we get

$$0 \le \sum_{n=1}^{N} x_n^2 - \frac{\left(\sum_{n=1}^{N} x_n y_n\right)^2}{\sum_{n=1}^{N} y_n^2}, \tag{1.3}$$

from which the Schwarz inequality follows easily. □

We conclude by looking at four elementary but useful geometric ideas.

First, notice that if $x,\ y$ are nonzero vectors and θ is that angle between them, then $\theta = \pi/2$ if and only if $\cos(\theta) = 0$. Therefore we shall agree that x and y are *orthogonal* precisely when $x \cdot y = 0$. Under this definition the zero vector, $0 = (0, 0, \ldots, 0)$, is orthogonal to all vectors.

Second, consider the concept of an $(N-1)$-dimensional hyperplane $\mathcal{P}$ in $\mathbb{R}^N$. This is the higher dimensional analog of a plane in $\mathbb{R}^3$. Arguing informally, one would expect that there must be some nonzero vector a and some point $b \in \mathcal{P}$ with the property that $\mathcal{P}$ consists of the points $x \in \mathbb{R}^N$ such that $x - b$ is orthogonal to a. Therefore we define $\mathcal{P}$ to be an $(N-1)$-dimensional hyperplane in $\mathbb{R}^N$ if and only if we can find $a,\ b \in \mathbb{R}^N$, $a \neq 0$, such that $\mathcal{P}$ is the set of x satisfying $a \cdot (x - b) = 0$. We say a is *normal* to $\mathcal{P}$.

Third, consider how one would describe lines in $\mathbb{R}^N$. Let us first agree that we will consider two nonzero vectors, x and y, to be *parallel* to one another if and only if one is a scalar multiple of the other; i.e., $x = \lambda y$ for some $\lambda \in \mathbb{R}$. Then, informally speaking, a line $\mathcal{L}$ in $\mathbb{R}^N$ is specified by giving a point p that lies on $\mathcal{L}$ and a nonzero vector a that is parallel to $\mathcal{L}$. Taking this as our cue, we *define* the line passing through p and parallel to a to be the set of points x satisfying $x - p = \lambda a$ for some $\lambda \in \mathbb{R}$.

Finally, consider the line segment $\mathcal{S}$ in $\mathbb{R}^N$ that runs from p to q where p and q are distinct points. This line segment should be a subset of the line which runs through both p and q, and every point x on it must satisfy $x - p = \lambda(q - p)$ for some $\lambda \in \mathbb{R}$. Formally we define $\mathcal{S}$ to be the set of x satisfying $x = p + \lambda(q - p)$ where $0 \le \lambda \le 1$.

Exercises

1. Verify (a)–(d) of Theorem 1.2.1.

2. Show that two nonzero vectors x and y are parallel if and only if the angle between them is either 0 or π.

3. Reprove the Schwarz inequality using the dot product notation and compare this with the proof previously given.

1.3 Transformations and Linear Transformations

Calculus of functions of a single variable (at least over the field of real numbers) deals with functions $f\colon \mathbb{R} \to \mathbb{R}$. Multivariable calculus (again over the reals) deals with functions $f\colon \mathbb{R}^M \to \mathbb{R}^N$. Sometimes they are described as transformations of $\mathbb{R}^M$ into $\mathbb{R}^N$. What can be said about such functions?

Notice that if $x \in \mathbb{R}^M$ and $f(x) \in \mathbb{R}^N$, then we must be able to find functions $f_1, f_2, \ldots, f_N\colon \mathbb{R}^M \to \mathbb{R}$ such that $f(x) = (f_1(x), f_2(x), \ldots, f_N(x))$. These f_i functions — which one might call the coordinate or component functions of f — completely determine f.

Transformations $f\colon \mathbb{R}^2 \to \mathbb{R}^2$ are particularly nice to consider because they can often be visualized. Take, for example, $f(x, y) = (x^2 - y^2, 2xy)$. (That is, $f_1(x, y) = x^2 - y^2$ and $f_2(x, y) = 2xy$.) This has the property that it transforms hyperbolas of the form $x^2 - y^2 = a$ onto vertical lines of the form $x = a$ and hyperbolas of the form $xy = b$ onto horizontal lines of the form $y = 2b$.

Another easily visualized example is provided by transformations of the form $f\colon \mathbb{R} \to \mathbb{R}^2$ or $f\colon \mathbb{R} \to \mathbb{R}^3$. Think of t as denoting time and $f(t)$ as being the position of a point or particle at time t. As time progresses, $f(t)$ traces a curve, the path the particle follows through the plane or 3-dimensional space. For instance $f(t) = (\cos(t), \sin(t), t)$ traces a helix which winds its way upward about the z-axis.

A very important class of transformations is the class of *linear transformations*. We say that $f\colon \mathbb{R}^M \to \mathbb{R}^N$ is linear provided

$$f(x + y) = f(x) + f(y) \quad \text{for all} \quad x, y \in \mathbb{R}^M$$

and

$$f(\lambda x) = \lambda f(x) \quad \text{for all} \quad \lambda \in \mathbb{R} \quad \text{and all} \quad x \in \mathbb{R}^M.$$

There is a convenient notation for linear transformations, the *matrix* notation. To see how this works, we first introduce the following vectors from $\mathbb{R}^M$:

$$\begin{aligned} e_1 &= (1, 0, 0, \ldots, 0), \\ e_2 &= (0, 1, 0, \ldots, 0), \\ &\ldots \\ e_M &= (0, 0, \ldots, 0, 1). \end{aligned}$$

We shall take this as standard notation from now on and call $e_1, e_2, \ldots, e_M$ the *standard basis* for $\mathbb{R}^M$. Notice that if $x = (x_1, x_2, \ldots, x_M) \in \mathbb{R}^M$, we can write

$$x = x_1 e_1 + x_2 e_2 + \cdots + x_M e_M.$$

Then if $f\colon \mathbb{R}^M \to \mathbb{R}^N$ is a linear transformation, we see that

$$f(x) = x_1 f(e_1) + x_2 f(e_2) + \cdots + x_M f(e_M).$$

This means that the linear transformation is completely determined by $f(e_1)$, $f(e_2)$, $\ldots$, $f(e_M)$. Now each $f(e_i)$ is a vector in $\mathbb{R}^N$, say $f(e_i) = (a_{1i}, a_{2i}, \ldots, a_{Ni})$.

Then we can represent the linear transformation f by the $N \times M$ matrix

$$\begin{pmatrix} a_{11} & a_{12} & \dots & a_{1M} \\ a_{21} & a_{22} & \dots & a_{2M} \\ \dots \\ a_{N1} & a_{N2} & \dots & a_{NM} \end{pmatrix}.$$

Sometimes we will more briefly indicate such a matrix by the symbol (a_{ij}). Alternatively we will also sometimes use the symbol $[f]$ for the matrix which represents f.

For example, for the identity transformation $f(x) = x$, we see that

$$\begin{aligned} f(e_1) &= (1, 0, \dots, 0), \\ f(e_2) &= (0, 1, \dots, 0), \\ &\dots \\ f(e_M) &= (0, \dots, 0, 1) \end{aligned}$$

so that the matrix is

$$\begin{pmatrix} 1 & 0 & \dots & 0 \\ 0 & 1 & \dots & 0 \\ \dots \\ 0 & \dots & 0 & 1 \end{pmatrix}.$$

As another example consider the reflection of the xy-plane through the y-axis given by $g(x, y) = (-x, y)$. The matrix in this case is

$$\begin{pmatrix} -1 & 0 \\ 0 & 1 \end{pmatrix}.$$

Exercises

1. Show that for every linear transformation f we must have $f(0) = 0$.
2. Prove that $f(x, y) = (x^2 - y^2, 2xy)$ maps hyperbolas of the form $x^2 - y^2 = a$ *onto* vertical lines $x = a$ and hyperbolas of the form $xy = b/2$ *onto* horizontal lines $y = b$.
3. Show that a linear transformation f is one-to-one if and only if $f(x) = 0$ implies $x = 0$.
4. True or false: $f(x) = ax + b$ defines a linear transformation of $\mathbb{R}$ to $\mathbb{R}$.
5. Prove that $f\colon \mathbb{R}^M \to \mathbb{R}^N$ is a linear transformation if and only if there exist vectors $a_1, a_2, \dots, a_N \in \mathbb{R}^M$ such that $f(x) = (a_1 \cdot x, a_2 \cdot x, \dots, a_N \cdot x)$.
6. (a) Consider the map $f(x, y)= (y, x)$ which switches the x- and y-axes in the plane. Show that this is a linear transformation and find its matrix.

(b) Find a linear transformation $f\colon \mathbb{R}^N \to \mathbb{R}^N$ which switches the x_i- and x_j-axes but leaves all the others fixed. Write the formula for $f(x)$, prove f is a linear transformation, and find its matrix.

7. Find the linear transformation which rotates vectors in the xy-plane by a fixed amount θ and give the matrix of this transformation.

8. Find the linear transformation $f\colon \mathbb{R}^N \to \mathbb{R}^N$ which reflects points on the x_i-axis through the origin and leaves points on all the other x_j-axes fixed. Give the matrix of f.

9. Given scalars $\lambda_1, \lambda_2, \ldots, \lambda_N$, find the linear transformation $f\colon \mathbb{R}^N \to \mathbb{R}^N$ which maps each x_i-axis to itself and changes the scale on that axis by a factor of λ_i. Give the matrix of f.

1.4 A Little Matrix Algebra

It is useful to introduce some algebraic operations with matrices. Given a scalar λ and a matrix (a_{ij}), their *scalar product* is defined by $\lambda(a_{ij}) = (\lambda a_{ij})$. The *sum* of two matrices is defined by $(a_{ij}) + (b_{ij}) = (a_{ij} + b_{ij})$. It is trivial to see that if α and β are scalars and A, B, and C are all matrices of the same size, then

$$\begin{aligned} A + B &= B + A, \\ A + (B + C) &= (A + B) + C, \\ \alpha(A + B) &= \alpha A + \alpha B, \end{aligned}$$

and

$$(\alpha + \beta)A = \alpha A + \beta A.$$

Of course $(-1)A$ is the additive inverse of A and is written $-A$, and the matrix with all zeros is the additive identity for all matrices of the same size.

It should be stressed that these operations are not simply made up in order to do algebra with matrices but are determined by corresponding operations with the linear transformations which the matrices represent. Recall that $[f]$ is the matrix of the linear transformation f with respect to the standard bases.

Theorem 1.4.1 *Let $f, g\colon \mathbb{R}^M \to \mathbb{R}^N$ be linear transformations and λ a scalar. Then $[f + g] = [f] + [g]$ and $[\lambda f] = \lambda[f]$.*

The most interesting of the elementary algebraic operations with matrices is the *product* of two matrices. Let A be a $P \times N$ matrix and B a $N \times M$ matrix. There are uniquely determined linear transformations $g\colon \mathbb{R}^M \to \mathbb{R}^N$ and $f\colon \mathbb{R}^N \to \mathbb{R}^P$ such that $A = [f]$ and $B = [g]$. Then the product AB is defined by the equation $[f][g] = [f \circ g]$. Of course this product is not usually commutative; that is, in general, $AB \neq BA$ even when both products are defined. On the other hand, matrix multiplication is associative, a fact whose demonstration we leave as an exercise.

Theorem 1.4.2 *If (a_{ij}) is a $P \times N$ matrix and (b_{jk}) is an $N \times M$ matrix, then*

$$(a_{ij})(b_{jk}) = \left(\sum_{j=1}^{N} a_{ij}b_{jk}\right)$$

which is a $P \times M$ matrix.

If A is an $N \times N$ matrix and I is the $N \times N$ identity matrix (i.e., the matrix of the identity map from $\mathbb{R}^N$ to $\mathbb{R}^N$), then it is easy to see that $AI = IA = A$. That is, I acts as a multiplicative identity. If A and B are both $N \times N$ matrices and $AB = BA = I$, then B is the multiplicative inverse of A and we write $B = A^{-1}$. (It can be shown that in this last case we need only one of the equations $AB = I$ or $BA = I$ for the other to hold, but we need to first introduce the notions of basis and dimension to prove this.)

We denote the *transpose* of a matrix A by the symbol A^T. We mean by this that $(a_{ij})^T = (a_{ji})$, or, perhaps a little more clearly that,

$$\begin{pmatrix} a_{11} & a_{12} & \dots & a_{1N} \\ a_{21} & a_{22} & \dots & a_{2N} \\ \dots & & & \\ a_{M1} & a_{M2} & \dots & a_{MN} \end{pmatrix}^T = \begin{pmatrix} a_{11} & a_{21} & \dots & a_{M1} \\ a_{12} & a_{22} & \dots & a_{M2} \\ \dots & & & \\ a_{1N} & a_{2N} & \dots & a_{MN} \end{pmatrix}.$$

If we think of $x = (x_1, x_2, \dots, x_N)$ as a $1 \times N$ matrix, then we have

$$x^T = \begin{pmatrix} x_1 \\ x_2 \\ \vdots \\ x_N \end{pmatrix}.$$

We note that x and x^T are examples of a *row vector* and a *column vector*, respectively. If f is a linear transformation, we can use the idea of the transpose to indicate the evaluation of f at x, namely $f(x)$, entirely in terms of matrix operations: $f(x)^T = [f]x^T$. For instance, consider the linear transformation $f: \mathbb{R}^2 \to \mathbb{R}^2$ given by $f(x, y) = (x + y, -2x)$. We see that

$$[f] = \begin{pmatrix} 1 & 1 \\ -2 & 0 \end{pmatrix}$$

and

$$\left(f(x, y)\right)^T = \begin{pmatrix} 1 & 1 \\ -2 & 0 \end{pmatrix} \begin{pmatrix} x \\ y \end{pmatrix} = \begin{pmatrix} x + y \\ -2x \end{pmatrix}.$$

Exercises

1. Show that if A is an $N \times M$ matrix, then there is a uniquely determined linear transformation $f: \mathbb{R}^M \to \mathbb{R}^N$ such that $A = [f]$.

2. Prove 1.4.1.

3. Show that the composition of two linear transformations is again a linear transformation.

4. Prove 1.4.2.

5. If A and B are $N \times N$ matrices with inverses, show that $(AB)^{-1} = B^{-1}A^{-1}$.

6. If A and B are matrices of the right size so that the product AB is defined, then show that $(AB)^T = B^T A^T$.

7. Show that for a linear transformation f one must have $(f(x))^T = [f]x^T$.

8. (a) Show that the set of $N \times M$ matrices is a vector space over the reals. (b) Show that the set of linear transformations $f\colon \mathbb{R}^M \to \mathbb{R}^N$ is a vector space over the reals.

9. Show that for matrices we have $A(BC) = (AB)C$, assuming the products are defined. (Hint: Replace the matrices by the corresponding linear transformations and recall that the composition of functions is associative.)

1.5 Oriented Volume and Determinants

When we develop the theory of integrals of real-valued functions on $\mathbb{R}^N$ and of line integrals and surface integrals and their higher dimensional generalizations, it will be very useful to be able to compute the "volume" of certain higher dimensional sets, namely the K-dimensional parallelepipeds. In studying such higher dimensional "volume," it turns out to be convenient to first study "oriented volume." This leads to a very basic idea in linear algebra, the concept of a determinant. We will also find that studying determinants and "volumes" yields a nice characterization of when a differentiable transformation is locally one-to-one.

Consider the parallelogram determined by two vectors a_1 and a_2. (See Figure 1.5.1.) We may think of this parallelogram as consisting of points of the form $\lambda_1 a_1 + \lambda_2 a_2$ where $0 \le \lambda_1, \lambda_2 \le 1$. Notice that the vertices are the points obtained by setting λ_1, λ_2 equal to 0 or 1. The idea of the parallelogram determined by two vectors makes sense whether we think of a_1 and a_2 as lying in $\mathbb{R}^2$, in $\mathbb{R}^3$, or in $\mathbb{R}^N$ where

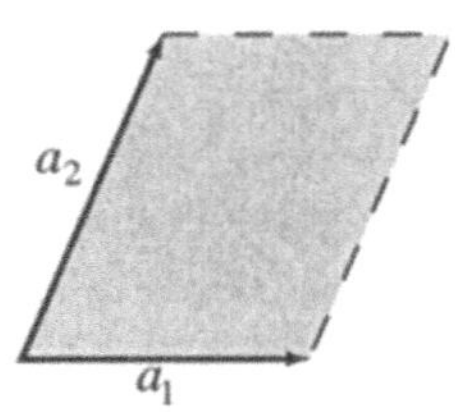

FIGURE 1.5.1.

$N \geq 2$. However if the parallelogram lies in $\mathbb{R}^2$, we can think of it as having an *orientation*, and if it lies in a higher dimensional space, we cannot. Let us try to make this clearer by a fanciful word picture.

We specify an oriented parallelogram lying in $\mathbb{R}^2$ by giving our two vectors as an ordered pair, either (a_1, a_2) or (a_2, a_1). Think of this oriented parallelogram as being a small living creature with two arms, a_1 and a_2. The *first* vector occurring in the ordered pair (a_i, a_j) is the arm which the creature prefers to use. To write the vectors in a particular order is simply to specify whether the creature is right handed or left handed. This is what the idea of orientation amounts to in this case.

(Notice we cannot say the creature is right handed or left handed if one arm lies on top of the other, that is, if a_1 and a_2 are collinear, or if one of the two vectors is the zero vector. Our discussion only makes sense if both of these conditions are avoided.)

Now we go to a more complex construction. Three vectors a_1, a_2, and a_3 in $\mathbb{R}^N$, where $N \geq 3$, will, in general, determine a parallelepiped. (See Figure 1.5.2.) If one of the vectors is coplanar with the other two or if one of them is the zero vector, then the parallelepiped will be degenerate. We may identify the parallelepiped with the set of points of the form $\lambda_1 a_1 + \lambda_2 a_2 + \lambda_3 a_3$ where $0 \leq \lambda_1, \lambda_2, \lambda_3 \leq 1$. If a_1, a_2, a_3 are chosen to lie in $\mathbb{R}^3$, we may think of the resulting parallelepiped as being oriented. The orientation is specified by writing the vectors as an ordered triple such as (a_2, a_1, a_3) or (a_3, a_1, a_2), etc. Think of (a_i, a_j, a_k) as representing some sort of living creature in 3-dimensional space. Its head is a_i and a_j and a_k are two arms with a_j being the arm the creature prefers to use. To specify an orientation is simply to decide whether the creature is right handed or left handed. Analogous to what happened with the parallelograms, if we choose a_1, a_2, a_3 to lie in $\mathbb{R}^N$ where $N > 3$, then it still makes sense to talk of a parallelepiped determined by the vectors but not of its orientation.

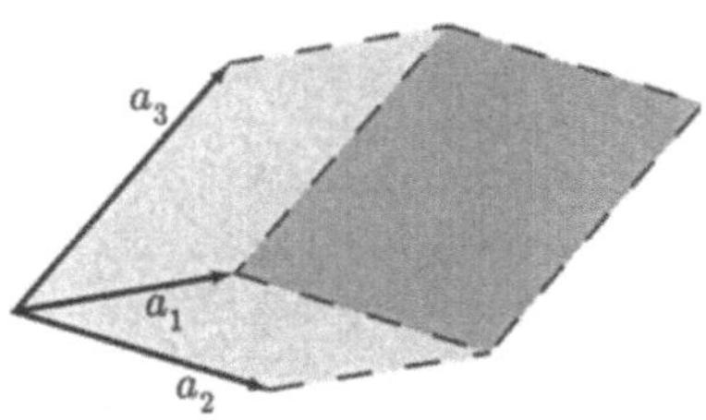

FIGURE 1.5.2.

More generally, if we choose $a_1, a_2, \ldots, a_K \in \mathbb{R}^N$ where $N \geq K$, then the K vectors determine a K-dimensional *parallelepiped* in $\mathbb{R}^N$. We may think of this parallelepiped as consisting of all points of the form $\lambda_1 a_1 + \lambda_2 a_2 + \cdots + \lambda_K a_K$ where $0 \leq \lambda_i \leq 1$ for each i. Under some circumstances this parallelepiped will be degenerate. We may hope to endow this parallelepiped with an orientation in the case where $N = K$ and in no other. In that case the orientation is determined by writing the vectors as an ordered K-tuple, $(a_{i_1}, a_{i_2}, \ldots, a_{i_K})$. (However that does not mean that different orders are necessarily associated with different orientations.)

If $a_1, a_2, \ldots, a_K$ are chosen from $\mathbb{R}^K$, let $V(a_1, a_2, \ldots, a_K)$ stand for the oriented volume of the K-dimensional parallelepiped determined by the vectors. We write down some properties we would expect V to have:

(V1) $V(e_1, e_2, \ldots, e_K) = 1$.

(V2) $V(a_1, \ldots, \lambda a_i, \ldots, a_K) = \lambda V(a_1, \ldots, a_i, \ldots, a_K)$
where λ is a scalar (possibly negative).

(V3) $V(a_1, \ldots, a_{i-1}, a_i + b, a_{i+1}, \ldots, a_K)$
$= V(a_1, \ldots, a_{i-1}, a_i, a_{i+1}, \ldots, a_K)$
$+V(a_1, \ldots, a_{i-1}, b, a_{i+1}, \ldots, a_K)$.

(V4) $V(a_1, \ldots, a_i, \ldots, a_j, \ldots, a_K) = -V(a_1, \ldots, a_j, \ldots, a_i, \ldots, a_K)$
where $i \neq j$.

(V1) in effect establishes a unit of measure for volumes; without this property, the value of V is not uniquely determined. Properties (V2) and (V3) can be summed up by saying that the oriented volume function should be linear in each variable. Of course (V4) simply expresses the fact that we are dealing with *oriented* volume. The reasonableness of (V3) can be seen in Figure 1.5.3 which shows what this property amounts to for the area of a parallelogram.

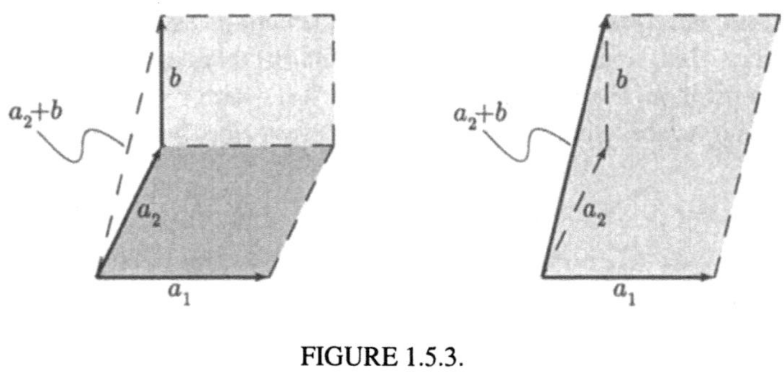

FIGURE 1.5.3.

It is not hard to see that these properties completely determine V. Notice that it follows from (V4) that any time we have $a_i = a_j$ for $i \neq j$, we must have $V(a_1, \ldots, a_K) = 0$. This fact plus diligent use of (V2)–(V4) permits us to reduce any $V(a_1, \ldots, a_K)$ to $\eta V(e_1, \ldots, e_K)$ and hence to η where η is a scalar. For example, if we are given the vectors $a_1 = (\alpha_{11}, \alpha_{12})$ and $a_2 = (\alpha_{21}, \alpha_{22})$ from $\mathbb{R}^2$, then we can show $V(a_1, a_2) = \alpha_{11}\alpha_{22} - \alpha_{12}\alpha_{21}$.

The reader with exposure to linear algebra or matrix theory may recognize that (V1)–(V4) are properties associated with determinants. The *determinant* is a function that associates with a square matrix A a real (or complex) number $\det(A)$. If $a_1^T, a_2^T, \ldots, a_M^T$ are the column vectors of A or if A is the matrix of a linear transformation $f: \mathbb{R}^M \to \mathbb{R}^M$, then we shall feel free to write $\det(A) = \det(a_1^T, \ldots, a_M^T) = \det(f)$. We shall assume both the existence of det and that it satisfies the following axioms:

(D1) $\det(e_1^T, e_2^T, \ldots, e_M^T) = 1$.

(D2) $\det(a_1^T, \ldots, \lambda a_i^T, \ldots, a_M^T) = \lambda \det(a_1^T, \ldots, a_i^T, \ldots, a_M^T)$ where λ is a scalar.

(D3) $\det(a_1^T, \ldots, a_{i-1}^T, a_i^T + b^T, a_{i+1}^T, \ldots, a_M^T)$
$= \det(a_1^T, \ldots, a_{i-1}^T, a_i^T, a_{i+1}^T, \ldots, a_M^T)$
$+ \det(a_1^T, \ldots, a_{i-1}^T, b^T, a_{i+1}^T, \ldots, a_M^T)$.

(D4) $\det(a_1^T, \ldots, a_i^T, \ldots, a_j^T, \ldots, a_M^T)$
$= -\det(a_1^T, \ldots, a_j^T, \ldots, a_i^T, \ldots, a_M^T)$ for $i \neq j$.

We have not proved the existence of det. We leave that to a linear algebra or matrix theory course. We have, however, seen that $\det(a_1^T, \ldots, a_M^T)$ has a geometric interpretation: It may be thought of as the oriented M-dimensional volume of the M-dimensional parallelepiped determined by $a_1, \ldots, a_M$.

In a later section we will want to come back to this collection of ideas and consider the "volume" of a K-dimensional parallelepiped lying in $\mathbb{R}^M$ where $K < M$.

Exercises

1. Using (D1)–(D4), prove that

$$\det(a_1^T, a_2^T) = a_{11}a_{22} - a_{12}a_{21} \quad \text{where} \quad a_i = (a_{i1}, a_{i2}).$$

2. Use (D1)–(D4) to compute

$$\det \begin{pmatrix} \lambda_1 & \alpha & \beta \\ 0 & \lambda_2 & \gamma \\ 0 & 0 & \lambda_3 \end{pmatrix}.$$

1.6 Properties of Determinants

Determinants are very important in the study of multivariable calculus, so we need to learn something about them. In doing this it turns out to be helpful to develop some of the properties of the *sign* of a permutation.

By a *permutation* on $\{1, 2, \ldots, N\}$ we mean a one-to-one map

$$\sigma : \{1, 2, \ldots, N\} \to \{1, 2, \ldots, N\}.$$

More generally a permutation on the set $\{p_1, p_2, \ldots, p_N\}$ is a one-to-one map

$$\sigma : \{p_1, p_2, \ldots, p_N\} \to \{p_1, p_2, \ldots, p_N\}.$$

Let $\mathcal{P}_N$ stand for the set of all permutations on the first N natural numbers. If we write the first N natural numbers in the order $1, 2, \ldots, N$, then we may think of a permutation as being a reordering in which we write $\sigma(1), \sigma(2), \ldots, \sigma(N)$. It is not hard to see that every such reordering can be obtained by a sequence of operations in which, at each step, one natural number is switched with another. For example, to obtain the permutation 3, 2, 4, 1 from 1, 2, 3, 4, we can carry out the switches

$$1234 \to 1324 \to 3124 \to 3214 \to 3241.$$

We want a way of deciding when σ can be carried out by an odd or an even number of such switches. One way to do this is to define

$$\operatorname{sgn}(\sigma) = \det(e^T_{\sigma(1)}, e^T_{\sigma(2)}, \ldots, e^T_{\sigma(N)}).$$

For example, if we consider the permutation defined by the sequence of switches

$$1234 \to 1324 \to 3124 \to 3214 \to 3241,$$

we see that

$$\begin{aligned}
&\det(e_1^T, e_2^T, e_3^T, e_4^T) = 1,\\
&\det(e_1^T, e_3^T, e_2^T, e_4^T) = -1,\\
&\det(e_3^T, e_1^T, e_2^T, e_4^T) = (-1)^2,\\
&\det(e_3^T, e_2^T, e_1^T, e_4^T) = (-1)^3,\\
&\det(e_3^T, e_2^T, e_4^T, e_1^T) = \det(e^T_{\sigma(1)}, e^T_{\sigma(2)}, e^T_{\sigma(3)}, e^T_{\sigma(4)}) = (-1)^4.
\end{aligned}$$

Notice that the exponent 4 on -1 is the number of switches carried out on 1, 2, 3, 4. For an arbitrary permutation σ, if $\operatorname{sgn}(\sigma) = 1$, we say σ is even, and if $\operatorname{sgn}(\sigma) = -1$, then we say it is odd. Every permutation is either even or odd. A transpose, the kind of permutation which amounts to a single switch of two elements, is always odd.

Theorem 1.6.1 *If $A = (\alpha_{ij})$ is an $N \times N$ matrix, then*

$$\det(A) = \sum_{\sigma \in \mathcal{P}_N} \operatorname{sgn}(\sigma)\alpha_{\sigma(1)1}\alpha_{\sigma(2)2}\ldots\alpha_{\sigma(N)N}.$$

(One should understand from the symbolism that the sum is taken over all permutations in $\mathcal{P}_N$.)

Proof. The column vectors of A have the form

$$a_j^T = \alpha_{1j}e_1^T + \alpha_{2j}e_2^T + \cdots + \alpha_{Nj}e_N^T.$$

Then we must have

$$\begin{aligned}\det(A) &= \det(a_1^T, \ldots, a_N^T)\\ &= \det\left(\sum_{j_1=1}^{N} \alpha_{j_1 1} e_{j_1}^T, \sum_{j_2=1}^{N} \alpha_{j_2 2} e_{j_2}^T, \ldots, \sum_{j_N=1}^{N} \alpha_{j_N N} e_{j_N}^T\right)\\ &= \sum_{j_1=1}^{N}\sum_{j_2=1}^{N} \ldots, \sum_{j_N=1}^{N} \alpha_{j_1 1}\alpha_{j_2 2}\ldots\alpha_{j_N N} \det(e_{j_1}^T, e_{j_2}^T, \ldots, e_{j_N}^T).\end{aligned}$$

Whenever we have distinct p and q such that $j_p = j_q$, then $\det(e_{j_1}^T, e_{j_2}^T, \ldots, e_{j_N}^T) = 0$, so we need consider only terms in the last sum for which $j_1, j_2, \ldots, j_N$ are distinct. Suppose that we have given distinct $j_1, j_2, \ldots, j_N$. There must be some $\sigma \in \mathcal{P}_N$ such that $j_k = \sigma(k)$ for all $k \in \{1, \ldots, N\}$. Then

$$\det(e_{j_1}^T, \ldots, e_{j_N}^T) = \det(e_{\sigma(1)}^T, \ldots, e_{\sigma(N)}^T) = \operatorname{sgn}(\sigma).$$

Since in the sum for $\det(A)$ every possible arrangement of $j_1, j_2, \ldots, j_N$ occurs exactly once, every possible permutation of $\{1, \ldots, N\}$ must occur exactly once. Therefore

$$\det(A) = \sum_{\sigma\in\mathcal{P}_N} \operatorname{sgn}(\sigma)\alpha_{\sigma(1)1}\alpha_{\sigma(2)2}\ldots\alpha_{\sigma(N)N}$$

and we are done. □

Recall that among $N \times N$ matrices, I is the identity matrix, that is, it is the matrix whose kth column vector is e_k^T. For every $\sigma \in \mathcal{P}_N$ let I_σ be the $N \times N$ matrix whose kth column vector is $e_{\sigma(k)}^T$. For example, if $N = 3$ and we have $\sigma(1) = 3$, $\sigma(2) = 1$, and $\sigma(3) = 2$, then

$$I_\sigma = \begin{pmatrix} 0 & 1 & 0\\ 0 & 0 & 1\\ 1 & 0 & 0\end{pmatrix}.$$

Note that by definition $\operatorname{sgn}(\sigma) = \det(I_\sigma)$.

Theorem 1.6.2 *If A is an $N \times N$ matrix and $\sigma \in \mathcal{P}_N$, then* $\det(AI_\sigma) = \operatorname{sgn}(\sigma)\det(A)$.

Proof. If the kth column vector of A is a_k^T, it is straightforward to see that the kth column vector of AI_σ must be $a_{\sigma(k)}^T$. Then it must be possible, by a sequence of column interchanges, to write

$$\begin{aligned}\det(AI_\sigma) &= \det(a_{\sigma(1)}^T, a_{\sigma(2)}^T, \ldots, a_{\sigma(N)}^T)\\ &= (-1)^p \det(a_1^T, a_2^T, \ldots, a_N^T)\\ &= (-1)^p \det(A)\end{aligned}$$

for some nonnegative integer p. Notice that the same sequence of column interchanges may be used to bring about this result for *every* A. We might, for example, find i

such that $\sigma(i) = 1$ and switch $a^T_{\sigma(i)}$ with $a^T_{\sigma(1)}$. Then we could find j such that $\sigma(j) = 2$ and switch $a^T_{\sigma(j)}$ with $a^T_{\sigma(2)}$, and so on. Therefore, given σ, there is some nonnegative integer p with the property that $\det(AI_\sigma) = (-1)^p \det(A)$ for every A. In particular, if $A = I$, we have $\det(II_\sigma) = (-1)^p \det(I)$, which amounts to saying that $\operatorname{sgn}(\sigma) = (-1)^p$. □

Corollary 1.6.1 *For every* $\sigma, \phi \in \mathcal{P}_N$ *we have* $\operatorname{sgn}(\sigma \circ \phi) = \operatorname{sgn}(\sigma)\operatorname{sgn}(\phi)$.

Proof. It is straightforward to show that $I_{\sigma\circ\phi} = I_\sigma I_\phi$. Taking the determinant of each side of the last equation yields the desired result. □

Corollary 1.6.2 *For every* $\sigma \in \mathcal{P}_N$ *we have* $\operatorname{sgn}(\sigma^{-1}) = \operatorname{sgn}(\sigma)$.

Proof. We see that

$$1 = \det(I) = \det(I_{\sigma^{-1}} I_\sigma) = \operatorname{sgn}(\sigma^{-1})\operatorname{sgn}(\sigma).$$

Since the sign of a permutation can only be 1 or -1, we are done. □

Theorem 1.6.3 *If A is an $N \times N$ matrix, then* $\det(A^T) = \det(A)$.

Proof. If $A = (\alpha_{ij})$, then $A^T = (\alpha_{ji})$. By Theorem 1.6.1 we have

$$\det(A^T) = \sum_{\sigma\in\mathcal{P}_N} \operatorname{sgn}(\sigma)\alpha_{1\sigma(1)}\alpha_{2\sigma(2)}\cdots\alpha_{N\sigma(N)}.$$

Now choose $\phi \in \mathcal{P}_N$. There is a unique $\sigma \in \mathcal{P}_N$ such that $\phi = \sigma^{-1}$. Consider the expression $\alpha_{1\sigma(1)}\alpha_{2\sigma(2)}\cdots\alpha_{N\sigma(N)}$. The factors of this expression have form α_{ij} where $j = \sigma(i)$. But this is the same as saying that $i = \phi(j)$. This means that

$$\alpha_{1\sigma(1)}\alpha_{2\sigma(2)}\cdots\alpha_{N\sigma(N)} = \alpha_{\phi(1)1}\alpha_{\phi(2)2}\cdots\alpha_{\phi(N)N}.$$

We also have $\operatorname{sgn}(\phi) = \operatorname{sgn}(\sigma^{-1}) = \operatorname{sgn}(\sigma)$. This means

$$\det(A^T) = \sum_{\phi\in\mathcal{P}_N} \operatorname{sgn}(\phi)\alpha_{\phi(1)1}\alpha_{\phi(2)2}\cdots\alpha_{\phi(N)N} = \det(A).$$ □

We have only discussed the sign of a permutation of the first N natural numbers. It is convenient to be able to talk about the sign of a permutation of *any* finite set. Let σ be a permutation of $\{p_1, p_2, \ldots, p_N\}$, that is, σ is a one-to-one map of this set onto itself. We can turn it into a permutation of the first N natural numbers by taking any one-to-one map $\phi\colon \{p_1, \ldots, p_N\} \to \{1, \ldots, N\}$ and introducing the permutation $\psi = \phi \circ \sigma \circ \phi^{-1}$ on $\{1, \ldots, N\}$. This is the same as saying that the following diagram of maps is *commutative*, that is, $\psi \circ \phi = \phi \circ \sigma$:

$$\begin{array}{ccc} \{p_1, p_2, \ldots, p_N\} & \overset{\sigma}{\to} & \{p_1, p_2, \ldots, p_N\} \\ \downarrow\phi & & \downarrow\phi \\ \{1, 2, \ldots, N\} & \overset{\psi}{\to} & \{1, 2, \ldots, N\}. \end{array} \tag{1.4}$$

We would like to define $\operatorname{sgn}(\sigma)$ to be $\operatorname{sgn}(\psi)$. But we cannot do this until we are sure that $\operatorname{sgn}(\psi)$ is independent of which particular map ϕ we use.

Therefore let $\phi_1, \phi_2 : \{p_1, \ldots, p_N\} \to \{1, \ldots, N\}$ be one-to-one maps and let us define $\psi_1 = \phi_1 \circ \sigma \circ \phi_1^{-1}$ and $\psi_2 = \phi_2 \circ \sigma \circ \phi_2^{-1}$. Notice that

$$\psi_2 = (\phi_2 \circ \phi_1^{-1}) \circ \psi_1 \circ (\phi_1 \circ \phi_2^{-1})$$

and that $\phi_2 \circ \phi_1^{-1}$ and $\phi_1 \circ \phi_2^{-1}$ are permutations of $\{1, \ldots, N\}$ and are inverses of one another. This means that $\operatorname{sgn}(\phi_2 \circ \phi_1^{-1}) = \operatorname{sgn}(\phi_1 \circ \phi_2^{-1})$. From this we deduce

$$\begin{aligned}\operatorname{sgn}(\psi_2) &= \operatorname{sgn}(\phi_2 \circ \phi_1^{-1})\operatorname{sgn}(\psi_1)\operatorname{sgn}(\phi_1 \circ \phi_2^{-1}) \\ &= [\operatorname{sgn}(\phi_2 \circ \phi_1^{-1})]^2 \operatorname{sgn}(\psi_1) \\ &= \operatorname{sgn}(\psi_1),\end{aligned}$$

and thus we can see that $\operatorname{sgn}(\sigma)$ is a well-defined quantity. □

We leave it to the exercises to show that in general $\operatorname{sgn}(\sigma^{-1}) = \operatorname{sgn}(\sigma)$ and $\operatorname{sgn}(\sigma \circ \phi) = \operatorname{sgn}(\sigma)\operatorname{sgn}(\phi)$.

We include here one other useful property of permutations.

Theorem 1.6.4 *Suppose σ is a permutation of the finite set T and S is a nonempty subset of T with the property that $\sigma(i) = i$ for all $i \notin S$. Then the restriction of σ to S, call it ψ, is a permutation of S and* $\operatorname{sgn}(\psi) = \operatorname{sgn}(\sigma)$.

Proof. We may write $\sigma = \sigma_1 \circ \sigma_2 \circ \cdots \circ \sigma_p$ where each σ_i is permutation of T which interchanges two elements of S and $\operatorname{sgn}(\sigma_i) = -1$. Then $\operatorname{sgn}(\sigma) = (-1)^p$. Let ψ be the restriction of σ to S and let ψ_i be the restriction of σ_i to S for each i. Then ψ and each ψ_i is a permutation of S and $\psi = \psi_1 \circ \psi_2 \circ \cdots \circ \psi_p$. Clearly $\operatorname{sgn}(\psi) = (-1)^p$. □

This result can be used to establish a very useful property for the evaluation of determinants.

Theorem 1.6.5 *If $N \geq 2$, then*

$$\det \begin{pmatrix} 1 & 0 & \ldots & 0 \\ \alpha_{21} & \alpha_{22} & \ldots & \alpha_{2N} \\ \ldots & & & \\ \alpha_{N1} & \alpha_{N2} & \ldots & \alpha_{NN} \end{pmatrix} = \det \begin{pmatrix} \alpha_{22} & \ldots & \alpha_{2N} \\ \ldots & & \\ \alpha_{N2} & \ldots & \alpha_{NN} \end{pmatrix}$$

and

$$\det \begin{pmatrix} 1 & \beta_{12} & \ldots & \beta_{1N} \\ 0 & \beta_{22} & \ldots & \beta_{2N} \\ \ldots & & & \ldots \\ 0 & \beta_{N2} & \ldots & \beta_{NN} \end{pmatrix} = \det \begin{pmatrix} \beta_{22} & \ldots & \beta_{2N} \\ \ldots & & \\ \beta_{N2} & \ldots & \beta_{NN} \end{pmatrix}.$$

Proof. We establish only the first half. Let $A = (\alpha_{ij})$ be an $N \times N$ matrix whose first row vector is $(1, 0, 0, \ldots, 0)$. Then

$$\det(A) = \sum_{\sigma \in \mathcal{P}_N} \operatorname{sgn}(\sigma)\alpha_{\sigma(1)1}\alpha_{\sigma(2)2}\ldots\alpha_{\sigma(N)N}.$$

Notice that if $\sigma(i) = 1$ for any $i \neq 1$, then $\alpha_{\sigma(1)1}\alpha_{\sigma(2)2}\ldots\alpha_{\sigma(N)N} = 0$, but if $\sigma(1) = 1$, then $\alpha_{\sigma(1)1}\alpha_{\sigma(2)2}\ldots\alpha_{\sigma(N)N} = \alpha_{\sigma(2)2}\ldots\alpha_{\sigma(N)N}$. Therefore

$$\det(A) = \sum_{\substack{\sigma \in \mathcal{P}_N \\ \sigma(1)=1}} \operatorname{sgn}(\sigma)\alpha_{\sigma(2)2}\ldots\alpha_{\sigma(N)N}.$$

This amounts to

$$\det(A) = \sum_{\psi \in \mathcal{S}} \operatorname{sgn}(\psi)\, \alpha_{\psi(2)2}\ldots\alpha_{\psi(N)N}$$

where $\mathcal{S}$ is the set of all permutations of $\{2, 3, \ldots, N\}$. But this last sum is clearly

$$\det\begin{pmatrix} \alpha_{22} & \ldots & \alpha_{2N} \\ \ldots & & \\ \alpha_{N2} & \ldots & \alpha_{NN} \end{pmatrix}. \qquad \square$$

We conclude this section with a very useful result which is connected with the study of K-dimensional volume.

Theorem 1.6.6 (Binet–Cauchy formula.) *If $A = (\alpha_{ij})$ is an $M \times N$ matrix and $B = (\beta_{ij})$ is an $N \times M$ matrix where $M \leq N$, then*

$$\det(AB) = \sum_{i_1<i_2<\cdots<i_M} \det\begin{pmatrix} \alpha_{1i_1} & \ldots & \alpha_{1i_M} \\ \ldots & & \\ \alpha_{Mi_1} & \ldots & \alpha_{Mi_M} \end{pmatrix} \det\begin{pmatrix} \beta_{i_11} & \ldots & \beta_{i_1M} \\ \ldots & & \\ \beta_{i_M1} & \ldots & \beta_{i_MM} \end{pmatrix}.$$

The summation symbolism here is to be understood as denoting the sum over all possible M-tuples $(i_1, i_2, \ldots, i_M)$ such that each i_k is an element of $\{1, 2, \ldots, N\}$ and $i_1 < i_2 < \cdots < i_M$. Also, in the proof below we may write

$$\sum_{k_1,\ldots,k_M=1}^{N}$$

by which we mean

$$\sum_{k_1=1}^{N}\sum_{k_2=1}^{N}\cdots\sum_{k_M=1}^{N}.$$

Notice that the order of the summations makes no difference. For example,

$$\sum_{k_1=1}^{N}\sum_{k_2=1}^{N} \text{ and } \sum_{k_2=1}^{N}\sum_{k_1=1}^{N}$$

both amount to the same summation.

Example 1.6.1 Consider

$$A = \begin{pmatrix} 1 & 0 & 2 \\ 3 & 1 & 4 \end{pmatrix} \quad \text{and} \quad B = \begin{pmatrix} -2 & 1 \\ 1 & 3 \\ 1 & 0 \end{pmatrix}.$$

We see that

$$AB = \begin{pmatrix} 0 & 1 \\ -1 & 6 \end{pmatrix},$$

and hence $\det(AB) = 1$. According to the Binet–Cauchy formula we should also have

$$\begin{aligned} \det(AB) &= \det\left\{\begin{pmatrix} 1 & 0 \\ 3 & 1 \end{pmatrix}\begin{pmatrix} -2 & 1 \\ 1 & 3 \end{pmatrix}\right\} \\ &\quad + \det\left\{\begin{pmatrix} 1 & 2 \\ 3 & 4 \end{pmatrix}\begin{pmatrix} -2 & 1 \\ 1 & 0 \end{pmatrix}\right\} + \det\left\{\begin{pmatrix} 0 & 2 \\ 1 & 4 \end{pmatrix}\begin{pmatrix} 1 & 3 \\ 1 & 0 \end{pmatrix}\right\} \\ &= \det\begin{pmatrix} -2 & 1 \\ -5 & 6 \end{pmatrix} + \det\begin{pmatrix} 0 & 1 \\ -2 & 3 \end{pmatrix} + \det\begin{pmatrix} 2 & 0 \\ 5 & 3 \end{pmatrix}, \end{aligned}$$

and it is easily checked that this indeed reduces to 1.

Proof of Theorem 1.6.6. Since $AB = \left(\sum_{j=1}^{N} \alpha_{ij}\beta_{jk}\right)$, we must have

$$\begin{aligned} \det(AB) &= \sum_{\sigma \in \mathcal{P}_M} \operatorname{sgn}(\sigma) \left(\sum_{k_1=1}^{N} \alpha_{\sigma(1)k_1}\beta_{k_1 1}\right)\left(\sum_{k_2=1}^{N} \alpha_{\sigma(2)k_2}\beta_{k_2 2}\right) \\ &\qquad \cdots \left(\sum_{k_M=1}^{N} \alpha_{\sigma(M)k_M}\beta_{k_M M}\right) \\ &= \sum_{k_1,\dots,k_M=1}^{N} \sum_{\sigma \in \mathcal{P}_M} \operatorname{sgn}(\sigma)\alpha_{\sigma(1)k_1}\beta_{k_1 1}\alpha_{\sigma(2)k_2}\beta_{k_2 2}\dots\alpha_{\sigma(M)k_M}\beta_{k_M M} \\ &= \sum_{k_1,\dots,k_M=1}^{N} \beta_{k_1 1}\beta_{k_2 2}\dots\beta_{k_M M} \sum_{\sigma \in \mathcal{P}_M} \operatorname{sgn}(\sigma)\alpha_{\sigma(1)k_1}\alpha_{\sigma(2)k_2}\dots\alpha_{\sigma(M)k_M} \\ &= \sum_{k_1,\dots,k_M=1}^{N} \beta_{k_1 1}\beta_{k_2 2}\dots\beta_{k_M M} \det\begin{pmatrix} \alpha_{1k_1} & \dots & \alpha_{1k_M} \\ \dots & & \\ \alpha_{Mk_1} & \dots & \alpha_{Mk_M} \end{pmatrix}. \end{aligned}$$

In this last line the summation is taken over all possible choices of values of $k_1, k_2, \dots, k_M$ from the set $\{1, 2, \dots, N\}$, each such choice occurring exactly once. Whenever we choose in this way that $i \neq j$ but $k_i = k_j$, then

$$\det\begin{pmatrix} \alpha_{1k_1} & \dots & \alpha_{1k_M} \\ \dots & & \\ \alpha_{Mk_1} & \dots & \alpha_{Mk_M} \end{pmatrix} = 0$$

because two column vectors in the matrix are identical. Consequently we can restrict our attention to choices for which $k_1, k_2, \ldots, k_M$ are distinct.

Choose $i_1, i_2, \ldots, i_M$ from $\{1, 2, \ldots, N\}$ such that $i_1 < i_2 < \cdots < i_M$. Let $\mathcal{Q}$ be the set of all permutations of $\{i_1, i_2, \ldots, i_M\}$. Suppose we have a term in the sum for $\det(AB)$ such that $\{k_1, k_2, \ldots, k_M\} = \{i_1, i_2, \ldots, i_M\}$. There must be an element ϕ of $\mathcal{Q}$ such that $\phi(i_r) = k_r$ for all r. The term we are considering can then be rewritten as follows:

$$\beta_{k_1 1}\beta_{k_2 2}\ldots\beta_{k_M M}\det\begin{pmatrix}\alpha_{1k_1} & \ldots & \alpha_{1k_M}\\ \ldots & & \\ \alpha_{Mk_1} & \ldots & \alpha_{Mk_M}\end{pmatrix}$$

$$= \beta_{\phi(i_1)1}\beta_{\phi(i_2)2}\ldots\beta_{\phi(i_M)M}\det\begin{pmatrix}\alpha_{1\phi(i_1)} & \ldots & \alpha_{1\phi(i_M)}\\ \ldots & & \\ \alpha_{M\phi(i_1)} & \ldots & \alpha_{M\phi(i_M)}\end{pmatrix}$$

$$= \operatorname{sgn}(\phi)\ \beta_{\phi(i_1)1}\beta_{\phi(i_2)2}\ldots\beta_{\phi(i_M)M}\det\begin{pmatrix}\alpha_{1i_1} & \ldots & \alpha_{1i_M}\\ \ldots & & \\ \alpha_{Mi_1} & \ldots & \alpha_{Mi_M}\end{pmatrix}.$$

Every selection of distinct $k_1, k_2, \ldots, k_M$ such that $\{k_1, k_2, \ldots, k_M\} = \{i_1, i_2, \ldots, i_M\}$ corresponds to the selection of a unique element ϕ of $\mathcal{Q}$. Therefore we may write

$$\det(AB) = \sum_{i_1<i_2<\cdots<i_M}\ \sum_{\phi\in\mathcal{P}_M}\operatorname{sgn}(\phi)\beta_{\phi(i_1)1}\beta_{\phi(i_2)2}$$
$$\ldots\beta_{\phi(i_M)M}\det\begin{pmatrix}\alpha_{1i_1} & \ldots & \alpha_{1i_M}\\ \ldots & & \\ \alpha_{Mi_1} & \ldots & \alpha_{Mi_M}\end{pmatrix}$$

which is the same as

$$\sum_{i_1<i_2<\cdots<i_M}\left(\sum_{\phi\in\mathcal{P}_M}\operatorname{sgn}(\phi)\ \beta_{\phi(i_1)1}\beta_{\phi(i_2)2}\ldots\beta_{\phi(i_M)M}\right)$$
$$\det\begin{pmatrix}\alpha_{1i_1} & \ldots & \alpha_{1i_M}\\ \ldots & & \\ \alpha_{Mi_1} & \ldots & \alpha_{Mi_M}\end{pmatrix}.$$

But $\sum_{\phi\in\mathcal{P}_M}\operatorname{sgn}(\phi)\beta_{\phi(i_1)1}\beta_{\phi(i_2)2}\ldots\beta_{\phi(i_M)M}$ can be shown in a straightforward way to be

$$\det\begin{pmatrix}\beta_{i_1 1} & \ldots & \beta_{i_1 M}\\ \ldots & & \\ \beta_{i_M 1} & \ldots & \beta_{i_M M}\end{pmatrix},$$

and we are done. □

Corollary 1.6.3 *If A and B are $M \times M$ matrices, then* $\det(AB) = \det(A)\det(B)$.

Proof. Note that the only $M \times M$ submatrices of A and B are A and B themselves. □

Exercises

1. Let $\sigma, \phi \in \mathcal{P}_N$ and let A be an $N \times N$ matrix. Prove the following:
   (a) $I_{\sigma^{-1}} = I_\sigma^T = (I_\sigma)^{-1}$.
   (b) $I_\sigma I_\phi = I_{\sigma\circ\phi}$.
   (c) $\det(I_\sigma A) = \mathrm{sgn}(\sigma)\det(A)$.

2. Show that permutations of the finite set $\{p_1, p_2, \ldots, p_M\}$ satisfy the relations $\mathrm{sgn}(\sigma \circ \phi) = \mathrm{sgn}(\sigma)\mathrm{sgn}(\phi)$ and $\mathrm{sgn}(\sigma^{-1}) = \mathrm{sgn}(\sigma)$.

3. Show that if $A = (\alpha_{ij})$ is an $N \times N$ matrix, then
$$\det(A) = \sum_{\sigma\in\mathcal{P}_N} \mathrm{sgn}(\sigma)\alpha_{1\sigma(1)}\alpha_{2\sigma(2)}\ldots\alpha_{N\sigma(N)}.$$

4. Write out a statement, without the summation symbol $\sum$, of the Binet–Cauchy theorem for
   (a) $M = 2$ and $N = 3$.
   (b) $M = 2$ and $N = 4$.

5. Let $\{p_1, p_2, \ldots, p_M\}$ be a given set of M objects and let $\mathcal{Q}$ be the set of permutations of this set. Define $\psi\colon \{1, 2, \ldots, M\} \to \{p_1, p_2, \ldots, p_M\}$ by $\psi(k) = p_k$. Show that the map $\sigma \mapsto \phi$ defined by $\phi = \psi \circ \sigma \circ \psi^{-1}$ is a one-to-one map of $\mathcal{P}_M$ onto $\mathcal{Q}$ which satisfies $\phi(p_k) = p_{\sigma(k)}$.

6. Let $B = (\beta_{ij})$ be an $N \times M$ matrix where $M < N$ and choose $i_1, i_2, \ldots, i_M$ distinct elements of $\{1, 2, \ldots, N\}$. Let $\mathcal{Q}$ be the set of permutations of $\{i_1, i_2, \ldots, i_M\}$. Then show that
$$\det\begin{pmatrix} \beta_{i_1 1} & \cdots & \beta_{i_1 M} \\ \cdots & & \\ \beta_{i_M 1} & \cdots & \beta_{i_M M}\end{pmatrix} = \sum_{\phi\in\mathcal{Q}} \mathrm{sgn}(\phi)\, \beta_{\phi(1)1}\cdots\, \beta_{\phi(M)M}.$$

7. (a) Show that for $i \neq j$ we have
$$\det(a_1^T, \ldots, a_i^T + \lambda a_j^T, \ldots, a_j^T, \ldots, a_N^T)$$
$$= \det(a_1^T, \ldots, a_i^T, \ldots, a_j^T, \ldots, a_N^T)$$
   where $a_1, \ldots, a_N \in \mathbb{R}^N$ and $\lambda \in \mathbb{R}$.
   (b) Show that a similar result holds for row vectors of a square matrix.

8. If I is the $M \times M$ identity matrix, D is an $N \times N$ matrix, B is an $M \times N$ matrix, and C is an $N \times M$ matrix, then show that

$$\det\begin{pmatrix} I & B \\ O & D \end{pmatrix} = \det\begin{pmatrix} I & O \\ C & D \end{pmatrix} = \det(D)$$

where O stands for zero matrices.

9. We shall show later that a linear transformation $f\colon \mathbb{R}^N \to \mathbb{R}^N$ has an inverse precisely when $\det(f) \neq 0$. Let us assume that fact for the time being and show a way to construct inverses of square matrices with nonzero determinants. Let

$$A = \begin{pmatrix} \alpha_{11} & \alpha_{12} & \dots & \alpha_{1N} \\ \alpha_{21} & \alpha_{22} & \dots & \alpha_{2N} \\ \dots & & & \\ \alpha_{N1} & \alpha_{N2} & \dots & \alpha_{NN} \end{pmatrix} \quad \text{and} \quad a_i^T = \begin{pmatrix} \alpha_{1i} \\ \alpha_{2i} \\ \dots \\ \alpha_{Ni} \end{pmatrix}$$

for $i = 1, 2, \dots, N$. Now for $i, j = 1, 2, \dots, N$ set

$$\beta_{ij} = \frac{1}{\det(A)} \det(a_1^T, \dots, a_{i-1}^T, e_j^T, a_{i+1}^T, \dots, a_N^T)$$

and

$$B = \begin{pmatrix} \beta_{11} & \beta_{12} & \dots & \beta_{1N} \\ \beta_{21} & \beta_{22} & \dots & \beta_{2N} \\ \dots & & & \\ \beta_{N1} & \beta_{N2} & \dots & \beta_{NN} \end{pmatrix}.$$

Using the fact that det is linear in each place, show that $BA = I$. Then show that $B = A^{-1}$.

10. Use the construction of the last exercise.

(a) Find the inverse for

$$A = \begin{pmatrix} \alpha & \beta \\ \gamma & \delta \end{pmatrix}$$

assuming $\det(A) \neq 0$.

(b) Find the inverse matrix for

$$B = \begin{pmatrix} 1 & x & x^2 \\ 0 & 1 & x \\ 0 & 0 & 1 \end{pmatrix}.$$

1.7 Linear Independence, Linear Subspaces, and Bases

Lurking behind the last several sections has been the unifying idea of *volume*, not just in 3-dimensional space but generalized to Euclidean spaces of any finite dimension.

Volume, or more properly K-dimensional volume, is a crucial idea in understanding multivariable calculus, particularly integration over subsets of higher dimensional spaces. We now have to dig down to an idea which is more basic than that of volume. Once we have properly unearthed this idea, we shall return to the consideration of K-dimensional volume.

To see what our prize is, think of a single vector x in some $\mathbb{R}^N$. If we picture it as a directed line segment, we see that it determines a line in $\mathbb{R}^N$, that this line is a 1-dimensional object, and that it might be proper in this setting to talk about *length* (the 1-dimensional version of volume), unless x is a zero vector, in which case only a point is determined in $\mathbb{R}^N$, not a line. Now think of a pair of vectors x and y. In general they will determine a plane, and this is the setting in which to discuss *area* (the 2-dimensional version of volume), unless the two vectors are collinear, in which case we have determined at most a line. Three vectors, x, y, and z, usually determine a 3-dimensional space, which is the proper setting for discussing *volume*, unless the vectors turn out to be coplanar, in which case we have determined at most a plane. And so on to higher dimensional versions of these ideas. Whenever we discuss volume in any of its various incarnations, we usually start with some set of vectors $x_1, x_2, \ldots, x_K$. These vectors can be thought of as determining a line, a plane, or some sort of higher dimensional analog. Before we can decide whether we are discussing length, area, or volume, we have to know the *dimension* of the space determined by the vectors. This is our prize, the dimension of the geometrical "object" determined by a given set of vectors. The key concept here turns out to be that of *linear independence*.

By a *linear combination* of the vectors $x_1, x_2, \ldots, x_K$ in the vector space V we mean a vector of the form $\alpha_1 x_1 + \alpha_2 x_2 + \cdots + \alpha_K x_K$ where $\alpha_1, \ldots, \alpha_K$ are scalars. We say that $x_1, x_2, \ldots, x_K$ are *linearly independent* provided the equation $\alpha_1 x_1 + \cdots + \alpha_K x_K = 0$ is valid only for $\alpha_1 = \cdots = \alpha_K = 0$. If, on the other hand, we can find scalars $\alpha_1, \ldots, \alpha_K$, at least some of which are nonzero, that satisfy $\alpha_1 x_1 + \cdots + \alpha_K x_k = 0$, then we say $x_1, \ldots, x_K$ are *linearly dependent*.

The standard example of linear independence is the set of vectors $e_1, e_2, \ldots, e_N$ from $\mathbb{R}^N$: It is trivial that $\alpha_1 e_1 + \cdots + \alpha_N e_N = 0$ holds only for $\alpha_1 = \cdots = \alpha_N = 0$. On the other hand, $x, e_1, e_2, \ldots, e_N$, where $x \in \mathbb{R}^N$, is always a linearly dependent set since if $x = (\chi_1, \ldots, \chi_N)$, we can write $\chi_1 e_1 + \cdots + \chi_N e_N + (-1)x = 0$.

Suppose we have $\alpha_1 x_1 + \alpha_2 x_2 + \cdots + \alpha_N x_N = 0$ where at least one of the coefficients α_i is nonzero. We may, without loss of generality, suppose $\alpha_1 \neq 0$. Then $x_1 = \beta_2 x_2 + \cdots + \beta_N x_N$ for some scalars $\beta_2, \ldots, \beta_N$. We see from this that to say $x_1, \ldots, x_N$ is a linearly dependent set of vectors amounts to saying that there must be some x_i which can be written as a linear combination of the others.

Note that a set of vectors containing 0 is automatically a linearly dependent set. Thus any set of linearly independent vectors does not contain 0.

If V is a vector space and W is a subset of V which happens to also be a vector space using the restrictions of vector addition and scalar multiplication from V to W, then we say W is a *vector subspace* or *linear subspace* of V. For instance the set of ordered 3-tuples $(\alpha, \beta, 0)$ constitutes a linear subspace of $\mathbb{R}^3$ under the usual definitions of vector addition and scalar multiplication.

The proof of the following result is left as an exercise.

Theorem 1.7.1 *Let W be a subset of the vector space V. If $x + y \in W$ whenever $x, y \in W$ and if $\alpha x \in W$ whenever $x \in W$ and $\alpha \in \mathbb{R}$, then W is a linear subspace of V.*

Example 1.7.1 Any line $\mathcal{L}$ through the origin in $\mathbb{R}^N$ may be regarded as a linear subspace of $\mathbb{R}^N$. This follows from the fact that there must be a nonzero vector a in $\mathbb{R}^N$ such that

$$\mathcal{L} = \{x \in \mathbb{R}^N : \ x = \lambda a \quad \text{for some } \lambda \in \mathbb{R}\}.$$

Since $\lambda_1 a + \lambda_2 a = (\lambda_1 + \lambda_2)a$ and $\gamma(\lambda a) = (\gamma\lambda)a$, we can then invoke the last theorem.

Example 1.7.2 Similarly we may regard the plane $x + y + z = 0$ in $\mathbb{R}^3$ as a linear subspace of $\mathbb{R}^3$. Suppose (x_1, y_1, z_1) and (x_2, y_2, z_2) both lie on this plane. From

$$x_1 + y_1 + z_1 = 0 \quad \text{and} \quad x_2 + y_2 + z_2 = 0$$

we deduce

$$(x_1 + x_2) + (y_1 + y_2) + (z_1 + z_2) = 0,$$

so that we know $(x_1 + x_2, \ y_1 + y_2, \ z_1 + z_2)$ also lies on the plane. Next, if (x, y, z) lies on the plane and α is a real number, then from $x + y + z = 0$ we deduce that $(\alpha x) + (\alpha y) + (\alpha z) = 0$, and hence $(\alpha x, \ \alpha y, \ \alpha z)$ also lies on the plane.

Theorem 1.7.2 *Let V be a vector space and $x_1, x_2, \ldots, x_K \in V$. Then the set of all linear combinations of $x_1, x_2, \ldots, x_K$ is a linear subspace of V.*

Proof. This is left as an exercise. □

Note: The particular subspace described in this theorem is called the *span* of $x_1, x_2, \ldots, x_K$ and is sometimes denoted by a special symbol such as $\text{span}\{x_1, \ldots, x_K\}$. If A is subset of V such that $\text{span}\, A = V$, then we call A a *spanning set* of V. We shall be concerned here only with finite spanning sets, but the concept can be profitably generalized to infinite sets.

Definition 1.7.1 If V is a vector space, then a maximal collection of linearly independent vectors of V is called a *basis* for V. This means that if A is a basis for V and A is a proper subset of B, which is in turn also a subset of V, then B must contain linearly dependent vectors.

Example 1.7.3 Given previous remarks, we see that $\{e_1, e_2, \ldots, e_N\}$ is a basis for $\mathbb{R}^N$. If S is any collection of vectors in $\mathbb{R}^N$ that contains $e_1, \ldots, e_N$ and any other vector x, then S must be a linearly dependent set.

Example 1.7.4 Let $\mathcal{L}$ be the line through the origin which consists of all points of the form λa where a is a fixed nonzero vector and λ is an arbitrary real number. Then the single vector a is a basis for the vector space $\mathcal{L}$.

Example 1.7.5 Let $\mathcal{P}$ be the plane $x + y + z = 0$ in $\mathbb{R}^3$. We know this is a linear subspace of $\mathbb{R}^3$. We claim that the two vectors $b_1 = (1, -1, 0)$ and $b_2 = (1, 0, -1)$ constitute a basis for $\mathcal{P}$. To see this, note that b_1 and b_2 are linearly independent and are in $\mathcal{P}$. Also, $\{b_1, b_2\}$ must be a maximal set of linearly independent vectors of $\mathcal{P}$, for if $c = (x, y, z)$ is any vector in $\mathcal{P}$, we must have $x + y + z = 0$, and thus

$$c = (-y - z, y, z) = (-y, y, 0) + (-z, 0, z) = -yb_1 - zb_2.$$

The next theorem connects the notions of span and basis; its proof is left as an exercise.

Theorem 1.7.3 *If $x_1, x_2, \dots, x_K$ is a basis for V, then $V = \text{span}\{x_1, x_2, \dots, x_K\}$. Furthermore, for every $x \in V$, the scalars $\alpha_1, \alpha_2, \dots, \alpha_K$ such that $x = \alpha_1 x_1 + \alpha_2 x_2 + \cdots + \alpha_K x_K$ are uniquely determined.*

We say A is a *minimal spanning set* for the vector space V provided it is a spanning set for V but has no proper subset that is also a spanning set.

Theorem 1.7.4 *A finite subset A of V is a basis for V if and only if it is a minimal spanning set for V.*

Proof. If A is a basis for V and $A = \{x_1, \dots, x_K\}$, then trivially every element of A can be written as a linear combination of elements of A. Suppose x is an element of V that is not in A. By the maximality of A, we must be able to find scalars $\alpha, \alpha_1, \dots, \alpha_K$, not all zero, such that $\alpha x + \alpha_1 x_1 + \cdots + \alpha_K x_K = 0$. It cannot be that $\alpha = 0$, since this would contradict the linear independence of $x_1, \dots, x_K$. Thus we can write $x = \beta_1 x_1 + \cdots + \beta_K x_K$. Since this is true for any $x \in V$, we have $V = \text{span}\, A$. Now suppose B is a nonempty, proper subset of A and $x \in A$ but $x \notin B$. We may suppose that $x = x_1$. If B is a spanning set, then it must be possible to write x_1 as a linear combination of the elements of B. But this implies that it is possible to write $x_1 + \alpha_2 x_2 + \cdots + \alpha_K x_K = 0$, which contradicts the linear independence of the elements of A. Thus A is minimal.

Suppose A is a minimal spanning set and $A = \{x_1, \dots, x_K\}$. If A is a set of linearly dependent vectors, then we can write $\alpha_1 x_1 + \cdots + \alpha_K x_K = 0$ where at least one of the α_i coefficients is nonzero. Without loss of generality, we may suppose $\alpha_1 \neq 0$. Then we can write x_1 as a linear combination of $x_2, \dots, x_K$. Since every vector $x \in V$ can be written as a linear combination of $x_1, x_2, \dots, x_K$, it follows that x can be written as a linear combination of $x_2, \dots, x_K$. This leads us to conclude A is not a minimal spanning set, which we know to be false. Hence $x_1, \dots, x_K$ must be linearly independent. To see that A is maximal with respect to being a set of linearly independent vectors, let $B \subseteq V$ be a set that properly contains A as a subset. For any $x \in B$ where $x \notin A$, we must be able to find scalars $\alpha_1, \dots, \alpha_K$ such that $x = \alpha_1 x_1 + \cdots + \alpha_K x_K$. But this means $x, x_1, \dots, x_K$ are linearly dependent. □

Theorem 1.7.5 *If $x_1, x_2, \dots, x_K$ is a basis for the vector space V and $y_1, y_2, \dots, y_L$ are linearly independent vectors in V, then $L \leq K$.*

Proof. Either $L < K$ or $L \geq K$. We need only suppose that $L \geq K$, show that this implies $L = K$, and we are done.

We illustrate the steps in an inductive procedure which constructs a sequence of spanning sets for V.

The first set in our sequence is $\{x_1, \ldots, x_K\}$, trivially a spanning set. To get to the next step in our sequence, consider the set $\{y_1, x_1, x_2, \ldots, x_K\}$. This is clearly a spanning set. Since $\{x_1, \ldots, x_K\}$ is a basis, we must be able to find scalars, not all zero, such that $\beta_1 y_1 + \alpha_1 x_1 + \alpha_2 x_2 + \cdots + \alpha_K x_K = 0$. We cannot have $\alpha_1 = \alpha_2 = \cdots = \alpha_K = 0$ since this would force $y_1 = 0$, so we may suppose $\alpha_1 \neq 0$. This means we may write x_1 as a linear combination of $y_1, x_2, \ldots, x_K$. Since any vector $x \in V$ can be written as a linear combination of $x_1, x_2, \ldots, x_K$, we see that x can also be written as a linear combination of $y_1, x_2, \ldots, x_K$. We take the second set in our sequence to be $\{y_1, x_2, \ldots, x_K\}$.

Now consider $\{y_1, y_2, x_2, \ldots, x_K\}$. This is a spanning set but not a minimal one. So the elements of this set must be linearly dependent and we can find scalars, not all zero, such that $\beta_1 y_1 + \beta_2 y_2 + \alpha_2 x_2 + \cdots + \alpha_K x_K = 0$. We cannot have $\alpha_2 = \alpha_3 = \cdots = \alpha_K = 0$ since this would contradict the linear independence of y_1, y_2. We may therefore suppose $\alpha_2 \neq 0$. So we may write x_2 as a linear combination of $y_1, y_2, x_3, \ldots, x_K$. Since $\{y_1, x_2, x_3, \ldots, x_K\}$ is a spanning set, this implies $\{y_1, y_2, x_3, \ldots, x_K\}$ is also a spanning set for V. We take the third set in our sequence to be $\{y_1, y_2, x_3, \ldots, x_K\}$.

Continuing in this fashion we thus construct the sequence of spanning sets

$$\begin{gathered}
\{x_1, x_2, x_3, \ldots, x_K\} \\
\{y_1, x_2, x_3, \ldots, x_K\} \\
\{y_1, y_2, x_3, \ldots, x_K\} \\
\ldots \\
\{y_1, y_2, y_3, \ldots, y_K\}.
\end{gathered}$$

(Note that as we do this, we may have to switch the indices on the x_i vectors, but this does not affect the validity of the argument.) Notice that if $L > K$, then it must be possible to write y_L as a linear combination of $y_1, \ldots, y_K$, a contradiction of the linear independence of $y_1, \ldots, y_L$. So we are forced to conclude that $L = K$. □

Corollary 1.7.1 *If $x_1, x_2, \ldots, x_K$ is a basis for the vector space V, then every basis for V contains exactly K elements.*

Definition 1.7.2 We say that a vector space V has *dimension* K provided it has a basis with K elements. We may, if we wish, indicate this fact by writing $\dim V = K$.

Corollary 1.7.2 $\dim \mathbb{R}^N = N$.

Proof. We know that $e_1, e_2, \ldots, e_N$ is a basis for $\mathbb{R}^N$. □

Corollary 1.7.3 *If $x_1, \ldots, x_K$ are linearly independent vectors of the vector space V, then* $\operatorname{span}\{x_1, \ldots, x_K\}$ *is a K-dimensional linear subspace of V.*

Corollary 1.7.4 *If V is a linear subspace of a finite-dimensional vector space W, then* $\dim V \leq \dim W$.

Proof. Let $K = \dim W$. If $V = \{0\}$, the conclusion is trivial. Assuming V is not the trivial subspace $\{0\}$, we must be able to find a nonzero $x_1 \in V$. We manufacture a sequence of linearly independent vectors $x_1, x_2, \ldots, x_L$ in V in the following way: If $\operatorname{span}\{x_1\} = V$, we are done. If not, we must be able to choose $x_2 \in V$ such that x_1 and x_2 are linearly independent. If $\operatorname{span}\{x_1, x_2\} = V$, then we stop. If not, we can find $x_3 \in V$ such that x_1, x_2, x_3 are linearly independent, etc. The process must eventually terminate with a set $x_1, \ldots, x_L$ which is a basis for V, since if it did not, we would find it is possible to have $L > K$ in contradiction of Theorem 1.7.5. Since $L \leq K$, we are done. □

Corollary 1.7.5 *Suppose V is a linear subspace of a finite-dimensional vector space W. If $x_1, \ldots, x_K$ is a basis for V, then it is possible to extend this to a basis $x_1, \ldots, x_K, x_{K+1}, \ldots, x_{K+L}$ for W.*

Proof. We suppose $\dim W = K + L$. Then we start with the sequence of vectors $x_1, \ldots, x_K$ and use the procedure employed in the proof of the last corollary to extend this to a basis for W. We know by Theorem 1.7.5 that the procedure will terminate with a sequence of $K + L$ vectors. □

From now on the only vector spaces with which we shall deal are $\mathbb{R}^N$ and its linear subspaces. A very important property of such spaces is that the dot product $x \cdot y$ is defined on them.

A particularly nice basis for a vector space is $x_1, x_2, \ldots, x_K$ in which the vectors are mutually orthogonal, that is, $x_i \cdot x_j = 0$ whenever $i \neq j$. It is easy to see that any collection of nonzero, mutually orthogonal vectors $x_1, x_2, \ldots, x_K$ must be a linearly independent collection. Suppose

$$\alpha_1 x_1 + \alpha_2 x_2 + \cdots + \alpha_K x_K = 0.$$

Then

$$\begin{aligned} x_i \cdot (\alpha_1 x_1 + \alpha_2 x_2 + \cdots + \alpha_K x_K) &= x_i \cdot 0, \\ \alpha_i |x_i|^2 &= 0, \\ \alpha_i &= 0, \end{aligned}$$

the last line following from the fact that x_i is nonzero. An *orthonormal basis* is one in which $x_1, x_2, \ldots, x_K$ are mutually orthogonal and each x_i is a *unit* vector, that is, $|x_i| = 1$. Any merely orthogonal basis may be converted to an orthonormal one by multiplying each x_i by the scalar $1/|x_i|$.

More generally, there is a sense in which any set of linearly independent vectors may be replaced by an orthonormal set spanning the same vector space. One may apply the *Gram–Schmidt orthogonalization process*, which we now describe.

Let $x_1, x_2, \ldots, x_K$ be linearly independent vectors in $\mathbb{R}^N$ and let $V = \text{span}\{x_1, \ldots, x_K\}$. We inductively construct an orthonormal basis $y_1, y_2, \ldots, y_K$ for V having the property that for $p = 1, 2, \ldots, K$ we have

$$\text{span}\{y_1, \ldots, y_p\} = \text{span}\{x_1, \ldots, x_p\}.$$

If we set $y_1 = x_1/|x_1|$, this condition is trivially fulfilled for $p = 1$. Suppose we have constructed $y_1, \ldots, y_p$ (where $p < K$) to specifications and now we wish to construct y_{p+1}. We set

$$u = x_{p+1} - \sum_{j=1}^{p} (y_j \cdot x_{p+1}) y_j.$$

If $u = 0$, then x_{p+1} can be written as a linear combination of $y_1, \ldots, y_p$. Since $\text{span}\{y_1, \ldots, y_p\} = \text{span}\{x_1, \ldots, x_p\}$, this means x_{p+1} can be written as a linear combination of $x_1, \ldots, x_p$. But this contradicts the linear independence of $x_1, \ldots, x_p$, x_{p+1}. We must have $u \neq 0$.

Using the orthonormality of $y_1, \ldots, y_p$, we have for $i = 1, \ldots, p$ that

$$\begin{aligned} y_i \cdot u &= (y_i \cdot x_{p+1}) - \sum_{j=1}^{p} (y_j \cdot x_{p+1})(y_i \cdot y_j) \\ &= (y_i \cdot x_{p+1}) - (y_i \cdot x_{p+1}) \\ &= 0. \end{aligned}$$

That is, u is orthogonal to $y_1, \ldots, y_p$.

Set $y_{p+1} = u/|u|$. Since $\text{span}\{y_1, \ldots, y_p\} = \text{span}\{x_1, \ldots, x_p\}$, we know each of $y_1, \ldots, y_p$ can be written as a linear combination of $x_1, \ldots, x_p$. It follows that u, and hence y_{p+1}, can be written as a linear combination of $x_1, \ldots, x_{p+1}$. Therefore $\text{span}\{y_1, \ldots, y_{p+1}\} \subseteq \text{span}\{x_1, \ldots, x_{p+1}\}$. One can clearly give an argument to show containment in the other direction. Thus $\text{span}\{y_1, \ldots, y_{p+1}\} = \text{span}\{x_1, \ldots, x_{p+1}\}$.

The desired orthonormal basis for V is $y_1, \ldots, y_K$.

Example 1.7.6 Suppose in $\mathbb{R}^3$ we apply the Gram–Schmidt process to the basis

$$\begin{aligned} x_1 &= e_1, \\ x_2 &= e_1 + e_2, \\ x_3 &= e_1 + e_2 + e_3. \end{aligned}$$

To begin, we set $y_1 = e_1$. Next we set

$$u = x_2 - (y_1 \cdot x_2) y_1.$$

This gives us $u = e_2$, and since this is already a unit vector, we set $y_2 = e_2$.

For the last step we set

$$u = x_3 - (y_1 \cdot x_3) y_1 - (y_2 \cdot x_3) y_2.$$

This reduces to $u = e_3$, and since this is a unit vector, we set $y_3 = e_3$. Thus in this case the Gram–Schmidt orthogonalization process produces the usual orthonormal basis, $e_1,\ e_2,\ e_3$.

Example 1.7.7 The Gram–Schmidt process is sensitive to the order in which one takes the original basis. Suppose we consider the same basis as the last example but this time in the order

$$\begin{aligned} x_1 &= e_1 + e_2 + e_3, \\ x_2 &= e_1 + e_2, \\ x_3 &= e_1. \end{aligned}$$

Normalizing x_1 yields $y_1 = \frac{1}{\sqrt{3}}\,(e_1 + e_2 + e_3)$.

We set

$$u = x_2 - (y_1 \cdot x_2)y_1.$$

This yields

$$u = \frac{1}{3}e_1 + \frac{1}{3}e_2 - \frac{2}{3}e_3,$$

which, upon normalization, gives us $y_2 = \frac{1}{\sqrt{6}}(e_1 + e_2 - 2e_3)$.

For the last step we set

$$u = x_3 - (y_1 \cdot x_3)y_1 - (y_2 \cdot x_3)y_2.$$

This reduces to

$$u = \frac{1}{2}e_1 - \frac{1}{2}e_2,$$

which we normalize to obtain $y_3 = \frac{1}{\sqrt{2}}(e_1 - e_2)$. It is easily seen that y_1, y_2, y_3 are orthonormal vectors.

One useful consequence of the Gram-Schmidt construction is the following:

Theorem 1.7.6 *If W is a linear subspace of $\mathbb{R}^N$ and V is a linear subspace of W, then every orthonormal basis $x_1, \dots, x_K$ of V can be extended to an orthonormal basis $x_1, \dots, x_K, x_{K+1}, \dots, x_{K+L}$ for W.*

Proof. One can extend the sequence $x_1, \dots, x_K$ to a basis $x_1, \dots, x_K, y_1, \dots, y_L$ for W. Then holding $x_1, \dots, x_K$ fixed, one starts at y_1 and applies the Gram–Schmidt process. □

Exercises

1. Show that $V = \{(\alpha, \alpha, \alpha) : \alpha \in \mathbb{R}\}$ is a linear subspace of $\mathbb{R}^3$ and find a basis for it.

2. Show that a line in $\mathbb{R}^N$ is a linear subspace of $\mathbb{R}^N$ if and only if the line passes through the origin.

3. Show that a plane in $\mathbb{R}^3$ is a linear subspace of $\mathbb{R}^3$ if and only if the plane passes through the origin.

4. If $x_1, x_2, \ldots, x_K$ belong to a finite-dimensional vector space V and W is the span of $x_1, x_2, \ldots, x_K$, show that $\dim W \leq K$.

5. Show that the set of vectors having $x_1, \ldots, x_K$ as its members is a basis for the vector space V if and only if

   (1) $x_1, \ldots, x_K$ are linearly independent and

   (2) $\operatorname{span}\{x_1, \ldots, x_K\} = V$.

6. Let $x_1, \ldots, x_K$ be linearly independent vectors in $\mathbb{R}^N$. Show that if y is a nonzero vector in $\mathbb{R}^N$ which is orthogonal to each one of $x_1, \ldots, x_K$, then $y, x_1, \ldots, x_K$ is also a linearly independent set of vectors.

7. Apply the Gram–Schmidt process to $b_1 = (1, -1, 0)$ and $b_2 = (1, 0, -1)$ to obtain an orthonormal basis for the plane $x + y + z = 0$.

8. (a) Let x_0 be a nonzero vector of $\mathbb{R}^N$ and show that $V = \{x \in \mathbb{R}^N : x_0 \cdot x = 0\}$ is a linear subspace of $\mathbb{R}^N$.

   (b) Show that $\dim V = N - 1$.

9. Apply the Gram–Schmidt orthogonalization process to $e_1 + e_2, e_2$. Plot the resulting vectors on the plane.

10. Let $a_1, \ldots, a_N$ be a basis for the vector space V and let $b_1, \ldots, b_N$ be a basis for the vector space W. Define $f\colon V \to W$ by

$$f\left(\sum_{i=1}^{N} \lambda_i a_i\right) = \sum_{i=1}^{N} \lambda_i b_i.$$

    Show that f is well defined, is a one-to-one onto linear transformation, and has an inverse that is also a linear transformation. (Such a linear transformation is called an isomorphism of vector spaces.)

11. Prove Theorem 1.7.1.

12. Prove Theorem 1.7.2.

13. Prove Theorem 1.7.3.

1.8 Orthogonal Transformations

An important property of area and volume in dimensions 2 and 3 is their invariance under reflections and rotations. We want to generalize this idea, so in our study of higher dimensional analogs of volume, it will be important to consider what kind of linear transformations correspond to reflections and rotations.

The kind of transformations we want would preserve lengths and the angles between vectors. A simple way to ensure a linear transformation f did this would be to require it to preserve the values of dot products.

Definition 1.8.1 If $f : \mathbb{R}^N \to \mathbb{R}^N$ is a linear transformation such that $f(x) \cdot f(y) = x \cdot y$ for all $x,\ y$, we call it an *orthogonal transformation*.

Notice that if f is orthogonal, we must have

$$|f(x)| = \sqrt{f(x) \cdot f(x)} = \sqrt{x \cdot x} = |x|$$

so that lengths are preserved. Furthermore, if θ is the angle between x and y and ψ is that between $f(x)$ and $f(y)$, then

$$|x||y|\cos(\psi) = |f(x)||f(y)|\cos(\psi) = f(x) \cdot f(y) = x \cdot y = |x||y|\cos(\theta),$$

so that angles are preserved.

Example 1.8.1 Suppose $f\colon \mathbb{R}^2 \to \mathbb{R}^2$ is the rotation of the xy-plane by an angle of α. We must have

$$\begin{aligned} f(e_1) &= (\ \ \cos(\alpha), \sin(\alpha)), \\ f(e_2) &= (-\sin(\alpha), \cos(\alpha)). \end{aligned}$$

See Figure 1.8.1. Then the matrix of f is

$$\begin{pmatrix} \cos(\alpha) & -\sin(\alpha) \\ \sin(\alpha) & \cos(\alpha) \end{pmatrix}.$$

We leave to the reader the task of showing that f is indeed orthogonal, that is, that it satisfies $f(x) \cdot f(y) = x \cdot y$.

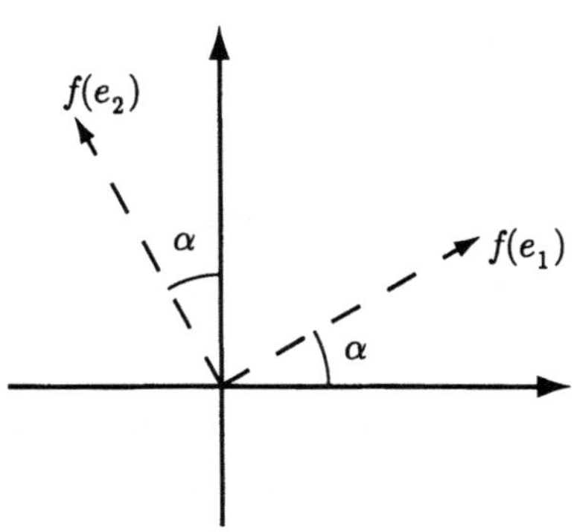

FIGURE 1.8.1.

Example 1.8.2 Rotations in $\mathbb{R}^3$ are orthogonal transformations. It should be intuitively reasonable that every rotation will leave some line through the origin fixed. Consequently every rotation can be specified by choosing an orthonormal basis v_1, v_2, v_3 for $\mathbb{R}^3$ (usually not the standard one) and defining $f: \mathbb{R}^3 \to \mathbb{R}^3$ by

$$\begin{aligned} f(v_1) &= \cos(\alpha)v_1 + \sin(\alpha)v_2, \\ f(v_2) &= -\sin(\alpha)v_1 + \cos(\alpha)v_2, \end{aligned}$$

and

$$f(v_3) = v_3$$

for some given α.

Example 1.8.3 Reflections through a plane in $\mathbb{R}^3$ are also orthogonal transformations. Here is the matrix of the reflection of $\mathbb{R}^3$ through the yz-plane:

$$R = \begin{pmatrix} -1 & 0 & 0 \\ 0 & 1 & 0 \\ 0 & 0 & 1 \end{pmatrix}.$$

We will need some properties of orthogonal transformations when we return to the study of K-dimensional volume in the next section.

Theorem 1.8.1 *A linear transformation $f: \mathbb{R}^N \to \mathbb{R}^N$ is orthogonal if and only if $f(e_1), f(e_2), \ldots, f(e_N)$ are orthonormal vectors.*

Proof. If f is orthogonal, then since f preserves lengths and angles, we see that $f(e_1), \ldots, f(e_N)$ must be unit vectors and must be mutually orthogonal.

Suppose $f(e_1), \ldots, f(e_N)$ are orthonormal vectors. Let $a = \alpha_1 e_1 + \cdots + \alpha_N e_N$ and $b = \beta_1 e_1 + \cdots + \beta_N e_N$. Then

$$f(a) \cdot f(b) = \sum_{i,j=1}^{N} \alpha_i \beta_j f(e_i) \cdot f(e_j) = \sum_{i=1}^{N} \alpha_i \beta_i = a \cdot b. \qquad \square$$

Note: Remember in connection with this result, that if a_i^T is the ith column vector of $[f]$, then $a_i = f(e_i)$. So the orthogonality of a linear transformation may be revealed by simply inspecting the column vectors of its matrix for orthonormality.

Corollary 1.8.1 *Orthogonal transformations are onto.*

For this next result it is useful to notice that to every linear transformation $f: \mathbb{R}^M \to \mathbb{R}^N$ we may associate a *dual* linear transformation $f^\diamond: \mathbb{R}^N \to \mathbb{R}^M$, the one whose matrix is given by $[f^\diamond] = [f]^T$. These two linear transformations are related to one another by the equation $f(x) \cdot y = x \cdot f^\diamond(y)$ for all $x \in \mathbb{R}^M$ and all $y \in \mathbb{R}^N$.

(**Note**: The usual notation for the dual of the linear transformation f is f^*. We have chosen to use instead $f^\diamond$ because we make use of the notation f^* later, for more general functions, for a different concept, the pullback, in such a way that there is danger of confusing the two concepts.)

Example 1.8.4 Suppose $x = (x_1, x_2)$, $y = (y_1, y_2)$, and f is the linear transformation with matrix

$$\begin{pmatrix} a_{11} & a_{12} \\ a_{21} & a_{22} \end{pmatrix}.$$

Then note that the equation $f(x) \cdot y = x \cdot f^{\diamond}(y)$ amounts to simply

$$(y_1, y_2) \begin{pmatrix} a_{11} & a_{12} \\ a_{21} & a_{22} \end{pmatrix} \begin{pmatrix} x_1 \\ x_2 \end{pmatrix} = (x_1, x_2) \begin{pmatrix} a_{11} & a_{12} \\ a_{21} & a_{22} \end{pmatrix}^T \begin{pmatrix} y_1 \\ y_2 \end{pmatrix}.$$

Theorem 1.8.2 *A linear transformation $f : \mathbb{R}^N \to \mathbb{R}^N$ is orthogonal if and only if f has an inverse and $f^{-1} = f^{\diamond}$.*

Proof. Suppose f is orthogonal. Then for every $x \in \mathbb{R}^N$ and $i = 1, 2, \ldots, N$, we have

$$(f^{\diamond} \circ f)(x) \cdot e_i = f(x) \cdot f(e_i) = x \cdot e_i.$$

But this can be true only if $f^{\diamond} \circ f =$ the identity transformation on $\mathbb{R}^N$. Next note that for any $x \in \mathbb{R}^N$ we can find $y \in \mathbb{R}^N$ such that $x = f(y)$ and hence

$$(f \circ f^{\diamond})(x) = (f \circ f^{\diamond} \circ f)(y) = f(y) = x.$$

So $f \circ f^{\diamond} =$ the identity transformation also, and thus $f^{\diamond} = f^{-1}$.

Suppose $f^{-1} = f^{\diamond}$. Then for every $x, y \in \mathbb{R}^N$ we have

$$f(x) \cdot f(y) = x \cdot (f^{\diamond} \circ f)(y) = x \cdot (f^{-1} \circ f)(y) = x \cdot y.$$

Thus a matrix A represents an orthogonal transformation if and only if $A^T = A^{-1}$.

□

Exercises

1. Show directly that the linear transformation f with matrix

$$\begin{pmatrix} \cos(\alpha) & -\sin(\alpha) \\ \sin(\alpha) & \cos(\alpha) \end{pmatrix},$$

is orthogonal. That is, show that

$$f(x) \cdot f(y) = x \cdot y.$$

2. Show directly that

$$R = \begin{pmatrix} -1 & 0 & 0 \\ 0 & 1 & 1 \\ 0 & 0 & 1 \end{pmatrix}$$

is the matrix of an orthogonal transformation f. That is, show that $f(x) \cdot f(y) = x \cdot y$.

3. (a) Show that the composition of two orthogonal transformations is orthogonal.

   (b) What does this tell us about the linear transformation with matrix

$$\begin{pmatrix} \cos(\alpha) & -\sin(\alpha) & 0 \\ \cos(\beta)\sin(\alpha) & \cos(\beta)\cos(\alpha) & -\sin(\beta) \\ \sin(\beta)\sin(\alpha) & \sin(\beta)\cos(\alpha) & \cos(\beta) \end{pmatrix}?$$

4. If an orthogonal transformation has the matrix

$$\begin{pmatrix} \frac{\sqrt{2}}{2} & -\frac{\sqrt{2}}{2} & \alpha \\ \frac{\sqrt{5}}{4} & \frac{\sqrt{5}}{4} & \beta \\ \frac{\sqrt{3}}{4} & \frac{\sqrt{3}}{4} & \gamma \end{pmatrix},$$

then solve for α, β, γ.

5. Show that for every linear transformation $f : \mathbb{R}^M \to \mathbb{R}^N$ we must have $f(x) \cdot y = x \cdot f^\diamond(y)$ whenever $x \in \mathbb{R}^M$ and $y \in \mathbb{R}^N$.

6. Prove Corollary 1.8.1.

7. Prove that every orthogonal transformation has determinant ± 1.

8. The orthogonal transformations which correspond to rotations are those whose determinant is 1 while the ones which correspond to reflections are those whose determinant is -1. If $f : \mathbb{R}^N \to \mathbb{R}^N$ is the identity transformation, show that $-f$ is a rotation if N is even and a reflection if N is odd.

9. Show that if $f : \mathbb{R}^2 \to \mathbb{R}^2$ is an orthogonal transformation with determinant 1, then there is a unique $\theta \in [0, 2\pi)$ with the property that the matrix of f is

$$\begin{pmatrix} \cos(\theta) & -\sin(\theta) \\ \sin(\theta) & \cos(\theta) \end{pmatrix}.$$

10. Let V and W be K-dimensional linear subspaces of $\mathbb{R}^M$ and $\mathbb{R}^N$, respectively. Let $a_1, \ldots, a_K$ and $b_1, \ldots, b_K$ be orthonormal bases of V and W respectively. Show that if $f : V \to W$ is the linear transformation which satisfies $f(a_i) = b_i$, then it must be orthogonal.

11. Show that for linear transformations f and g we have $(g \circ f)^\diamond = f^\diamond \circ g^\diamond$.

1.9 K-dimensional Volume of Parallelepipeds in $\mathbb{R}^N$

Recall that a K-dimensional parallelepiped in $\mathbb{R}^N$ is determined by giving K vectors $a_1, \ldots, a_K \in \mathbb{R}^N$. (We usually require these vectors to be independent, but here it is convenient to permit them to be dependent.) It may be thought of as the set of points $\lambda_1 a_1 + \cdots + \lambda_K a_K$ where $0 \le \lambda_i \le 1$. We can represent this parallelepiped by an $N \times K$ matrix A whose column vectors are $a_1^T, \ldots, a_K^T$ or by the linear transformation $f\colon \mathbb{R}^K \to \mathbb{R}^N$ whose matrix is A (that is, $f(e_i) = a_i$ for $i = 1, 2, \ldots, K$).

We want to attach to this parallelepiped a number $\mathcal{D}(A)$ or $\mathcal{D}(f)$ which may be thought of as its *K-dimensional volume*. We will stipulate first that only the case $K \le N$ is of interest to us, because if $K > N$, then we would expect the parallelepiped to be degenerate (like a 3-dimensional box squashed into a plane) and to have $\mathcal{D}(A) = 0$.

Let us list some desirable properties of $\mathcal{D}(f)$.

(KV1) If $f\colon \mathbb{R}^K \to \mathbb{R}^N$ is a linear transformation and $K \le N$ and $g\colon \mathbb{R}^N \to \mathbb{R}^N$ is an orthogonal transformation, then $\mathcal{D}(g \circ f) = \mathcal{D}(f)$.

Suppose $a_i = f(e_i)$, so f corresponds to the K-dimensional parallelepiped determined by $a_1, \ldots, a_K$. We see that $g \circ f$ corresponds to the K-dimensional parallelepiped determined by $g(a_1), \ldots, g(a_K)$. So (KV1) amounts to saying that the volume of a K-dimensional parallelepiped ought to remain unchanged when it is subjected to an orthogonal transformation.

(KV2) Suppose $K \le N$ and we are given K vectors $a_i = (\alpha_{i1}, \alpha_{i2}, \ldots, \alpha_{iK}, 0, \ldots, 0)$ in $\mathbb{R}^N$. Then for the volume of the parallelepiped determined by $a_1, \ldots, a_K$ we have

$$\mathcal{D}\left(\begin{pmatrix} \alpha_{11} & \alpha_{21} & \ldots & \alpha_{K1} \\ \ldots & \ldots & & \ldots \\ \alpha_{1K} & \alpha_{2K} & \ldots & \alpha_{KK} \\ 0 & 0 & \ldots & 0 \\ \ldots & \ldots & & \ldots \\ 0 & 0 & \ldots & 0 \end{pmatrix}\right) = \left|\det\begin{pmatrix} \alpha_{11} & \ldots & \alpha_{K1} \\ \ldots & & \\ \alpha_{1K} & \ldots & \alpha_{KK} \end{pmatrix}\right|.$$

This says that if $a_1, \ldots, a_K$ lie in a linear subspace of $\mathbb{R}^N$ which can be identified in a natural way with $\mathbb{R}^K$, then the volume of the K-dimensional parallelepiped should amount to computing the determinant of the K vectors in $\mathbb{R}^K$. This makes sense in view of our earlier discussion of how the determinant can be interpreted in terms of volume.

We will now show that (KV1) and (KV2) completely determine the value of $\mathcal{D}(f)$. This means that if we believe in the existence of a function $f \mapsto \mathcal{D}(f)$, then there can be at most one such function. Then we shall exhibit a function satisfying (KV1) and (KV2), and it must clearly be the function having these properties.

Theorem 1.9.1 *If $K \le N$ and $a_1, \ldots, a_K \in \mathbb{R}^N$, then there is an orthogonal transformation $g\colon \mathbb{R}^N \to \mathbb{R}^N$ with the property that each $g(a_i)$ has the form $(\beta_1, \ldots, \beta_K, 0, \ldots, 0)$.*

Proof. Let V be the span of $a_1, \ldots, a_K$. Then $\dim V$ is some L which is less than or equal to K. We can find an orthonormal basis $b_1, \ldots, b_L$ for V and extend it to an orthonormal basis $b_1, \ldots, b_N$ for $\mathbb{R}^N$. Define $g\colon \mathbb{R}^N \to \mathbb{R}^N$ to be the unique linear transformation that satisfies $g(b_i) = e_i$ for all i. Since it takes an orthonormal basis to an orthonormal basis, g must be orthogonal, and it must carry everything in V (which is the same thing as $\operatorname{span}\{b_1, \ldots, b_L\}$) to something of the form $(\beta_1, \ldots, \beta_L, 0, \ldots, 0)$.

This means that given $a_1, \ldots, a_K$ in $\mathbb{R}^N$ (where $K \leq N$), we can replace them by K vectors of the form $(\beta_1, \ldots, \beta_K, 0, \ldots, 0)$ and we can be certain, by virtue of (KV1), that the volume of the new parallelepiped is the same as that of the old one. We then invoke (KV2) to compute the volume of the new parallelepiped. Thus our two properties are sufficient to determine K-dimensional volume. □

Now we need to exhibit the promised function. Suppose we are given $a_1, \ldots, a_K \in \mathbb{R}^N$ where $K \leq N$ and each $a_i = (\alpha_{i1}, \ldots, \alpha_{iN})$. Consider the matrix

$$A = \begin{pmatrix} \alpha_{11} & \ldots & \alpha_{K1} \\ \ldots & & \\ \alpha_{1N} & \ldots & \alpha_{KN} \end{pmatrix}$$

or, equivalently, the linear transformation $f : \mathbb{R}^K \to \mathbb{R}^N$ which carries each e_i to a_i. From the Binet–Cauchy theorem we deduce that

$$\det(A^T A) = \sum_{i_1 < i_2 < \cdots < i_K} \left(\det \begin{pmatrix} \alpha_{1i_1} & \ldots & \alpha_{1i_K} \\ \ldots & & \\ \alpha_{Ki_1} & \ldots & \alpha_{Ki_K} \end{pmatrix} \right)^2 .$$

This is nonnegative and permits us to make the following definition: We set

$$\mathcal{D}(A) = \sqrt{\det(A^T A)},$$

or, equivalently,

$$\mathcal{D}(f) = \sqrt{\det(f^\diamond \circ f)}.$$

To see that (KV1) holds, let $g : \mathbb{R}^N \to \mathbb{R}^N$ be an orthogonal transformation. Then

$$\begin{aligned} \mathcal{D}(g \circ f) &= \sqrt{\det((g \circ f)^\diamond \circ (g \circ f))} \\ &= \sqrt{\det(f^\diamond \circ g^\diamond \circ g \circ f)} \\ &= \sqrt{\det(f^\diamond \circ g^{-1} \circ g \circ f)} \\ &= \sqrt{\det(f^\diamond \circ f)} \\ &= \mathcal{D}(f). \end{aligned}$$

Verification of (KV2) is almost equally easy. Suppose for $i = 1, 2, \ldots, K$ that $a_i = (\alpha_{i1}, \ldots, \alpha_{iK}, 0, \ldots, 0)$. By appealing again to the Binet–Cauchy theorem we obtain

$$\det(A^T A) = \left(\det \begin{pmatrix} \alpha_{11} & \ldots & \alpha_{K1} \\ \ldots & & \\ \alpha_{1K} & \ldots & \alpha_{KK} \end{pmatrix} \right)^2 ,$$

and we are done.

Example 1.9.1 Consider the parallelogram in $\mathbb{R}^3$ formed by $a = (1, 1, 0)$ and $b = (1, -1, 3)$. Since the two vectors are orthogonal, the area of the parallelogram must be $|a|\ |b|$, that is, $\sqrt{22}$. It is easy to check that the use of $\mathcal{D}$ gives the same figure: Set

$$A = \begin{pmatrix} 1 & 1 \\ 1 & -1 \\ 0 & 3 \end{pmatrix}.$$

Then

$$\mathcal{D}(A) = \sqrt{\det(A^T A)} = \sqrt{\det \begin{pmatrix} 2 & 0 \\ 0 & 11 \end{pmatrix}} = \sqrt{22}.$$

We see, via the Binet–Cauchy theorem, that there is an interesting geometric fact contained in the definition of $\mathcal{D}(A)$.

Consider the 1-dimensional parallelepiped determined by a single vector $a = (\alpha_1, \ldots, \alpha_N) \in \mathbb{R}^N$. We construct the matrix $A = (a^T)$ and find that $\mathcal{D}(A) = \sqrt{\alpha_1^2 + \cdots + \alpha_N^2}$ which is simply the magnitude of the vector. Notice that we find $\mathcal{D}(A)$ by taking all orthogonal projections of a onto the coordinate axes, taking the sum of the squares of the lengths of these projections, and then taking the square root of this sum.

Now think of two vectors, $a = (\alpha_1, \alpha_2, \alpha_3)$ and $b = (\beta_1, \beta_2, \beta_3)$, in $\mathbb{R}^3$. These determine a 2-dimensional parallelepiped (a parallelogram). To find its area we form the matrix

$$A = \begin{pmatrix} \alpha_1 & \beta_1 \\ \alpha_2 & \beta_2 \\ \alpha_3 & \beta_3 \end{pmatrix}$$

and compute

$$\mathcal{D}(A) = \sqrt{\det \begin{pmatrix} \alpha_1 & \beta_1 \\ \alpha_2 & \beta_2 \end{pmatrix}^2 + \det \begin{pmatrix} \alpha_1 & \beta_1 \\ \alpha_3 & \beta_3 \end{pmatrix}^2 + \det \begin{pmatrix} \alpha_2 & \beta_2 \\ \alpha_3 & \beta_3 \end{pmatrix}^2}.$$

Note that in this computation we first take the orthogonal projections of a and b onto all the coordinate planes. For instance, the orthogonal projections of a and b onto the x_1x_2-plane are $(\alpha_1, \alpha_2, 0)$ and $(\beta_1, \beta_2, 0)$ respectively. Let us ignore the zeros and just think of them as (α_1, α_2) and (β_1, β_2). Then the area of the orthogonal projection onto the x_1x_2-plane must be

$$\left| \det \begin{pmatrix} \alpha_1 & \beta_1 \\ \alpha_2 & \beta_2 \end{pmatrix} \right|.$$

Similar remarks may be made about the other two determinants in the formula for $\mathcal{D}(A)$. So we see that the area of the parallelogram in $\mathbb{R}^3$ is the square root of the sum of the squares of the areas of the orthogonal projections of the parallelogram onto the coordinate planes.

This description can be carried over to K-dimensional parallelepipeds in $\mathbb{R}^N$ for arbitrary K and N, $K \leq N$, and it tells us there is a generalization of the theorem of Pythagoras which applies not only to lengths but also to areas, volumes, and K-dimensional volumes.

We have until now emphasized the interpretation of $\mathcal{D}(A)$ as K-dimensional volume. However one can also think of $\mathcal{D}(g)$ as a sort of distortion factor.

Theorem 1.9.2 *Let $f\colon \mathbb{R}^K \to \mathbb{R}^K$ and $g\colon \mathbb{R}^K \to \mathbb{R}^N$ be linear transformations where $K \leq N$. Then $\mathcal{D}(g \circ f) = \mathcal{D}(g)\,\mathcal{D}(f)$.*

Proof. For $i = 1, 2, \ldots, K$, let $(\alpha_{i1}, \alpha_{i2}, \ldots, \alpha_{iK}) = f(e_i)$. We can find an orthogonal transformation $h\colon \mathbb{R}^N \to \mathbb{R}^N$ with the property that $(h \circ g)(e_i) = (\beta_{i1}, \ldots, \beta_{iK}, 0, \ldots, 0)$ for each i. Note that each $(h \circ g \circ f)(e_i)$ must have the form $(\gamma_{i1}, \ldots, \gamma_{iK}, \ldots, 0, \ldots, 0)$ and we must have

$$\begin{pmatrix} \gamma_{11} & \gamma_{21} & \cdots & \gamma_{K1} \\ \cdots & \cdots & & \cdots \\ \gamma_{1K} & \gamma_{2K} & \cdots & \gamma_{KK} \\ 0 & 0 & \cdots & 0 \\ \cdots & \cdots & & \cdots \\ 0 & 0 & \cdots & 0 \end{pmatrix} = \begin{pmatrix} \beta_{11} & \beta_{21} & \cdots & \beta_{K1} \\ \cdots & \cdots & & \cdots \\ \beta_{1K} & \beta_{2K} & \cdots & \beta_{KK} \\ 0 & 0 & \cdots & 0 \\ \cdots & \cdots & & \cdots \\ 0 & 0 & \cdots & 0 \end{pmatrix} \begin{pmatrix} \alpha_{11} & \cdots & \alpha_{K1} \\ \cdots & & \\ \alpha_{1K} & \cdots & \alpha_{KK} \end{pmatrix}$$

and hence

$$\begin{pmatrix} \gamma_{11} & \cdots & \gamma_{K1} \\ \cdots & & \\ \gamma_{1K} & \cdots & \gamma_{KK} \end{pmatrix} = \begin{pmatrix} \beta_{11} & \cdots & \beta_{K1} \\ \cdots & & \\ \beta_{1K} & \cdots & \beta_{KK} \end{pmatrix} \begin{pmatrix} \alpha_{11} & \cdots & \alpha_{K1} \\ \cdots & & \\ \alpha_{1K} & \cdots & \alpha_{KK} \end{pmatrix}.$$

Therefore

$$\begin{aligned}
\mathcal{D}(g \circ f) = \mathcal{D}(h \circ g \circ f) &= \mathcal{D}\left(\begin{pmatrix} \gamma_{11} & \gamma_{21} & \cdots & \gamma_{K1} \\ \cdots & \cdots & & \cdots \\ \gamma_{1K} & \gamma_{2K} & \cdots & \gamma_{KK} \\ 0 & 0 & \cdots & 0 \\ \cdots & \cdots & & \cdots \\ 0 & 0 & \cdots & 0 \end{pmatrix}\right) \\
&= \left|\det\begin{pmatrix} \gamma_{11} & \cdots & \gamma_{K1} \\ \cdots & & \\ \gamma_{1K} & \cdots & \gamma_{KK} \end{pmatrix}\right| \\
&= \left|\det\begin{pmatrix} \beta_{11} & \cdots & \beta_{K1} \\ \cdots & & \\ \beta_{1K} & \cdots & \beta_{KK} \end{pmatrix}\right| \left|\det\begin{pmatrix} \alpha_{11} & \cdots & \alpha_{K1} \\ \cdots & & \\ \alpha_{1K} & \cdots & \alpha_{KK} \end{pmatrix}\right| \\
&= \mathcal{D}\left(\begin{pmatrix} \beta_{11} & \beta_{21} & \cdots & \beta_{K1} \\ \cdots & \cdots & & \cdots \\ \beta_{1K} & \beta_{2K} & \cdots & \beta_{KK} \\ 0 & 0 & \cdots & 0 \\ \cdots & \cdots & & \cdots \\ 0 & 0 & \cdots & 0 \end{pmatrix}\right) \left|\det\begin{pmatrix} \alpha_{11} & \cdots & \alpha_{K1} \\ \cdots & & \\ \alpha_{1K} & \cdots & \alpha_{KK} \end{pmatrix}\right| \\
&= \mathcal{D}(h \circ g)\ \mathcal{D}(f) = \mathcal{D}(g)\ \mathcal{D}(f).
\end{aligned}$$

□

Think in this theorem of g as being fixed and f as varying. Each new choice of f in effect specifies a different parallelepiped. $g \circ f$ amounts to the image of the parallelepiped f under the linear transformation g. The volume of the image $g \circ f$ is $\mathcal{D}(g)$ times the volume of f, so $\mathcal{D}(g)$ is the factor by which K-dimensional volume is altered under the transformation g. This idea will be useful when we study transformations of integrals.

Here is another application of $\mathcal{D}(f)$, one which will be helpful when we study differentiability in higher dimensions.

Theorem 1.9.3 *If $f: \mathbb{R}^K \to \mathbb{R}^N$ is a linear transformation and $K \leq N$, then the following are equivalent:*

(a) f is one-to-one.

(b) $f(e_1),\ f(e_2),\ \ldots,\ f(e_K)$ are linearly independent.

(c) $\mathcal{D}(f) > 0$.

Proof. Suppose (a) holds and $f(e_1), \ldots, f(e_K)$ are linearly dependent. Then there are scalars $\alpha_1, \ldots, \alpha_K$, not all zero, such that $\alpha_1 f(e_1) + \cdots + \alpha_K f(e_K) = 0$. This means $f(\alpha_1 e_1 + \cdots + \alpha_K e_K) = 0$ where $\alpha_1 e_1 + \cdots + \alpha_K e_K \neq 0$. But this means that f is not one-to-one, a contradiction. Hence (a) implies (b).

Now suppose (b) holds. Set $x_i = f(e_i)$ for $i = 1, 2, \ldots, K$ and let us apply the Gram–Schmidt orthogonalization process to $x_1, \ldots, x_K$. Let A_0 be the $N \times K$ matrix having the transposes of $x_1, \ldots, x_K$ as its column vectors. Note that $\mathcal{D}(f) = \mathcal{D}(A_0)$. Let $y_1 = x_1/|x_1|$ and let A_1 be the matrix having the transposes of $y_1, x_2, \ldots, x_K$ as its column vectors. Then $\mathcal{D}(A_0) = |x_1|\mathcal{D}(A_1)$. Note $|x_1| > 0$. For the sake of notational uniformity in the remainder of the proof, we denote x_1 by the symbol u_1. We now suppose we have constructed orthonormal vectors $y_1, \ldots, y_p$, using the Gram–Schmidt process, where $1 \leq p < K$. We further suppose A_p is the $N \times K$ matrix having the transposes of $y_1, \ldots, y_p, x_{p+1}, \ldots, x_K$ as its column vectors and satisfying $\mathcal{D}(A_0) = |u_1|\,|u_2| \ldots |u_p| \mathcal{D}\left(A_p\right)$ where $u_1, \ldots, u_p$ are nonzero vectors. Now set

$$u_{p+1} = x_{p+1} - \sum_{j=1}^{p} (y_j \cdot x_{p+1}) y_j$$

and $y_{p+1} = u_{p+1}/|u_{p+1}|$. (Note that we know, from having gone through this construction before, that u_{p+1} is nonzero and $y_1, \ldots, y_{p+1}$ are orthonormal.) Let B be the matrix having as its column vectors the transposes of $y_1, \ldots, y_p, u_{p+1}, x_{p+2}, \ldots, x_K$. Since u_{p+1} is a linear combination of $y_1, \ldots, y_p, x_{p+1}$, we have $\mathcal{D}(B) = \mathcal{D}\left(A_p\right)$. Let A_{p+1} be the matrix having the transposes of $y_1, \ldots, y_{p+1}, x_{p+2}, \ldots, x_K$ as its column vectors. Then $\mathcal{D}\left(A_p\right) = |u_{p+1}| \mathcal{D}\left(A_{p+1}\right)$ and hence $\mathcal{D}(A_0) = |u_1| \ldots |u_{p+1}| \mathcal{D}\left(A_{p+1}\right)$. At the conclusion of the Gram–Schmidt process we obtain $\mathcal{D}(A_0) = |u_1| \ldots |u_K|\, \mathcal{D}(A_K)$. The vectors whose transposes occur as the columns of A_K are $y_1, \ldots, y_K$. Since this is an orthonormal set, we must have $\mathcal{D}(A_K) = 1$. (The verification of this last statement is an exercise.) Hence $\mathcal{D}(f) = \mathcal{D}(A_0) > 0$. Thus (b) implies (c).

Finally suppose (c) holds and f is not one-to-one. There must be some $x \neq 0$ such that $f(x) = 0$. Then there must exist scalars $\alpha_1, \dots, \alpha_K$, not all zero, such that

$$\alpha_1 f(e_1) + \cdots + \alpha_K f(e_K) = f(\alpha_1 e_1 + \cdots + \alpha_K e_K) = 0.$$

Then $f(e_1), \dots, f(e_K)$ are linearly dependent and it must be possible to write one of them as a linear combination of the others. But if this is so, then (by another exercise) we have $\mathcal{D}(f) = 0$, a contradiction. Therefore (c) implies (a). □

Corollary 1.9.1 *A linear transformation* $f: \mathbb{R}^N \to \mathbb{R}^N$ *is one-to-one and onto if and only if* $\det(f) \neq 0$.

Proof. The equivalence of f being one-to-one and $\det(f) \neq 0$ follows immediately from Theorem 1.9.3. If $\det(f) \neq 0$, then the fact that f is onto follows from $f(e_1), \dots, f(e_N)$ being linearly independent and $N = \dim \mathbb{R}^N$. □

Exercises

1. Show that if A is an $N \times K$ matrix with $K \leq N$ and two column vectors which are identical, then $\mathcal{D}(A) = 0$.

2. Show that if A and B are $N \times K$ matrices with $K \leq N$ and B is obtained from A by multiplying one of the columns of A by the scalar λ, then $\mathcal{D}(B) = |\lambda| \mathcal{D}(A)$.

3. Show that if A and B are $N \times K$ matrices with $K \leq N$ and B is obtained from A by adding to one column of A a linear combination of the other column vectors, then $\mathcal{D}(B) = \mathcal{D}(A)$.

4. Show that if A is an $N \times K$ matrix with $K \leq N$ and the vectors whose transposes constitute its column vectors happen to form an orthonormal set, then $\mathcal{D}(A) = 1$.

5. If $K \leq N$ and $a_1, a_2, \dots, a_K$ are linearly dependent vectors in $\mathbb{R}^N$ and A is the matrix having $a_1^T, a_2^T, \dots, a_K^T$ as its column vectors, then show that $\mathcal{D}(A) = 0$. (Note that this is equivalent to considering a linear transformation $f: \mathbb{R}^K \to \mathbb{R}^N$ such that $f(x) = 0$ for some nonzero x.)

6. For $K \leq N$ compute the K-dimensional volume of the parallelepiped determined by

$$\begin{aligned}
&e_1, \\
&e_1 + e_2, \\
&e_1 + e_2 + e_3, \\
&\dots \\
&e_1 + e_2 + \cdots + e_K.
\end{aligned}$$

7. Show that for $K \leq N$ the K-dimensional volume of the parallelepiped determined by vectors $a_1, a_2, \ldots, a_K$ in $\mathbb{R}^N$ is given by

$$\sqrt{\det \begin{pmatrix} a_1 \cdot a_1 & \ldots & a_1 \cdot a_K \\ \ldots & & \\ a_K \cdot a_1 & \ldots & a_K \cdot a_K \end{pmatrix}}.$$

8. Show that for $a, b \in \mathbb{R}^N$, where $2 \leq N$, the area of the parallelogram determined by a and b is $|a||b|\sin(\theta)$ where θ is the angle between a and b.

9. Show that if A and B are both $N \times N$ matrices and $AB = I$, then $BA = I$.

2

METRIC SPACES

2.1 Metric Spaces

In the study of analysis in $\mathbb{R}^N$ (and later on manifolds) we are interested in such things as continuity, differentiability, and integrability. All these ideas depend on limit processes and convergence. Let us glance at some examples of convergence which may be familiar to the reader from a previous study of functions of a single variable. If some of the ideas — for example, Lebesgue integration or uniform convergence — are unfamiliar, this should not be cause for dismay. We are called not so much to appreciate the particular ideas as their variety.

If x_n converges to a, then $f(x_n)$ converges to $f(a)$.

Here "converges" means the usual convergence in $\mathbb{R}$. The statement is true if f is continuous at a.

If the sequence of differentiable functions f_n converges to a function f, then the sequence f_n' converges to f'.

If, in this example, convergence means pointwise convergence, the statement is false. A stronger type of convergence is necessary. It turns out that if the first convergence is interpreted as "f_n converges uniformly to f and f_n' converges uniformly" and the second convergence means uniform convergence, then the statement becomes true. One can say that differentiation is a continuous operation with respect to the two types of convergence described above.

If the sequence of functions f_n converges to a function f, then the sequence $\int_a^b f_n(x)\,dx$ converges to $\int_a^b f(x)\,dx$.

This statement is true for various types of convergence. Uniform convergence on $[a, b]$ is sufficient, but it is much too strong and thus makes the statement weak. Pointwise convergence is not sufficient. The most useful version assumes the so-called "dominated convergence." Then it is known as the Lebesgue dominated convergence theorem.

These few examples show that in solving problems in analysis one has to use many different types of convergence. For this reason it is worthwhile to study convergence and continuity in abstract spaces. The theory of metric spaces is one of the most important and easily describable settings for these concepts. Intuitively, a sequence converges to a limit if its terms are closer and closer to that limit. Thus, if we can measure the distance between points, we can tell whether or not a sequence converges to a point. The notion of distance is axiomatically described in the definition of metric spaces.

Definition 2.1.1 (Metric space) A *metric space* is a pair (X, d), where X is a non-empty set and d is a function $d\colon X \times X \to \mathbb{R}$, called a *metric*, such that

(a) $d(x, y) = 0$ if and only if $x = y$;

(b) $d(x, y) = d(y, x)$ for all $x, y \in X$;

(c) $d(x, y) \leq d(x, z) + d(z, y)$ for all $x, y, z \in X$.

Conditions (a), (b), and (c) are very natural if one thinks of distance between points. The first condition says that the distance between points x and y is 0 if and only if these two points coincide, that is $x = y$. Condition (b) says that the distance from x to y is the same as from y to x. Finally, (c) says that the distance measured from x to y cannot be greater than the distance from x to a third point z plus the distance from z to y. This property is usually called the triangle inequality.

In the definition of a metric space these three conditions are chosen as fundamental properties of the notion of distance. The definition is general enough to allow many important applications. At the same time it is not too general, so that we can prove many interesting properties of convergence and continuity in this abstract setting.

Let us consider some examples of metric spaces. Some of these examples are important in other areas of mathematics. Others are given to illustrate the possibilities allowed by the definition of metric spaces. The definition of metric spaces generalizes our intuition of distance in $\mathbb{R}^2$ or $\mathbb{R}^3$. One often tends to expect some properties of the distance in $\mathbb{R}^2$ or $\mathbb{R}^3$ to remain true in any metric space. Some examples given below are useful in testing those properties. They are important as counterexamples.

Example 2.1.1 The most important example of a metric space is the set $\mathbb{R}$ of all real numbers with the metric $d(x, y) = |x - y|$. It is relatively simple, yet rich enough to illustrate the basic concepts of metric spaces. Moreover, it can be called a model metric space since a metric on any metric space X translates problems in X to considerations in $\mathbb{R}$.

Example 2.1.2 Let X be an arbitrary nonempty set and let

$$d(x,y) = \begin{cases} 1 & \text{if } x \neq y \\ 0 & \text{if } x = y. \end{cases}$$

It is not difficult to verify that this is a metric space. This space is good for counterexamples. One should remember it when testing conjectures. Note that this example shows that any nonempty set can be made into a metric space. This metric is sometimes called the *discrete metric*.

In most examples verifying conditions (a)–(c) is routine. However, in some it is not easy at all. It may be unexpected but this is the case in the next example. It is possibly the most important example in this book.

Example 2.1.3 (Euclidean metric) Let $X = \mathbb{R}^N$. For points $x = (x_1, \dots, x_N)$, $y = (y_1, \dots, y_N)$ of $\mathbb{R}^N$, the so-called *Euclidean metric* is defined by

$$d(x,y) = \sqrt{(x_1 - y_1)^2 + \cdots + (x_N - y_N)^2};$$

this is often written as

$$d(x,y) = \sqrt{\sum_{n=1}^{N} (x_n - y_n)^2}.$$

Note that if $N = 3$, then the defined metric is the usual distance in $\mathbb{R}^3$. It is easy to check that (a) and (b) are satisfied. To prove that (c) is satisfied we shall use the Schwarz inequality (see Theorem 1.2.2):

$$\sum_{n=1}^{N} x_n y_n \leq \sqrt{\sum_{n=1}^{N} x_n^2} \sqrt{\sum_{n=1}^{N} y_n^2}.$$

Theorem 2.1.1 *For any real numbers $x_1, \dots, x_N, y_1, \dots, y_N$ we have*

$$\sqrt{\sum_{n=1}^{N} (x_n + y_n)^2} \leq \sqrt{\sum_{n=1}^{N} x_n^2} + \sqrt{\sum_{n=1}^{N} y_n^2}. \tag{2.1}$$

Proof. By the Schwarz inequality we have

$$\begin{aligned} \sum_{n=1}^{N} (x_n + y_n)^2 &= \sum_{n=1}^{N} x_n^2 + 2 \sum_{n=1}^{N} x_n y_n + \sum_{n=1}^{N} y_n^2 \\ &\leq \sum_{n=1}^{N} x_n^2 + 2 \sqrt{\sum_{n=1}^{N} x_n^2} \sqrt{\sum_{n=1}^{N} y_n^2} + \sum_{n=1}^{N} y_n^2 \\ &= \left(\sqrt{\sum_{n=1}^{N} x_n^2} + \sqrt{\sum_{n=1}^{N} y_n^2} \right)^2 \end{aligned}$$

proving the inequality. □

Now we can prove the triangle inequality for the Euclidean metric.

Corollary 2.1.1 *For any $x, y, z \in \mathbb{R}^N$ we have*

$$\sqrt{\sum_{n=1}^{N}(x_n - z_n)^2} \leq \sqrt{\sum_{n=1}^{N}(x_n - y_n)^2} + \sqrt{\sum_{n=1}^{N}(y_n - z_n)^2}$$

Proof. Note that $x_n - z_n = (x_n - y_n) + (y_n - z_n)$ and use (2.1). □

Example 2.1.4 ($\mathcal{L}(\mathbb{R})$) Let X be the set of all Lebesgue integrable functions on $\mathbb{R}$ and $d(f, g) = \int |f - g|$. Then

$$d(f, g) = \int |f - g| = \int |g - f| = d(g, f).$$

Moreover, since

$$|f - h| \leq |f - g| + |g - h|,$$

we have

$$\int |f - h| \leq \int |f - g| + \int |g - h|.$$

However, d is not a metric on X because $\int |f - g| = 0$ does not imply $f = g$. Indeed, if $f = 0$ and g is the characteristic function of $\mathbb{Q}$ (the set of all rational numbers), then

$$\int |f - g| \leq \int |f| + \int |g| = 0.$$

Consequently (X, d) is not a metric space. On the other hand, this is a very important space and we would like to be able to use the methods of metric spaces. It turns out that it is possible. This is how it can be done.

We introduce an equivalence relation in X:

$$f \sim g \quad \text{if} \quad \int |f - g| = 0.$$

One can also say that $f \sim g$ if $f = g$ almost everywhere. Then we identify functions which are equivalent. More precisely, we form a new space Y of equivalence classes of integrable functions:

$$[f] = \left\{g \in X : \int |f - g| = 0\right\} \quad \text{and} \quad Y = \{[f] : f \in X\}.$$

The space Y will be denoted by $\mathcal{L}^1(\mathbb{R})$. Using properties of the Lebesgue integral one can prove that $(\mathcal{L}^1(\mathbb{R}), d)$ is a metric space. (Note: The reader who is unfamiliar with Lebesgue integration should not feel at a loss. We shall develop the Lebesgue integral in the setting of $\mathbb{R}^N$ in Chapter 4. In the meantime, the reader should look at this example more to obtain a general feeling for what can happen in attempts to construct metric spaces than as a situation in which he or she should understand every detail.)

Let (X, d) be a metric space and let Y be a nonempty subset of X. Denote by δ the restriction of d to $Y \times Y$. The pair (Y, δ) is a metric space. (Y, δ) is called a subspace of (X, d). For example, $[0, 1]$ with d defined by $\delta(x, y) = |x - y|$ is a subspace of $\mathbb{R}$ with the metric defined the same way. Thus $[0, 1]$ with δ is a metric space.

The main topic of this book is calculus in $\mathbb{R}^N$. It is often important to think of $\mathbb{R}^N$ as the Cartesian product of N copies of the real line $\mathbb{R}$. Now we are going to define the Cartesian product of any finite collection of metric spaces. Throughout this chapter we prove properties of metric spaces which are defined as product spaces.

Example 2.1.5 (Cartesian product of metric spaces) Let $(X_1, d_1), \ldots, (X_m, d_m)$ be metric spaces and let $X = X_1 \times \cdots \times X_m$. For $x = (x_1, \ldots, x_m) \in X$ and $y = (y_1, \ldots, y_m) \in X$ define

$$d(x, y) = \max\{d_1(x_1, y_1), \ldots, d_m(x_m, y_m)\}.$$

It is not difficult to prove that d is a metric. Thus (X, d) is a metric space. Note that the Euclidean metric in $\mathbb{R}^N$ is not of the above form. If we wish to generalize that example, we should define the metric in X as

$$d(x, y) = \sqrt{(d_1(x_1, y_1))^2 + \cdots + (d_m(x_m, y_m))^2}.$$

As we will see later, in some sense, it does not matter which one is used. This question will be discussed in Section 2.3 (see Theorem 2.3.7).

Exercises

1. Prove that $d(x, y) \geq 0$ for all $x, y \in X$.
2. Prove that if $x_1, \ldots, x_n \in X$, then $d(x_1, x_n) \leq d(x_1, x_2) + \cdots + d(x_{n-1}, x_n)$.
3. Prove that $|d(x, z) - d(y, z)| \leq d(x, y)$ for all $x, y, z \in X$.
4. Prove that the following are metrics in $\mathbb{R}^N$:

   (a) $d(x, y) = |x_1 - y_1| + \cdots + |x_N - y_N|$,
   (b) $d(x, y) = \max\{|x_1 - y_1|, \ldots, |x_N - y_N|\}$,

   where $x = (x_1, \ldots, x_N)$ and $y = (y_1, \ldots, y_N)$.
5. Prove that $X = \mathbb{R}^2$ with

$$d_r((x_1, y_1), (x_2, y_2)) = \begin{cases} |y_1| + |y_2| + |x_1 - x_2| & \text{if } x_1 \neq x_2 \\ |y_1 - y_2| & \text{if } x_1 = x_2 \end{cases}$$

   is a metric space.

6. Let $\mathcal{C}([a, b])$ denote the set of all continuous real-valued functions on the interval $[a, b]$. Prove that $X = \mathcal{C}([a, b])$ with

$$d(f, g) = \max_{x \in [a,b]} |f(x) - g(x)|$$

is a metric space.

7. Let l^1 denote the set of all sequences $x = \{x_1, x_2, \dots\}$ of real numbers such that $\sum_{n=1}^{\infty} |x_n| < \infty$. Prove that l^1 with

$$d(x, y) = \sum_{n=1}^{\infty} |x_n - y_n|$$

is a metric space.

8. Let l^2 denote the set of all sequences $x = \{x_1, x_2, \dots\}$ of real numbers such that $\sum_{n=1}^{\infty} x_n^2 < \infty$. Prove that l^2 with

$$d(x, y) = \sqrt{\sum_{n=1}^{\infty} (x_n - y_n)^2}$$

is a metric space.

9. Let l^∞ denote the set of all bounded sequences $x = \{x_1, x_2, \dots\}$ of real numbers. Prove that l^∞ with

$$d(x, y) = \sup_{n \in \mathbb{N}} |x_n - y_n|$$

is a metric space.

10. Let (A, α) and (B, β) be metric spaces. Show that $(A \times B, d)$, where

$$d((a_1, b_1), (a_2, b_2)) = \alpha(a_1, a_2) + \beta(b_1, b_2),$$

is a metric space.

11. Let $X = \mathcal{C}^1([a, b])$, the set of all functions on the interval $[a, b]$ whose derivative is continuous.

    (a) Does $d(f, g) = \max_{x \in [a,b]} |f'(x) - g'(x)|$ define a metric in X?

    (b) Does $d(f, g) = \max_{x \in [a,b]} \left(|f(x) - g(x)| + |f'(x) - g'(x)|\right)$ define a metric in X?

12. Prove that both

$$d(x, y) = \max\{d_1(x_1, y_1), \dots, d_m(x_m, y_m)\}$$

and

$$d(x, y) = \sqrt{(d(x_1, y_1))^2 + \cdots + (d(x_m, y_m))^2}$$

define metrics in the Cartesian product space $X = X_1 \times \cdots \times X_m$.

2.2 Open and Closed Sets

In this section we discuss certain types of sets that play an important role in the study of continuity.

Definition 2.2.1 (Open balls, open sets, neighborhoods) For $x \in X$ and $\varepsilon > 0$, by the *open ball* at x of radius ε, we mean the set $B(x, \varepsilon) = \{y \in X \colon d(x, y) < \varepsilon\}$. A set $U \subseteq X$ is called *open* if for every $x \in U$ there exists $\varepsilon > 0$ such that $B(x, \varepsilon) \subseteq U$. By a *neighborhood of a point* x we mean any open set containing x.

It is not difficult to prove that, as one would expect, an open ball $B(x, \varepsilon)$ is an open set. Open intervals (a, b) are open sets in $\mathbb{R}$ with the usual metric. Sets of the form $\{(x_1, \dots, x_N) \in \mathbb{R}^N : \alpha_k < x_k < \beta_k \text{ for } k = 1, \dots, N\}$ are open in the Euclidean space $\mathbb{R}^N$. One can produce more examples of open sets by using (b) and (c) in the following theorem.

Theorem 2.2.1 *Let X be a metric space.*

(a) $\emptyset$ *and* X *are open.*

(b) If $U_1, U_2, \dots, U_n$ *are open in* X*, then* $U_1 \cap U_2 \cap \cdots \cap U_n$ *is open in* X.

(c) If U_α *is open in* X *for every* α *in some index set* $\mathcal{A}$*, then* $\bigcup_{\alpha \in \mathcal{A}} U_\alpha$ *is open in* X.

If U_n is open for every $n \in \mathbb{N}$, then $\bigcap_{n \in \mathbb{N}} U_n$ need not be open.

Definition 2.2.2 (Interior of a set) Let A be an arbitrary subset of a metric space X. The union of all open sets in X which are subsets of A is called the *interior* of A and denoted by A°. If $x \in A^\circ$, then x is called an *interior point* of A.

It is possible that a nonempty set has an empty interior. For example, the set of all irrational numbers, as a subset of $\mathbb{R}$, has empty interior. Note that it is necessary to add the phrase "as a subset of $\mathbb{R}$." If the set of all irrational numbers is the whole space X, then it is an open set and its interior is X.

The next theorem summarizes basic properties of the interior operation.

Theorem 2.2.2

(a) A° *is an open set.*

(b) $(A^\circ)^\circ = A^\circ$.

(c) $A^\circ \cup B^\circ \subseteq (A \cup B)^\circ$.

(d) $A^\circ \cap B^\circ = (A \cap B)^\circ$.

(e) $A^\circ = A$ *if and only if* A *is open.*

Definition 2.2.3 (Closed sets) A set $U \subseteq X$ is called *closed* if its complement $U^c = X - U$ is open.

It is important to remember that there are sets that are neither open nor closed. On the other hand, there are sets that are both open and closed.

Theorem 2.2.3

(a) *$\emptyset$ and X are closed.*

(b) *If $U_1, U_2, \ldots, U_n$ are closed, then $U_1 \cup U_2 \cup \cdots \cup U_n$ is closed.*

(c) *If U_α is closed for every α in some index set $\mathcal{A}$, then $\bigcap_{\alpha \in \mathcal{A}} U_\alpha$ is closed.*

Note that there is a certain symmetry between open sets and closed sets. The above theorem is similar to Theorem 2.2.2. (One should expect to be able to prove this theorem using Theorem 2.2.2 and DeMorgan's Law.) As open sets are used to define the interior, closed sets are used to define the closure.

Definition 2.2.4 (Closure) Let A be a subset of a metric space. The intersection of all closed subsets of X which contain A is called the *closure* of A and is denoted by $\overline{A}$.

The following theorem is very similar to Theorem 2.2.2. Note, however, the difference in parts (c) and (d).

Theorem 2.2.4

(a) *$\overline{A}$ is a closed set.*

(b) $\overline{\overline{A}} = \overline{A}$.

(c) $\overline{A \cup B} = \overline{A} \cup \overline{B}$.

(d) $\overline{A \cap B} \subseteq \overline{A} \cap \overline{B}$.

(e) *$\overline{A} = A$ if and only if A is closed.*

Consider a continuous real-valued function f on $\mathbb{R}$. If the values of f are known for all rational numbers, then the values at the remaining points are determined. Indeed, for any $x \in \mathbb{R}$ there exists a sequence $\{r_n\}$ of rational numbers convergent to x and, since f is continuous, we must have $f(x) = \lim_{n\to\infty} f(r_n)$. This is possible because the set of rational numbers is dense in $\mathbb{R}$. The notion of a dense subset can be easily generalized to any metric space. As we will see later, the property of continuous functions on $\mathbb{R}$ described here remains true in any metric space.

Definition 2.2.5 (Dense subset) Let A and B be subsets of X. The set A is said to be *dense* in B if $B \subseteq \overline{A}$.

The following characterization of dense subsets of a metric space is often useful.

Theorem 2.2.5 *A set A is dense in X if and only if every nonempty open set in X has a nonempty intersection with A.*

In Section 2.1, we noted that a nonempty subset Y of a metric space (X, d) is a metric space itself, with the metric defined as the restriction of d to Y. Although the metric in Y is the same as in X, open and closed sets in Y need not be open or closed in X. For example, let $X = \mathbb{R}$ and $d(x, y) = |x - y|$. If $Y = [0, 1]$, then both $[0, \frac{1}{2})$ and $[0, 1]$ are open sets in Y. On the other hand, if $Y = (0, 1)$, then both $(0, \frac{1}{2}]$ and $(0, 1)$ are closed subsets of Y. One can prove that if X is a metric space and $Y \subseteq X$, then $A \subseteq Y$ is open in Y if and only if $A = U \cap Y$ for some U open in X. Similarly, $A \subseteq Y$ is closed in Y if and only if $A = U \cap Y$ for some U closed in X.

Exercises

1. Prove that an open ball $B(x, \varepsilon)$ is an open set.

2. Sketch the balls $B((0, 0), 1)$ in $\mathbb{R}^2$ with the following metrics:

   (a) $d((x_1, y_1), (x_2, y_2)) = \sqrt{(x_1 - y_1)^2 + (x_2 - y_2)^2}$,

   (b) $d((x_1, y_1), (x_2, y_2)) = \max\{|x_1 - y_1|, |x_2 - y_2|\}$,

   (c) $d((x_1, y_1), (x_2, y_2)) = |x_1 - y_1| + |x_2 - y_2|$.

3. Describe the open ball $B(f, 1)$ in the space $C([0, 1])$ of all continuous real-valued functions on the interval $[0, 1]$ with the metric defined by

$$d(f, g) = \max_{x \in [a,b]} |f(x) - g(x)|$$

   if f is the function defined by $f(x) = x^2$.

4. Consider $\mathbb{R}^2$ with the Euclidean metric:

   (a) Prove $\{(x, y) \in \mathbb{R}^2 : y > e^x\}$ is open in $\mathbb{R}^2$.

   (b) Prove $\{(x, y) \in \mathbb{R}^2 : xy = 1\}$ is closed in $\mathbb{R}^2$.

5. Prove $\{(x, y) \in \mathbb{R}^2 : y = 1, x > 0\}$ is not closed in $\mathbb{R}^2$ with the usual metric but it is closed in $\mathbb{R}^2$ with the metric d_r. (See problem 5 on page 47.)

6. Find the interior and the closure of $\{(x, y) \in \mathbb{R}^2 : 0 < x < 1 \text{ and } 0 \le y \le 1\}$ in $\mathbb{R}^2$ with the usual metric and with the metric d_r. (See problem 5 on page 47.)

7. Consider the space $C([a, b])$ with the metric defined by

$$d(f, g) = \max_{x \in [a,b]} |f(x) - g(x)|.$$

   Find the closure of the set of all polynomials on $[a, b]$. (Hint: This requires some background material not in this book; locate a statement of the Weierstrass approximation theorem.)

8. Prove Theorem 2.2.1.

9. Prove that every open set is a union of open balls.

10. Give an example of a sequence of open sets $U_1, U_2, \ldots$ such that $\bigcap_{k=1}^{\infty} U_k$ is not open. Give an example of a sequence of open sets $U_1, U_2, \ldots$ such that $\bigcap_{k=1}^{\infty} U_k$ is open.

11. Prove Theorem 2.2.2. Give an example showing that the inclusion in (c) cannot be replaced by equality.

12. Prove that A is open if and only if every point of A is an interior point of A.

13. Prove Theorem 2.2.3.

14. Prove Theorem 2.2.4. Give an example showing that the inclusion in (d) cannot be replaced by equality.

15. Give examples to disprove:

    (a) If U_α is closed for every α in some index set $\mathcal{A}$, then $\bigcup_{\alpha \in \mathcal{A}} U_\alpha$ is closed.

    (b) Every set is either open or closed.

    (c) A set cannot be both open and closed.

16. Prove Theorem 2.2.5.

17. Prove that a set A is dense in B if and only if A has a nonempty intersection with every neighborhood of every point in B.

18. A metric space X is called separable if it has a dense countable subset.

    (a) Prove that $\mathbb{R}^N$ is a separable space.

    (b) Is l^1 a separable space?

    (c) Is l^2 a separable space?

19. Show that the closure of the unit ball in l^∞ contains an uncountable subset S such that $d(x, y) = 1$ for all $x, y \in S$. Is l^∞ a separable space?

20. Let (X, d) be the Cartesian product of metric spaces $(X_1, d_1), \ldots, (X_m, d_m)$, that is, let $X = X_1 \times \cdots \times X_m$ and $d(x, y) = \max\{d_1(x_1, y_1), \ldots, d_m(x_m, y_m)\}$, where $x = (x_1, \ldots, x_m)$ and $y = (y_1, \ldots, y_m)$.

    (a) Prove that $B(x, \varepsilon) = B(x_1, \varepsilon) \times \cdots \times B(x_m, \varepsilon)$.

    (b) Prove that if, for every $k = 1, \ldots, m$, S_k is an open subset of X_k, then $S = S_1 \times \cdots \times S_m$ is an open subset of X.

    (c) Prove that if, for every $k = 1, \ldots, m$, S_k is a closed subset of X_k, then $S = S_1 \times \cdots \times S_m$ is a closed subset of X.

(d) Prove that if, for every $k = 1, \dots, m$, S_k is a dense subset of X_k, then $S = S_1 \times \cdots \times S_m$ is a dense subset of X.

21. Let X be a metric space and let $Y \subseteq X$.

(a) Prove that $A \subseteq Y$ is open in Y if and only if $A = U \cap Y$ for some U open in X.

(b) Prove that $A \subseteq Y$ is closed in Y if and only if $A = U \cap Y$ for some U closed in X.

2.3 Convergence

As mentioned in Section 2.1, one of the main goals of introducing a metric in a set is to define convergence of sequences in that set.

Definition 2.3.1 (Convergent sequence) Let $\{x_n\}$ be a sequence in a metric space X. If there exists an $x \in X$ such that for every $\varepsilon > 0$ there exists $n_\varepsilon \in \mathbb{N}$ such that $d(x_n, x) < \varepsilon$ for every $n > n_\varepsilon$, then the sequence $\{x_n\}$ is called *convergent* in X. The point x is called the *limit* of $\{x_n\}$. We say that $\{x_n\}$ converges to x and write $\lim_{n\to\infty} x_n = x$ or just $x_n \to x$.

Note that the definition of convergence in a metric space can be simplified if it is expressed in terms of the usual convergence in $\mathbb{R}$:

$$\lim_{n\to\infty} x_n = x \quad \text{means} \quad \lim_{n\to\infty} d(x_n, x) = 0 \tag{2.2}$$

or

$$x_n \to x \quad \text{means} \quad d(x_n, x) \to 0. \tag{2.3}$$

This definition has a very simple intuitive meaning: the sequence $\{x_n\}$ converges to x if the distance between x_n and x converges to 0 as $n \to \infty$. Note that the symbol $\lim_{n\to\infty}$ in (2.2) (or $\to$ in (2.3)) has two different meanings. The one on the left denotes the convergence in X, while the one on the right denotes the convergence in $\mathbb{R}$. It will often be necessary to talk about more than one convergence at the same time. It is usually clear what is meant. When doubt can arise we will identify the convergence by saying something like "$x_n \to x$ in X." Some types of convergence have special names: pointwise convergence, uniform convergence, convergence almost everywhere, etc. To indicate that $\{x_n\}$ does not converge to x we will write $x_n \not\to x$.

It is important to remember that convergence is always defined relative to the space X and the metric d. For example, if $X = (0, 1)$ and $d(x, y) = |x - y|$, then the sequence $\{1/n\}$ is not convergent.

The following theorem lists basic properties of convergence in metric spaces.

Theorem 2.3.1

(a) If $x_n = x$ for all $n \in \mathbb{N}$, then $\lim_{n\to\infty} x_n = x$.

(b) *If* $\lim_{n\to\infty} x_n = x$ *and* $\{x_{p_n}\}$ *is a subsequence of* $\{x_n\}$, *then* $\lim_{n\to\infty} x_{p_n} = x$.

(c) *If* $\lim_{n\to\infty} x_n = x$ *and* $\lim_{n\to\infty} x_n = y$, *then* $x = y$.

The proofs of these properties are easy. They are left as exercises. The equivalence in the next theorem is less obvious; it is sometimes called the Urysohn property.

Theorem 2.3.2 $\lim_{n\to\infty} x_n = x$ *if and only if every subsequence of* $\{x_n\}$ *has a subsequence convergent to* x.

Proof. Since a subsequence of a subsequence of $\{x_n\}$ is a subsequence of $\{x_n\}$, $\lim_{n\to\infty} x_n = x$ implies that every subsequence of $\{x_n\}$ has a subsequence convergent to x by part (b) of Theorem 2.3.1.

Assume now that the sequence $\{x_n\}$ is not convergent to x. Then there exists $\varepsilon > 0$ such that $d(x_n, x) \geq \varepsilon$ for infinitely many $n \in \mathbb{N}$. Thus, there exists a subsequence $\{x_{p_n}\}$ such that $d(x_{p_n}, x) \geq \varepsilon$ for all $n \in \mathbb{N}$. But then $\{x_{p_n}\}$ cannot have a subsequence convergent to x. □

It is often more convenient to use the above theorem in the contrapositive version: $\{x_n\}$ does not converge to x if and only if $\{x_n\}$ has a subsequence such that none of its subsequences converges to x.

The next theorem shows that convergence can be defined by open sets without referring to the metric. This is important in generalizations of metric spaces, the so-called topological spaces. A topological space is a set where some subsets are designated as open such that conditions (a), (b), and (c) in Theorem 2.2.1 are satisfied.

Let $\{x_n\}$ be a sequence in X and let $U \subseteq X$. If there exists an index n_0 such that $x_n \in U$ for all $n > n_0$, then we say that $\{x_n\}$ is eventually in U.

Theorem 2.3.3 $\lim_{n\to\infty} x_n = x$ *if and only if* $\{x_n\}$ *is eventually in every open subset of* X *which contains* x.

Proof. Let $x_n \to x$ and let U be an open set such that $x \in U$. Since U is open, there exists $\varepsilon > 0$ such that $B(x, \varepsilon) \subseteq U$. Since $x_n \to x$, there exists $n_0 \in \mathbb{N}$ such that $d(x_n, x) < \varepsilon$ for every $n > n_0$. But this means that $x_n \in B(x, \varepsilon)$ and hence $x_n \in U$ for every $n > n_0$.

Now suppose that $x_n \not\to x$. Then there exists $\varepsilon > 0$ such that $d(x_n, x) \geq \varepsilon$ for infinitely many $n \in \mathbb{N}$. Consequently, $x_n \notin B(x, \varepsilon)$ for infinitely many $n \in \mathbb{N}$. Since $B(x, \varepsilon)$ is an open set which contains x, the proof is complete. □

As open sets define convergence, convergence defines open sets.

Theorem 2.3.4 *A set* A *is open if and only if for every* $x \in A$ *and every sequence* $\{x_n\}$ *convergent to* x, $\{x_n\}$ *is eventually in* A.

Closed sets can also be characterized by convergent sequences.

Theorem 2.3.5 *A set* A *is closed if and only if every convergent sequence of elements of* A *has its limit in* A, *that is,* $x_n \in A$ *and* $\lim_{n\to\infty} x_n = x$ *implies* $x \in A$.

Finally, dense sets can be characterized by convergent sequences.

Theorem 2.3.6 *A is dense in B if and only if every element of B is the limit of a sequence of elements of A.*

It seems that it is sufficient to know which sequences are convergent to define all the other concepts of a metric space. Is the metric uniquely determined by convergent sequences? The following simple example shows that it is not so: Consider $\mathbb{R}$ with $d_1(x, y) = |x - y|$ and $d_2(x, y) = \min\{|x - y|, 1\}$. These two metrics are obviously different, but they define the same convergence. In general, every convergence can be defined by infinitely many different metrics. On the other hand, since all these metrics define the same convergence, they define the same open, closed, dense sets, it seems that it really does not matter which metric is used. This is not quite true. In Section 2.6 we will discuss completeness of metric spaces. This is one of the most important properties of metric spaces. It is rather surprising that completeness depends on the metric chosen. If you only know which sequences are convergent, you cannot tell whether the space is complete.

Definition 2.3.2 (Equivalent metrics) Two metrics d_1 and d_2 defined on the same set X are called *equivalent* if they define the same convergence, that is,

$$d_1(x_n, x) \to 0 \quad \textit{if and only if} \quad d_2(x_n, x) \to 0.$$

The following are equivalent metrics in $\mathbb{R}^N$:

$$d_1(x, y) = \sqrt{\sum_{n=1}^{N}(x_n - y_n)^2},$$

$$d_2(x, y) = \sum_{n=1}^{N} |x_n - y_n|,$$

$$d_3(x, y) = \max\{|x_1 - y_1|, \dots, |x_N - y_N|\}.$$

Indeed, we have the following general result.

Theorem 2.3.7 *Let $(X_1, d_1), \dots, (X_m, d_m)$ be metric spaces and let $X = X_1 \times \cdots \times X_m$. The following metrics are equivalent:*

$$d(x, y) = \sqrt{\sum_{n=1}^{m}(d_n(x_n, y_n))^2}, \tag{2.4}$$

$$d'(x, y) = \sum_{n=1}^{m} d_n(x_n, y_n), \tag{2.5}$$

$$d^*(x, y) = \max\{d_1(x_1, y_1), \dots, d_m(x_m, y_m)\}. \tag{2.6}$$

Note that this theorem says that in the definition of the metric in the Cartesian product of metric spaces one could use (2.4), (2.5), or (2.6). Each one of these possibilities has its advantages. The metrics (2.5) and (2.6) are simple and easy to use. The metric (2.4) has the best geometric properties. We have seen indications of that in Chap ter 1.

From the theorems in this section it easily follows that equivalent metrics define the same open, closed, or dense sets. However, they need not define the same bounded sets.

Definition 2.3.3 (Bounded sets) A set A in a metric space is called *bounded* if there exists a constant $M > 0$ such that $d(x, y) \leq M$ for all $x, y \in A$.

Consider the real line $\mathbb{R}$ with $d_1(x, y) = |x - y|$ and $d_2(x, y) = \min\{|x - y|, 1\}$. It is easy to check that these two metrics are equivalent. Note that every subset of $\mathbb{R}$ is bounded with respect to d_2, which is obviously not true for d_1. The same can be done in an arbitrary metric space: d and $\min\{d, 1\}$ are equivalent for any metric d. In some sense, this shows that the property of being bounded does not mean much in metric spaces. The situation is completely different in the so-called normed spaces, which is an important class of metric spaces (see Section 2.7).

We close this section with a very useful property of convergence in Cartesian products of metric spaces.

Theorem 2.3.8 *Let (X, d) be the Cartesian product of metric spaces $(X_1, d_1), \dots, (X_m, d_m)$ and let $x_n = (x_{1,n}, \dots, x_{m,n})$ be a sequence of elements of X. Then the sequence $\{x_n\}$ converges to $x = (x_1, \dots, x_m) \in X$ if and only if for every $k \in \{1, \dots, m\}$ the sequence $\{x_{k,n}\}$ converges to x_k in X_k.*

Note that in the above theorem we do not specify the metric d. The theorem is true for any one of the metrics (2.3)–(2.5) or, as a matter of fact, any equivalent metric. In this book we are not going to consider metrics on the Cartesian product that are not equivalent with the above ones. Thus, when we say "the Cartesian product of metric spaces $(X_1, d_1), \dots, (X_m, d_m)$," we mean the set $X = X_1 \times \cdots \times X_m$ with one of the defined metrics. If for some reason using a specific metric is essential, the metric will be defined.

Exercises

1. Consider the space $C([0, 1])$ of all continuous, real-valued functions on the interval $[0, 1]$ and the sequence of functions $f_n(x) = x^n$. Show that the sequence $\{f_n\}$ is not convergent with respect to the metric $d_1(f, g) = \max_{x \in [0,1]} |f(x) - g(x)|$, but it is convergent with respect to the metric $d_2(f, g) = \int |f - g|$.

2. Consider the sequence of sequences $x_n = (x_{n,1}, x_{n,2}, x_{n,3}, \dots)$ where

$$x_{n,k} = \begin{cases} \frac{1}{k} & \text{if } k \leq n \\ 0 & \text{if } k > n. \end{cases}$$

Show that $\{x_n\}$ is convergent in l^2, but divergent in l^1. Is $\{x_n\}$ convergent in l^∞?

3. Consider the space $C^1([0, 2\pi])$ of all functions on the interval $[0, 2\pi]$ whose derivative is continuous. Let

$$d_1(f, g) = \max_{x \in [0,2\pi]} |f(x) - g(x)|$$

and

$$d_2(f, g) = \max_{x \in [0,2\pi]} \left(|f(x) - g(x)| + |f'(x) - g'(x)|\right).$$

Check convergence of the sequence $f_n(x) = \frac{1}{n}\sin(n^2 x)$ with respect to d_1 and d_2.

4. Prove Theorem 2.3.1.

5. Is it possible to define convergence using closed sets?

6. Prove Theorem 2.3.4.

7. Prove Theorem 2.3.5.

8. Prove Theorem 2.3.6.

9. Let (X, d) be a metric space and let $x_{n,k} \in X$ for $k, n \in \mathbb{N}$. Prove that if

$$\begin{array}{l} x_{1,1}, x_{1,2}, x_{1,3}, \ldots \to x \\ x_{2,1}, x_{2,2}, x_{2,3}, \ldots \to x \\ x_{3,1}, x_{3,2}, x_{3,3}, \ldots \to x \\ \ldots \\ x_{n,1}, x_{n,2}, x_{n,3}, \cdots \to x \\ \ldots, \end{array}$$

then there exists an increasing sequence of natural numbers p_n such that $x_{n,p_n} \to x$.

10. Let X be the space of all infinite sequences $\{x_n\}$ of real numbers such that $x_n = 0$ for all but a finite number of n. Define a convergence in X as follows:

A sequence $a_n = (\alpha_{n,1}, \alpha_{n,2}, \alpha_{n,3}, \ldots)$ converges to $a = (\alpha_1, \alpha_2, \alpha_3, \ldots)$ if the following two conditions are satisfied:

(1) $|\alpha_{n,k} - \alpha_k| \to 0$ as $n \to \infty$ for every $k \in \mathbb{N}$,

(2) there exists $k_0 \in \mathbb{N}$ such that $\alpha_{n,k} = 0$ for all $n \in \mathbb{N}$ and all $k \geq k_0$.

Prove that this convergence cannot be defined by a metric.

11. Let $\{x_n\}$ be a sequence in X. Show that if $\{x_n\}$ has no convergent subsequence, then for every $y \in X$ there exists an open neighborhood U_y of y such that $U_y \cap \{x_1, x_2, \dots\}$ is a finite or empty set.

12. Prove Theorem 2.3.7.

13. Prove that equivalent metrics define the same open, closed, or dense sets. Do equivalent metrics define the same closure and interior operations?

14. Let d be an arbitrary metric. Prove that d and $\min\{d, 1\}$ are equivalent.

15. Prove Theorem 2.3.8.

2.4 Continuous Mappings

Finally we are ready to define continuity of mappings between metric spaces.

Definition 2.4.1 (Continuity) Let X and Y be metric spaces and let $x_0 \in X$. A mapping $f: X \to Y$ is called *continuous at* x_0 if $x_n \to x_0$ implies $f(x_n) \to f(x_0)$. If f is continuous at every point of X, then we say that f is *continuous*.

If a mapping f is continuous, then we can interchange the order of evaluation of f and the limit operation:

$$f\left(\lim_{n\to\infty} x_n\right) = \lim_{n\to\infty} f(x_n).$$

There is also another way of interpreting continuity. One can say that if a function is continuous, then a small change in the input results in a small change in the output. The following theorem formulates this property more precisely.

Theorem 2.4.1 *Let f be a mapping from X to Y. The following conditions are equivalent:*

(a) f is continuous at $x_0 \in X$;

(b) For every $\varepsilon > 0$ there exists $\delta > 0$ such that $d(f(x), f(x_0)) < \varepsilon$ whenever $d(x, x_0) < \delta$.

Since the two conditions in this theorem are equivalent, the second condition can be used as the definition of continuity. This is actually a very common practice. There is no essential difference between choosing one or the other. As we will see, in some arguments it is more convenient to use convergent sequences, and in others it is better to use ε and δ.

In calculus textbooks, continuity of a function f at a point $x_0 \in \mathbb{R}$ is defined by the property $\lim_{x\to x_0} f(x) = f(x_0)$. The same approach can be used here. First we define the limit of a function at a point:

A function $f : X \to Y$, where X and Y are metric spaces, has a limit y_0 at x_0, denoted by $\lim_{x \to x_0} f(x) = y_0$, if for every $\varepsilon > 0$ there exists a $\delta > 0$ such that $d(f(x), y_0) < \varepsilon$ whenever $0 < d(x, x_0) < \delta$.

(Note the difference between condition (b) in Theorem 2.4.1 and the above definition. The reason for assuming that $0 < d(x, x_0)$ is to eliminate $x = x_0$ from consideration. This is important, because we want to consider limits of functions at points where the function is not defined.) Now, we can define continuity of f at $x_0 \in X$ by the familiar condition $\lim_{x \to x_0} f(x) = f(x_0)$, where it is understood that the limit exists and the function has a value at x_0. Equivalence of this definition and Definition 2.4.1 is an immediate consequence of Theorem 2.4.1.

Theorem 2.4.2 *The composition of two continuous functions is continuous.*

Proof. Let $f: X \to Y$ and $g: Y \to Z$. If $x_n \to x$ in X, then $f(x_n) \to f(x)$ in Y (by continuity of f) and thus $g(f(x_n)) \to g(f(x))$ in Z (by continuity of g). Thus $g \circ f$ is continuous. □

The next theorem describes a very useful property of the metric, namely, the metric is a continuous function.

Theorem 2.4.3 *If $x_n \to x$ and $y_n \to y$, then $d(x_n, y_n) \to d(x, y)$.*

Proof.

$$\begin{aligned} |d(x_n, y_n) - d(x, y)| &= |d(x_n, y_n) - d(x, y_n) + d(x, y_n) - d(x, y)| \\ &\le |d(x_n, y_n) - d(x, y_n)| + |d(x, y_n) - d(x, y)| \\ &\le d(x_n, x) + d(y_n, y) \to 0. \end{aligned}$$ □

In the next theorem we characterize continuous mappings in terms of open and closed sets. This possibility is essential in general topological spaces where the metric is not available.

Theorem 2.4.4 *Let f be a mapping from X to Y. The following conditions are equivalent:*

(a) f is continuous;

(b) $f^{-1}(U)$ is open in X for every U open in Y;

(c) $f^{-1}(U)$ is closed in X for every U closed in Y;

(d) $f(\overline{S}) \subseteq \overline{f(S)}$ for every subset S of X.

Let us look at the definition of continuity again. If a function f is continuous, then for every point x_0 and every $\varepsilon > 0$ there exists $\delta > 0$ such that $d(f(x), f(x_0)) < \varepsilon$ whenever $d(x, x_0) < \delta$. It is important to remember that the δ depends on both x_0 and ε. For example, if $X = Y = \mathbb{R}$ and $f : \mathbb{R} \to \mathbb{R}$ is defined by $f(x) = x^2$, then for $x_0 = 0$ and $\varepsilon = 0.01$ it suffices to take $\delta = 0.1$, but for $x_0 = 1$ and $\varepsilon = 0.01$ it is not sufficient to take $\delta = 0.1$. When we require that for every $\varepsilon > 0$ the same δ "works" for all $x \in X$, we obtain a stronger type of continuity, called uniform continuity.

Definition 2.4.2 (Uniform continuity) Let X and Y be metric spaces. A mapping $f\colon X \to Y$ is called *uniformly continuous* if for every $\varepsilon > 0$ there exists $\delta > 0$ such that $d(x_1, x_2) < \delta$ implies $d(f(x_1), f(x_2)) < \varepsilon$ for all $x_1, x_2 \in X$.

Example 2.4.1 Consider the function $f(x) = x^2$ on a bounded and closed interval $[a, b]$. If $|x - y| < \delta$, then

$$|x^2 - y^2| = |x - y||x + y| \le 2\max\{|a|, |b|\}\delta.$$

Thus, for any $\varepsilon > 0$ the choice $\delta = \frac{\varepsilon}{2\max\{|a|,|b|\}}$ will work for any pair of points $x, y \in [a, b]$. Consequently, f is uniformly continuous on $[a, b]$. One can show that f is not uniformly continuous on $\mathbb{R}$.

Example 2.4.2 Consider now the space $\mathcal{C}([a, b])$ of all continuous real-valued functions on the interval $[a, b]$ with the standard metric $d(f, g) = \max_{x\in[a,b]} |f(x) - g(x)|$. Let $\Lambda\colon \mathcal{C}([a, b]) \to \mathcal{C}([a, b])$ be a mapping defined by

$$\Lambda(f)(x) = \int_a^x f(t)\,dt.$$

We will show that Λ is uniformly continuous. Indeed, we have

$$\begin{aligned} d(\Lambda(f), \Lambda(g)) &= \max_{x\in[a,b]} |\Lambda(f)(x) - \Lambda(g)(x)| \\ &= \max_{x\in[a,b]} \left| \int_a^x (f(t) - g(t))\,dt \right| \\ &\le \max_{x\in[a,b]} \int_a^x |f(t) - g(t)|\,dt \\ &\le \max_{x\in[a,b]} |f(x) - g(x)|\,(b - a) \\ &= d(f, g)(b - a). \end{aligned}$$

Consequently, if for an arbitrary $\varepsilon > 0$ we take $\delta < \frac{\varepsilon}{2(b-a)}$, then $d(f, g) < \delta$ implies $d(\Lambda(f), \Lambda(g)) < \varepsilon$.

It is possible to define uniform continuity in terms of convergence.

Theorem 2.4.5 *Let $f\colon X \to Y$. The following conditions are equivalent:*

(a) f is uniformly continuous;

(b) $d(x_n, y_n) \to 0$ implies $d(f(x_n), f(y_n)) \to 0$.

Proof. The easier implication $(a) \Rightarrow (b)$ is left as an exercise. Now suppose that f is not uniformly continuous. This implies that there exists an $\varepsilon > 0$ such that for every $\delta > 0$ there exist points $x, y \in X$ such that $d(x, y) < \delta$ and $d(f(x), f(y)) \ge \varepsilon$. In particular, there exist sequences of points $x_n, y_n \in X$ such that $d(x_n, y_n) < 1/n$ and $d(f(x_n), f(y_n)) \ge \varepsilon$. But then (b) is not satisfied. □

Using condition (b) in the above theorem we could replace the final ε-δ argument in Example 2.4.2 by noting that $d\left(\Lambda(f_n), \Lambda(g_n)\right) \leq d(f_n, g_n)(b-a) \to 0$.

Let $(X_1, d_1), \ldots, (X_m, d_m)$ be metric spaces and let $X = X_1 \times \cdots \times X_m$. Let $k \in \{1, \ldots, m\}$ be fixed. Define a mapping Π_k from X into X_k by

$$\Pi_k((x_1, \ldots, x_m)) = x_k.$$

The defined mapping is called the projection map of X onto X_k. From Theorem 2.3.8 it immediately follows that projection maps are continuous. As in the case of Theorem 2.3.8, when we talk about continuity we mean the continuity with respect to any of the standard metrics in $X_1 \times \cdots \times X_m$. From the continuity of projection maps we obtain the following useful theorem.

Theorem 2.4.6 *Let $Y_1, \ldots, Y_m$ be metric spaces and let Π_k be the projection map from $Y_1 \times \cdots \times Y_m$ onto Y_k. A mapping $f : X \to Y_1 \times \cdots \times Y_m$ is continuous if and only if $\Pi_k \circ f$ is continuous for every $k \in \{1, \ldots, m\}$.*

Note that in the last theorem we might also write $f = (f_1, \ldots, f_m)$ where each $f_k = \Pi_k \circ f$. We would call each f_k the kth component of f. Note also that we can replace continuity by uniform continuity in the statement of this theorem.

Exercises

1. Show that $f(x) = x^2$ is not uniformly continuous on $\mathbb{R}$.
2. Prove Theorem 2.4.1.
3. True or false?
   (a) If $f\colon X \to Y$ is continuous, then for every open $U \subseteq X$ the set $f(U)$ is open in Y.
   (b) If $f\colon X \to Y$ is continuous, then for every closed $U \subseteq X$ the set $f(U)$ is closed in Y.
   (c) If $f\colon X \to Y$ and for every open $U \subseteq X$ the set $f(U)$ is open in Y, then f is continuous.
4. Prove that $\lim_{x \to x_0} f(x) = y_0$ if and only if $\lim_{n\to\infty} f(x_n) = y_0$ for every sequence $x_n \to x_0$ such that $x_n \neq x_0$ for all $n \in \mathbb{N}$.
5. Show that f is continuous at $x_0 \in X$ if and only if $\lim_{x \to x_0} f(x) = f(x_0)$.
6. Show that every uniformly continuous function is continuous.
7. Give an ε-δ proof for Theorem 2.4.2.
8. Prove Theorem 2.4.4.
9. Can you modify (a), (b), and (c) in Theorem 2.4.4 to characterize continuity at a point?
10. Prove that projection maps Π_k are uniformly continuous.
11. Prove Theorem 2.4.6.

2.5 Compact Sets

In this section we discuss one of the less intuitive concepts in metric spaces.

Definition 2.5.1 (Compact sets) A subset K of a metric space is called *compact* if every sequence in K has a subsequence convergent to an element of K. More precisely, if $x_1, x_2, \ldots \in K$ then there exists an increasing sequence of indices p_n and $x_0 \in K$ such that $x_{p_n} \to x_0$.

One of the fundamental properties of the real numbers is the property that every bounded sequence of real numbers contains a convergent subsequence. It is known as the Bolzano–Weierstrass Theorem. It easily follows from that theorem that every bounded, closed subset of $\mathbb{R}$ is compact. It turns out that these are the only compact subsets of $\mathbb{R}$. The same is true in $\mathbb{R}^N$.

Theorem 2.5.1 *Every bounded, closed subset of* $\mathbb{R}^N$ *is compact.*

Proof. Let K be a bounded, closed subset of $\mathbb{R}^N$ and let $x_n = (x_{n,1}, \ldots, x_{n,N}) \in K$, $n \in \mathbb{N}$. Since, for every $k = 1, \ldots, N$, the sequence $\{x_{n,k}\}$ is bounded, by the Bolzano–Weierstrass theorem (used N times), there exists an increasing sequence of indices $\{p_n\}$ such that the sequence $x_{p_n} = (x_{p_n,1}, \ldots, x_{p_n,N})$ converges to some $x \in \mathbb{R}^N$. Since K is closed, $x \in K$. □

To see that in an arbitrary metric space not every bounded, closed set is compact, consider the space $\mathcal{C}([0, 1])$ with $d(f, g) = \max_{x\in[0,1]} |f(x) - g(x)|$. The closed unit ball

$$\overline{B}(0, 1) = \{f \in \mathcal{C}([0, 1]) : \max_{x\in[0,1]} |f(x)| \le 1\}$$

is a bounded and closed subset of $\mathcal{C}([0, 1])$. However, the sequence of functions $f_n(x) = x^n$ does not have a convergent subsequence. Thus $\overline{B}(0, 1)$ is not compact.

On the other hand we have the following:

Theorem 2.5.2 *Every compact set is bounded and closed.*

Proof. Let K be a compact subset of a metric space X. If K is not bounded, then there exist sequences $x_n, y_n \in K$ such that

$$d(x_n, y_n) \ge n \quad \text{for all} \quad n \in \mathbb{N}. \tag{2.7}$$

Since K is compact, there exists an increasing sequence of indices $\{p_n\}$ such that $x_{p_n} \to x$ and $y_{p_n} \to y$ for some $x, y \in K$. But then, by Theorem 2.4.3, we have

$$d(x_{p_n}, y_{p_n}) \to d(x, y) \quad \text{as} \quad n \to \infty,$$

contradicting (2.7).

Suppose now that K is not closed. Then there exist $x_n \in K$ such that $x_n \to x$ and $x \notin K$. Since every subsequence of $\{x_n\}$ converges to x, the sequence cannot have a subsequence convergent to an element of K. But this contradicts the assumption that K is compact. □

Definition 2.6.2 (Complete space) A metric space X is called *complete* if every Cauchy sequence in X is convergent in X.

We know that $\mathbb{R}$ is a complete space. Using completeness of $\mathbb{R}$ we can easily prove completeness of $\mathbb{R}^N$. Here is a sketch of the proof: If $x_n = (x_{1,n}, \ldots, x_{N,n})$ and $\{x_n\}$ is a Cauchy sequence in $\mathbb{R}^N$, then $\{x_{k,n}\}$ is a Cauchy sequence in $\mathbb{R}$ for $k = 1, \ldots, N$. Thus there exist $x_{1,0}, \ldots, x_{N,0} \in \mathbb{R}$ such that $x_{1,n} \to x_{1,0}, x_{2,n} \to x_{2,0}, \ldots, x_{N,n} \to x_{N,0}$. But this means that $x_n \to (x_{1,0}, \ldots, x_{N,0})$ in $\mathbb{R}^N$. This argument can be easily generalized and used to prove the following theorem.

Theorem 2.6.2 *The Cartesian product of a finite number of complete spaces is complete.*

In the next theorem we prove completeness of a function space.

Theorem 2.6.3 *Let X be an arbitrary metric space. The space Y of all bounded continuous functions from X into $\mathbb{R}$ with the metric defined by*

$$d(f, g) = \sup_{x \in X} |f(x) - g(x)|$$

is complete.

Proof. Let $\{f_n\}$ be a Cauchy sequence in Y. Then, for every $x \in X$, the sequence $\{f_n(x)\}$ is a Cauchy sequence in $\mathbb{R}$. Since $\mathbb{R}$ is a complete space, for every $x \in X$ there exists a number $f(x)$ such that $f_n(x) \to f(x)$. This defines a function f from X into $\mathbb{R}$. We have to prove that $f \in Y$ and that $d(f_n, f) \to 0$. (Note that, in general, $f_n(x) \to f(x)$ does not imply $d(f_n, f) \to 0$).

Let ε be an arbitrary positive number. Since $\{f_n\}$ is a Cauchy sequence there exists an index n_0 such that

$$d(f_m, f_n) = \sup_{x \in X} |f_m(x) - f_n(x)| < \varepsilon \quad \text{for all} \quad m, n \geq n_0.$$

Consequently, for every $x \in X$ and every $n \geq n_0$ we have

$$|f(x) - f_n(x)| = \lim_{m \to \infty} |f_m(x) - f_n(x)| \leq \varepsilon.$$

This proves that $d(f, f_n) \to 0$ and, in fact, that f is a bounded function. It remains to prove that f is continuous. Let x be an arbitrary point in X and let ε be a positive number. Then, for n_0 defined as before, there exists $\delta > 0$ such that $|f_{n_0}(x) - f_{n_0}(y)| < \varepsilon$ for every $y \in X$ such that $d(x, y) < \delta$. Thus

$$\begin{aligned} |f(x) - f(y)| &\leq |f(x) - f_{n_0}(x)| + |f_{n_0}(x) - f_{n_0}(y)| + |f_{n_0}(y) - f(y)| \\ &< \varepsilon + \varepsilon + \varepsilon = 3\varepsilon \end{aligned}$$

for every $y \in X$ such that $d(x, y) < \delta$. □

The following two theorems describe connections between complete, closed, and compact sets.

Theorem 2.6.4 *A subspace of a complete space is complete if and only if it is closed.*

The easy proof is left as an exercise.

Theorem 2.6.5 *Compact spaces are complete.*

Proof. Let $\{x_n\}$ be a Cauchy sequence in a compact space X and let $\{x_{p_n}\}$ be a convergent subsequence of $\{x_n\}$. If $x_{p_n} \to x_0$, then $d(x_n, x_0) \leq d(x_n, x_{p_n}) + d(x_{p_n}, x_0) \to 0$. □

We close this section with the famous Banach fixed point theorem (called also the Contraction Mapping Theorem).

Definition 2.6.3 (Contraction mapping) Let (X, d) be a metric space. A mapping $f : X \to X$ is called a *contraction mapping* if there exists a constant $0 < a < 1$ such that

$$d(f(x), f(y)) \leq ad(x, y) \quad \textit{for all} \quad x, y \in X.$$

Note that every contraction mapping is uniformly continuous.

Theorem 2.6.6 (Banach fixed point theorem) *Let X be a complete metric space and let $f : X \to X$ be a contraction mapping. Then there exists a unique $x_0 \in X$ such that $f(x_0) = x_0$.*

Proof. Suppose $d(f(x), f(y)) \leq ad(x, y)$ for some $0 < a < 1$ and for all $x, y \in X$. Let x be an arbitrary point in X. Define a sequence of points $x_n \in X$ by the recursion

$$x_1 = x, \quad x_{n+1} = f(x_n).$$

We will show that $\{x_n\}$ is a Cauchy sequence. First note that

$$d(x_n, x_{n+1}) \leq a\, d(x_{n-1}, x_n) \leq a^2\, d(x_{n-2}, x_{n-1}) \leq \cdots \leq a^{n-1} d(x_1, x_2).$$

Consequently, if $m < n$ we have

$$\begin{aligned} d(x_m, x_n) &\leq d(x_m, x_{m+1}) + d(x_{m+1}, x_{m+2}) + \ldots + d(x_{n-1}, x_n) \\ &\leq (a^{m-1} + a^m + \ldots + a^{n-2})\, d(x_1, x_2) \\ &= \frac{a^{m-1}(1 - a^{n-m})}{1-a} d(x_1, x_2) \\ &\leq \frac{a^{m-1}}{1-a} d(x_1, x_2) \to 0 \quad \text{as} \quad m \to \infty. \end{aligned}$$

Thus, since $d(x_m, x_n) \to 0$ as $m, n \to \infty$, $\{x_n\}$ is a Cauchy sequence. Since X is complete, there exists an $x_0 \in X$ such that $x_n \to x_0$. The limit x_0 is the fixed point of the mapping f. Indeed, since f is continuous, we have

$$f(x_0) = f\left(\lim_{n\to\infty} x_n\right) = \lim_{n\to\infty} f(x_n) = \lim_{n\to\infty} x_{n+1} = x_0.$$

Finally, to prove uniqueness of the fixed point, suppose that x_0 and y_0 are such that $f(x_0) = x_0$ and $f(y_0) = y_0$. Then

$$d(x_0, y_0) = d(f(x_0), \ f(y_0)) \le a\, d(x_0, y_0).$$

Since $a < 1$, the above inequality is possible only if $d(x_0, y_0) = 0$, and consequently $x_0 = y_0$. □

Exercises

1. Prove that a Cauchy sequence is convergent if and only if it has a convergent subsequence.

2. Prove Theorem 2.6.2.

3. Prove Theorem 2.6.4.

4. Give an example of a complete space which is not compact.

5. A metric space X is called *totally bounded* if, for every $\varepsilon > 0$, X is the union of a finite number of closed balls of radius ε. Prove that the following conditions are equivalent.

   (a) Every sequence in X has a Cauchy subsequence.

   (b) X is totally bounded.

6. Prove that the following conditions are equivalent.

   (a) X is compact.

   (b) X is complete and totally bounded.

7. Prove Cantor's intersection property: Let $S_1 \supset S_2 \supset S_3 \supset \dots$. be a decreasing sequence of nonempty closed subsets of a complete metric space. If $\sigma(S_n) \to 0$, then $\bigcap_{n=1}^{\infty} S_n$ contains exactly one point ($\sigma(S_n)$ denotes the diameter of S_n).

8. Let $\{x_n\}$ be a sequence in X. Show that the following conditions are equivalent:

   (a) $\{x_n\}$ is a Cauchy sequence;

   (b) For every increasing sequence $p_n \in \mathbb{N}$ we have $d(x_{p_{n+1}}, x_{p_n}) \to 0$;

   (c) For every pair of increasing sequences $p_n, q_n \in \mathbb{N}$ we have $d(x_{p_n}, x_{q_n}) \to 0$.

9. Let $X = (0, \infty)$ and let $d_1(x, y) = |x - y|$, $d_2(x, y) = |\ln x - \ln y|$.

   (a) Show that d_1 and d_2 are equivalent metrics on X.

   (b) Show that (X, d_1) is not complete.

   (c) Show that (X, d_2) is complete.

10. Prove that every contraction mapping is uniformly continuous.

11. Let $X = \mathcal{C}([0, 1])$ be the space of all continuous real-valued functions on the interval $[1, 0]$ with the metric defined by

$$d(f, g) = \max_{x \in [0,1]} |f(x) - g(x)|.$$

Define a mapping $T : \mathcal{C}([0, 1]) \to \mathcal{C}([0, 1])$ by

$$(Tf)(x) = \int_0^1 e^{-tx} \cos(f(t)/2) dt.$$

Prove that T is a contraction mapping. Then use Theorem 2.6.6 to show that the nonlinear integral equation

$$f(x) = \int_0^1 e^{-tx} \cos(f(t)/2) dt$$

has a unique solution.

2.7 Normed Spaces

In this section we are going to define a norm on a vector space. As we will see, this concept is similar to the metric, but it takes advantage of the algebraic structure of the vector space. Note that in the definition of the metric space, the space X is just a set. In this section we assume that X is a vector space. In some sense every normed space is a metric space and thus it has all the properties of a metric space. On the other hand, there are important theorems in normed spaces that do not hold or simply do not make sense in a general metric space. In this section we prove some of them.

Definition 2.7.1 (Normed spaces) Let X be a vector space. A function which assigns a real number $\|x\|$ to every $x \in X$ is called a *norm* if

(a) $\|x\| = 0$ implies $x = 0$,

(b) $\|\lambda x\| = |\lambda| \|x\|$ for all $x \in X$ and all $\lambda \in \mathbb{R}$,

(c) $\|x + y\| \le \|x\| + \|y\|$ for all $x, y \in X$.

A vector space with a norm is called a *normed space*.

The set $\mathbb{R}$ with $\|x\| = |x|$ is an example of a normed space. More generally, the space $\mathbb{R}^N$ with $\|x\| = |x| = \sqrt{x_1^2 + \cdots + x_N^2}$ is a normed space. Some of the examples of metric spaces given in Section 2.1 can be easily changed to examples of normed spaces. For example,

$$\mathcal{L}^1(\mathbb{R}) \quad \text{with} \quad \|f\| = \int |f|$$

$$\mathcal{C}([a,b]) \quad \text{with} \quad \|f\| = \max_{x\in[a,b]} |f(x)|$$

$$l^1 \quad \text{with} \quad \|\{x_1, x_2, \dots\}\| = \sum_{n=1}^{\infty} |x_n|.$$

These are infinite dimensional normed spaces.

Every normed space becomes a metric space if we define $d(x, y) = \|x - y\|$. Since the change from a normed space to a metric space is so simple, we usually say that a normed space is a metric space, which formally is incorrect. It is convenient to do that because we can immediately use all the concepts of metric spaces. All the theorems proved for metric spaces automatically apply to normed spaces. Now we are going to discuss some properties of normed spaces that we do not have in general metric spaces.

Theorem 2.7.1 *Let $\|\cdot\|_1$ and $\|\cdot\|_2$ be norms on a vector space X. Then $\|\cdot\|_1$ and $\|\cdot\|_2$ are equivalent (that is, metrics defined by $\|\cdot\|_1$ and $\|\cdot\|_2$ are equivalent) if and only if there exist positive numbers α and β such that $\alpha\|x\|_1 \le \|x\|_2 \le \beta\|x\|_1$ for all $x \in X$.*

Proof. Assume that the norms are equivalent. Then $\|x_n\|_1 \to 0$ if and only if $\|x_n\|_2 \to 0$. Suppose there is no $\alpha > 0$ such that $\alpha\|x\|_1 \le \|x\|_2$ for every $x \in X$. Then, for each $n \in \mathbb{N}$, there exists $x_n \in X$ such that $\frac{1}{n}\|x_n\|_1 > \|x_n\|_2$. Define

$$y_n = \frac{1}{\sqrt{n}} \frac{x_n}{\|x_n\|_2}.$$

Then $\|y_n\|_2 = \frac{1}{\sqrt{n}} \to 0$. On the other hand, $\|y_n\|_1 \ge n\|y_n\|_2 = \sqrt{n}$. This contradiction shows that a number α with the required property exists. The existence of a number β can be proved in a similar way.

The other implication is obvious. □

Let $\{x_n\}$ be a sequence in a normed space X. The series $\sum_{k=1}^{\infty} x_k$ is called convergent if the sequence of partial sums converges in X. That is, there exists $x \in X$ such that $\|x_1 + x_2 + \cdots + x_n - x\| \to 0$ as $n \to \infty$. In this case we write $\sum_{n=1}^{\infty} x_n = x$. If $\sum_{n=1}^{\infty} \|x_n\| < \infty$, then the series is called *absolutely convergent*. We know that a convergent series need not be absolutely convergent; consider for example the series $\sum_{k=1}^{\infty} \frac{(-1)^k}{k}$. On the other hand, every absolutely convergent series of real numbers converges. In a general normed space, this need not be true (see Exercise 13). However, every complete normed space has the property. As with other concepts, completeness of a normed space can be defined as completeness with respect to the metric defined by the norm. Complete normed spaces are called Banach spaces. It turns out that the convergence of absolutely convergent series is equivalent to completeness.

Theorem 2.7.2 *A normed space is complete if and only if every absolutely convergent series converges.*

Proof. Let X be a Banach space and let $\sum_{k=1}^{\infty} x_k$ be an absolutely convergent series in X. Define $s_n = x_1 + \cdots + x_n$ for $n = 1, 2, \ldots$. We will show that the sequence $\{s_n\}$ is a Cauchy sequence. Let $\varepsilon > 0$ and let k be a positive integer such that $\sum_{n=k+1}^{\infty} \|x_n\| < \varepsilon$. Then, for every $n > m > k$, we have

$$\|s_n - s_m\| = \|x_{m+1} + \cdots + x_n\| \leq \sum_{r=m+1}^{\infty} \|x_r\| < \varepsilon.$$

Thus $\{s_n\}$ is a Cauchy sequence in X. Since X is complete, there exists $x \in X$ such that $s_n \to x$. But this means $\sum_{n=1}^{\infty} x_n = x$.

Assume now that X is a normed space in which every absolutely convergent series converges. Let $\{x_n\}$ be a Cauchy sequence in X. We have to show that $\{x_n\}$ converges in X. By the definition of Cauchy sequences, for every $k \in \mathbb{N}$ there exists $p_k \in \mathbb{N}$ such that

$$\|x_n - x_m\| < 2^{-k} \quad \text{for all} \quad m, n \geq p_k.$$

Without loss of generality, we can assume that the sequence $\{p_n\}$ is strictly increasing. Since the series $\sum_{k=1}^{\infty}(x_{p_{k+1}} - x_{p_k})$ is absolutely convergent, it is convergent and thus the sequence

$$x_{p_k} = x_{p_1} + (x_{p_2} - x_{p_1}) + \cdots + (x_{p_k} - x_{p_{k-1}})$$

converges to some $x \in X$. Consequently

$$\|x_n - x\| \leq \|x_n - x_{p_n}\| + \|x_{p_n} - x\| \to 0,$$

because $\{x_n\}$ is a Cauchy sequence. This shows that $x_n \to x$ and thus the proof is complete. □

Exercises

1. Show that $\mathcal{L}^1(\mathbb{R})$ with the norm $\|f\| = \int |f|$ is a normed space.

2. Show that $\mathcal{C}([a, b])$ with the norm $\|f\| = \max_{x \in [a,b]} |f(x)|$ is a normed space.

3. Show that l^1 with the norm $\|\{x_1, x_2, \ldots\}\| = \sum_{n=1}^{\infty} |x_n|$ is a normed space.

4. Show that l^∞ with the norm $\|\{x_1, x_2, \ldots\}\| = \sup_{n \in \mathbb{N}} |x_n|$ is a normed space.

5. Give definitions of the following concepts in terms of a norm:

   (a) open and closed balls, open and closed sets,

   (b) convergence, continuity, uniform continuity,

   (c) compact sets,

   (d) Cauchy sequences, completeness.

6. Show that in testing for equivalence of norms it suffices to consider only sequences which converge to 0.

7. Let $(X_1, \|\cdot\|_1)$ and $(X_2, \|\cdot\|_2)$ be normed spaces. Show that the space $X_1 \times X_2$ with the norm defined by $\|(x_1, x_2)\| = \|x_1\|_1 + \|x_2\|_2$ is a normed space.

8. These are equivalent norms in $\mathbb{R}^N$:

$$\|x\|_1 = \sqrt{\sum_{n=1}^{N} x_n^2}$$

$$\|x\|_2 = |x_1| + \ldots + |x_N|$$

$$\|x\|_3 = \max\{|x_1|, \ldots, |x_N|\}.$$

   For each pair of these norms find the numbers α and β in Theorem 2.7.1.

9. Show that the norms $\|f\|_1 = \int |f|$ and $\|f\|_2 = \max_{x\in[a,b]} |f(x)|$ are not equivalent in $\mathcal{C}([a, b])$.

10. Show that the norms $\|\{x_1, x_2, \ldots\}\|_1 = \sum_{n=1}^{\infty} |x_n|$ and $\|\{x_1, x_2, \ldots\}\|_2 = \sup_{n\in\mathbb{N}} |x_n|$ are not equivalent in l^1.

11. Prove that equivalent norms define the same bounded sets.

12. Prove that if $\{x_n\}$ is a Cauchy sequence, then the sequence $\{\|x_n\|\}$ converges.

13. Let $X = \mathcal{C}([0, 1])$ and let $\|f\| = \int |f|$. Show that the series $\sum_{k=1}^{\infty} \frac{x^k}{k}$ is absolutely convergent, but it is not convergent.

14. Let X be the space of all linear transformations from $\mathbb{R}^N$ into $\mathbb{R}^N$.

    (a) Is $\|f\| = \det(f)$ a norm in X?

    (b) Is $\|f\| = |\det(f)|$ a norm in X?

    (c) Can you define a norm in X?

15. Let X be the space of all linear transformations from $\mathbb{R}^K$ into $\mathbb{R}^N$ where $K \le N$.

    (a) Is $\|f\| = \mathcal{D}(f)$ a norm in X?

    (b) Is $\|f\| = \mathcal{D}(f)$ a norm in a subspace of X?

    (c) Can you define a norm in X?

    (The definition of $\mathcal{D}$ is given in Section 1.9).

16. Let $X = \mathcal{C}(\mathbb{R})$. Does there exist a norm on X which defines the following convergence in X:

$$f_n \to f \text{ if } \max_{x\in[a,b]} |f_n(x) - f(x)| \to 0 \text{ for every } -\infty < a < b < \infty?$$

    Does there exist a metric on X which defines this convergence?

3

DIFFERENTIATION

3.1 Rates of Change and Derivatives as Linear Transformations

Let $f : \mathbb{R}^N \to \mathbb{R}$. By the *partial derivative of f with respect to its ith variable* we mean the function

$$D_i f(x) = \lim_{\lambda \to 0} \frac{f(x + \lambda e_i) - f(x)}{\lambda}.$$

(Remember that e_i is the vector with 1 in the ith coordinate and 0 everywhere else.) This is also denoted by the symbol $\frac{\partial f}{\partial x_i}$. The domain of this function is, of course, the set of all x for which the limit exists. We recall from calculus that in terms of computing a partial derivative from a given function, we simply regard all variables except the ith one as constants and apply standard differentiation rules.

For repeated partial derivatives we use a symbolism such as $D_{ij} f$ or $\frac{\partial^2 f}{\partial x_j \partial x_i}$ for $D_j(D_i f)$. For $D_{ii} f$ we also use $\frac{\partial^2 f}{\partial x_i^2}$. We can extend this symbolism to derivatives of arbitrarily high order: For instance, $D_{j_1 j_2 \ldots j_k} f(x)$ means that $D_{j_k}(D_{j_{k-1}}(\ldots(D_{j_1} f(x))\ldots))$.

Example 3.1.1 Let $f(x_1, x_2, x_3) = x_1^2 + 4x_1x_2x_3$. Then

$$D_1 f(x_1, x_2, x_3) = 2x_1 + 4x_2x_3,$$
$$D_2 f(x_1, x_2, x_3) = 4x_1x_3,$$

$$\begin{aligned}
D_3 f(x_1, x_2, x_3) &= 4x_1x_2, \\
D_{11} f(x_1, x_2, x_3) &= 2, \\
D_{23} f(x_1, x_2, x_3) &= 4x_1, \\
D_{123} f(x_1, x_2, x_3) &= 4.
\end{aligned}$$

Partial derivatives are easy to compute, but unfortunately they do not capture all the information we would like to have about rates of change of a function. The difficulty is that they only tell us about rates of change of f in directions parallel to the coordinate axes. It is quite conceivable that the rates of change of f in *other* directions could be anything at all with absolutely no useful relation to the values of $D_i f(x)$. To really understand rates of change, we need a more subtle and more powerful idea than that of partial derivatives.

When we talk about the derivative $f'(x)$ of a function $f : \mathbb{R} \to \mathbb{R}$, we think of $f'(x)$ as being the slope of the line that is tangent to the graph of f at the point $(x, f(x))$, and it may be intuitively helpful to say that f "behaves like a straight line" in the vicinity of $(x, f(x))$. It would be consonant with this way of thinking to say that f is *differentiable* at x if and only if there exist a number m and a function $g(h)$, both dependent on our choice of x, having the properties that

$$f(x+h) - f(x) = mh + g(h)$$

and

$$\lim_{h \to 0} \frac{g(h)}{h} = 0.$$

(Of course it turns out that m is what we mean by $f'(x)$.) Thus differentiability amounts to saying that $f(x+h) - f(x) \approx mh$ where the function $h \mapsto mh$ is a linear transformation of $\mathbb{R}$ to $\mathbb{R}$.

We can take this same point of view when attempting to extend the idea of differentiability to higher dimensions.

Definition 3.1.1 Suppose $f : A \to \mathbb{R}^N$ where $A \subseteq \mathbb{R}^M$. We will say that f is *differentiable* at $x \in A^\circ$ if and only if we can find a linear transformation F and a function $g(h)$, dependent on our choice of x, such that

$$f(x+h) - f(x) = F(h) + g(h)$$

and

$$\lim_{h \to 0} \frac{g(h)}{|h|} = 0.$$

(In this case we must divide by $|h|$ rather than h because h is a vector.) Notice that this is the same as saying we can find a linear transformation F, dependent on x, which satisfies

$$\lim_{h \to 0} \frac{f(x+h) - f(x) - F(h)}{|h|} = 0.$$

It will follow from the proof of the next theorem that this linear transformation is unique. That is, if F and G are linear transformations that satisfy

$$\lim_{h \to 0} \frac{f(x+h) - f(x) - F(h)}{|h|} = 0 \quad \text{and} \quad \lim_{h \to 0} \frac{f(x+h) - f(x) - G(h)}{|h|} = 0,$$

then it must be true that $F = G$. We represent this F by the symbol $f'(x)$ or $Df(x)$ and refer to it as *the derivative of f at x*. If f is differentiable at every point of a set U, we say it is *differentiable over U*. If f is differentiable at every point of its domain, we simply say that f is *differentiable.*

At this point we introduce some new notation. Derivatives in our new sense require us to think about taking limits at a point x from all possible directions. It is easier to do this if we know there is some open set U containing x such that f is defined on U. At the same time it may be important to know that the domain of f is in $\mathbb{R}^M$. Therefore when we write

$$f : \mathbb{R}^M \rightarrowtail \mathbb{R}^N$$

we shall mean that the range of f is a subset of $\mathbb{R}^N$ and the domain of f is an *open* subset of $\mathbb{R}^M$.

Theorem 3.1.1 *If $f : \mathbb{R}^M \rightarrowtail \mathbb{R}^N$ is differentiable at x, then*

$$[f'(x)] = \begin{pmatrix} D_1 f_1(x) & D_2 f_1(x) & \dots & D_M f_1(x) \\ D_1 f_2(x) & D_2 f_2(x) & \dots & D_M f_2(x) \\ \dots & & & \\ D_1 f_N(x) & D_2 f_N(x) & \dots & D_M f_N(x) \end{pmatrix}$$

where $f_1, f_2, \dots, f_N$ are the coordinate functions of f.

Proof. Let x be a point in the domain of f. We know that

$$\lim_{|h| \to 0} \frac{|f(x+h) - f(x) - f'(x)(h)|}{|h|} = 0.$$

Notice that for $\lambda \neq 0$ we must have

$$\frac{f(x+\lambda e_j) - f(x) - f'(x)(\lambda e_j)}{\lambda} = \frac{f(x+\lambda e_j) - f(x)}{\lambda} - f'(x)(e_j).$$

Now let $\lambda \to 0$. Notice this expression goes to zero and $f'(x)(e_j)$ remains unchanged in the limit. At the same time

$$\frac{f(x+\lambda e_j) - f(x)}{\lambda}$$
$$= \left(\frac{f_1(x+\lambda e_j) - f_1(x)}{\lambda}, \frac{f_2(x+\lambda e_j) - f_2(x)}{\lambda}, \dots, \frac{f_N(x+\lambda e_j) - f_N(x)}{\lambda} \right)$$

which becomes $(D_j f_1(x), D_j f_2(x), \dots, D_j f_N(x))$ as $\lambda \to 0$. Therefore

$$f'(x)(e_j) = (D_j f_1(x), D_j f_2(x), \dots, D_j f_N(x)).$$

Since $f'(x)(e_j)$ is the jth column vector of the matrix of $f'(x)$, this completes the proof. □

The matrix $[D_j f_i(x)]$ is known as the *Jacobian matrix* of f.

Example 3.1.2 Consider the function $f : \mathbb{R}^2 \to \mathbb{R}^2$ defined by $f(r, \theta) = (r\cos\theta, r\sin\theta)$. If we set $x = r\cos\theta$ and $y = r\sin\theta$, we see this is our old friend, the transformation from polar to Cartesian coordinates. We have

$$[f'(r, \theta)] = \begin{pmatrix} \cos\theta & -r\sin\theta \\ \sin\theta & r\cos\theta \end{pmatrix}.$$

As indicated earlier, the existence of the Jacobian matrix is not enough to assure us that f is differentiable. For that we need a stronger condition, one involving the existence of continuous partial derivatives.

Example 3.1.3 Let

$$g(t) = \begin{cases} \frac{1}{6}t^3 & \text{if } t \geq 0, \\ -\frac{1}{6}t^3 & \text{if } t \leq 0. \end{cases}$$

Note that g' and g'' exist and are continuous functions. Indeed $g''(t) = |t|$. However g''' does not exist at $t = 0$. In accord with the definition we are about to state formally, g is a function of class C^2.

Example 3.1.4 Let g be the same function as in the last example. Define $f(x, y) = g(x)e^y$. We may form any partial derivative of f in which differentiation with respect to x occurs no more than twice. (For instance, $D_{11}f$ or $D_{21212}f$.) Furthermore, every such partial will be continuous. However partial derivatives such as $D_{12121}f$ will fail to exist because they require us to differentiate g more than twice. This function is also of class C^2.

Example 3.1.5 Consider once more the function $f : \mathbb{R}^2 \to \mathbb{R}^2$ defined by $f(r, \theta) = (r\cos\theta, r\sin\theta)$. Each of its coordinate functions can be differentiated as often as we wish with respect to both r and θ, and the partial derivatives are always continuous. We say this function is of class C^∞.

Definition 3.1.2 A function $f : \mathbb{R}^M \rightarrowtail \mathbb{R}$ is said to be of class C^r at x_0 provided all partial derivatives $D_{j_1 j_2 \ldots j_k} f(x)$, where $k \leq r$, exist and are continuous at all points x of some open set U containing x_0. In these circumstances we also say f is *continuously differentiable of order r at x_0*. If we simply say f is of class C^r, then we understand it to be of class C^r at all points of its domain. We may also say that f is a C^r function or, more briefly, f is C^r. If f is a C^r function for all r, then we say that it is a C^∞ function. In the case where f is a C^1 function, we say that it is *continuously differentiable.* If $f : \mathbb{R}^M \rightarrowtail \mathbb{R}^N$, then we say that it is a C^r function if this is true of all its coordinate functions. By a C^0 function we mean one that is continuous.

Theorem 3.1.2 *If $f : \mathbb{R}^M \rightarrowtail \mathbb{R}^N$ is a C^1 function, then it is differentiable.*

Proof. Choose x from the domain of f. Let F be the linear transformation with matrix $[D_j f_i(x)]$. We need to show that

$$\lim_{h\to 0} \frac{|f(x+h) - f(x) - F(h)|}{|h|} = 0.$$

Since the domain of f is an open set, we can find some open ball U centered at x which is contained in the domain of f. For $|h|$ sufficiently small, namely, less than the radius of U, we can be sure that $x + h \in U$. It will also be useful to us to realize that for any pair of points in U, the line segment between those points will also be in U (that is, U is *convex*).

We can write $h = \lambda_1 e_1 + \lambda_2 e_2 + \cdots + \lambda_M e_M$. Let us define

$$\begin{aligned}
h_1 &= \lambda_1 e_1 + \lambda_2 e_2 + \cdots + \lambda_M e_M,\\
h_2 &= \lambda_2 e_2 + \lambda_3 e_3 + \cdots + \lambda_M e_M,\\
h_3 &= \lambda_3 e_3 + \lambda_4 e_4 + \cdots + \lambda_M e_M,\\
&\ldots\\
h_M &= \lambda_M e_M,\\
h_{M+1} &= 0.
\end{aligned}$$

Note that for $j = 1, 2, \ldots, M$ the line segment from h_j to h_{j+1} lies in U and every point on that line segment has the form $\alpha e_j + h_{j+1}$ where α varies through all the values in the closed interval from 0 to λ_j. Next we see that

$$f(x+h) - f(x) = \sum_{j=1}^{M} \left(f(x+h_j) - f(x+h_{j+1})\right).$$

For each coordinate function f_i and each choice of j we may treat $f_i(x+h_j) - f_i(x+h_{j+1})$ as a real-valued function of the single variable α, namely the function $f_i(x+\alpha e_j + h_{j+1}) - f_i(x+h_{j+1})$. If $\lambda_j \neq 0$, then by the mean value theorem for real-valued functions of a single variable it is possible to find an α_{ij} in the interval from 0 to λ_j such that

$$\begin{aligned}
f_i(x+h_j) - f_i(x+h_{j+1}) &= f_i(x+\lambda_j e_j + h_{j+1}) - f_i(x+h_{j+1})\\
&= [D_j f_i(x+\alpha_{ij} e_j + h_{j+1})]\lambda_j.
\end{aligned}$$

If $\lambda_j = 0$, then we can set $\alpha_{ij} = 0$ and the equation is satisfied trivially. Then the ith component of

$$\frac{f(x+h) - f(x) - F(h)}{|h|}$$

can be rewritten as

$$\frac{\sum_{j=1}^{M} \left(f_i(x+h_j) - f_i(x+h_{j+1})\right) - \sum_{j=1}^{M} D_j f_i(x)\lambda_j}{|h|},$$

which in turn becomes

$$\frac{\sum_{j=1}^{M}\left(D_j f_i(x+\alpha_{ij}e_j+h_{j+1})-D_j f_i(x)\right)\lambda_j}{|h|}.$$

Note that

$$|\alpha_{ij}|\le|\lambda_i|\le\sqrt{\lambda_1^2+\lambda_2^2+\cdots+\lambda_j^2}=|h_j|\le|h|,$$

so we see that $\frac{|\lambda_j|}{|h|}\le 1$ and that letting $h\to 0$ forces λ_j, α_{ij}, $h_j\to 0$. Since $D_j f_i$ is continuous on U, as $h\to 0$, we have

$$D_j f_i(x+\alpha_{ij}e_j+h_{j+1})\to D_j f_i(x).$$

This tells us that

$$\frac{f(x+h)-f(x)-F(h)}{|h|}$$
$$=\frac{\sum_{i=1}^{N}\left(\sum_{j=1}^{M}\left(D_j f_i(x+\alpha_{ij}e_j+h_{j+1})-D_j f_i(x)\right)\lambda_j\right)e_i}{|h|}\to 0$$

as $h\to 0$, and we are done. □

We adopted our view of $f'(x)$ as a linear transformation because we wanted a piece of mathematics that would allow us to talk about the instantaneous rate of change of f in any direction. We need to see that we can do that now.

Suppose that $f:\mathbb{R}^M\rightarrowtail\mathbb{R}$. Think of it as denoting a physical quantity such as temperature in a planar or 3-dimensional region. Let us fix a point x at which we wish to calculate the instantaneous rate of change of f and specify a direction in which to do it by choosing a nonzero vector v. We may assume that v is a *unit vector*, that is, that $|v|=1$. (If $|v|\ne 1$, then we replace it by $(1/|v|)v$.) This restriction on v has the consequence that for any $\lambda\in\mathbb{R}$ we have $|\lambda v|=|\lambda|$. Then we define the *directional derivative of f at the point x in the direction v* to be

$$D_v f(x)=\lim_{\lambda\to 0}\frac{f(x+\lambda v)-f(x)}{\lambda}.$$

With this definition, partial derivatives turn out to be special cases of directional derivatives: $D_i f=D_{e_i}f$. It is easy to see that the following is true:

Theorem 3.1.3 *If $f:\mathbb{R}^M\rightarrowtail\mathbb{R}$ is differentiable at x, then*

$$D_v f(x)=(D_1 f(x),D_2 f(x),\ldots,D_M f(x))\cdot v.$$

It is convenient to introduce a special vector associated with a real-valued f, the *gradient* of f at x:

$$\nabla f(x)=(D_1 f(x),D_2 f(x),\ldots,D_M f(x)).$$

Note that $D_v f(x) = \nabla f(x) \cdot v = [f'(x)]v^T = f'(x)(v)$.

Therefore for a real-valued f, knowledge that f is differentiable *and* knowledge of all the partial derivatives of f is sufficient to determine the instantaneous rate of change of f in all directions. A similar statement holds for the components of a vector-valued function if, for example, we know that each component function has continuous first partials.

The following result is left as an exercise.

Theorem 3.1.4 *Compositions of C^r functions are C^r.*

Exercises

1. Explain why, in the definition of differentiability, we must require that

$$\lim_{h \to 0} \frac{g(h)}{|h|} = 0$$

rather than simply $\lim_{h \to 0} g(h) = 0$.

2. Compute the matrix of $f'(p)$ for the following, where p is a point in the appropriate domain:

   (a) $f(x, y) = (x^2 - y^2, 2xy)$.

   (b) $f(t) = (f_1(t), f_2(t), \ldots, f_N(t))$.

   (c) $f(x_1, x_2, \ldots, x_N) = x_1^2 + x_2^2 + \cdots + x_N^2$.

   (d) $f(x, y, z, w) = (x, xy, xyz, xyzw)$.

   (e) $f(x, y) = (e^x \cos(y), e^x \sin(y))$.

3. Show that an open ball in $\mathbb{R}^N$ is a convex set. That is, show that if p and q are points in the open ball, then the points on the line segment from p to q also lie in the open ball.

4. In the proof of Theorem 3.1.2 show that for each j the line segment from h_j to h_{j+1} lies in U.

5. Prove Theorem 3.1.3.

6. Show that

$$f(x, y) = \begin{cases} \frac{xy}{x^2+y^2} & \text{if } (x, y) \neq (0, 0) \\ 0 & \text{if } (x, y) = (0, 0) \end{cases}$$

is not differentiable at $(0, 0)$ even though both $\frac{\partial f}{\partial x}(0, 0)$ and $\frac{\partial f}{\partial y}(0, 0)$ are defined.

7. Show that when $v = -e_i$, we have $D_v f = -D_i f$.

8. Show that for vectors v and w in the appropriate space and α a scalar, we have $D_{v+w} f = D_v f + D_w f$ and $D_{\alpha v} f = \alpha D_v f$.

9. Keeping in mind that if f is differentiable, then $f'(x)$ is to be thought of as a linear transformation, show that the operation of this linear transformation on the vector v is given by

$$\left(f'(x)\right)(v) = \lim_{\lambda \to 0} \frac{f(x+\lambda v) - f(x)}{\lambda}.$$

10. Show that

$$f(x) = \begin{cases} x^2 \sin(1/x) \text{ if } x \neq 0 \\ 0 \ \text{ if } x = 0 \end{cases}$$

has a derivative for all real x but is not of class C^1.

11. *An alternate definition of differentiability*: Suppose $x \in A^\circ$ and $f : A \to \mathbb{R}^N$ where $A \subseteq \mathbb{R}^M$. Show that f is differentiable at x if and only if the following two conditions hold:

    (a) For every $h \in \mathbb{R}^M$,

    $$\lim_{\lambda \to 0} \frac{1}{\lambda}(f(x+\lambda h) - f(x))$$

    exists, where $\lambda \in \mathbb{R}$.

    (b) If we set

    $$F(h) = \lim_{\lambda \to 0} \frac{1}{\lambda}(f(x+\lambda h) - f(x)),$$

    then $F : \mathbb{R}^M \to \mathbb{R}^N$ is a linear transformation.

12. Use the definition of differentiability in Exercise 11 to decide whether or not the following functions $f : \mathbb{R}^2 \to \mathbb{R}$ are differentiable at (0, 0):

    (a) $$f(u, v) = \begin{cases} \frac{u^2 v^3}{u^4 + v^6} & \text{if } (u, v) \neq (0, 0), \\ 0 & \text{otherwise.} \end{cases}$$

    (b) $$f(u, v) = |uv|.$$

13. Establish the following by induction on r:

    (a) Products of real-valued, C^r functions are C^r.

    (b) Compositions of C^r functions are C^r. (Hint: Consider $h = f \circ g$ where f and g are C^r and compute $D_i h$.)

3.2 Some Elementary Properties of Differentiation

Most of the results we present in this section simply generalize known properties of functions of a single variable. We leave the first theorem as an exercise.

Theorem 3.2.1 *Suppose $f, g : \mathbb{R}^M \rightarrowtail \mathbb{R}^N$ and $h : \mathbb{R}^M \rightarrowtail \mathbb{R}$ are differentiable and $\alpha \in \mathbb{R}$.*

(a) If f is a constant function, then $f'(x) = 0$ whenever x belongs to the domain of f.

(b) If f is a linear transformation, then $f'(x) = f$ whenever x belongs to the domain of f.

(c) $(f+g)'(x) = f'(x) + g'(x)$ whenever x belongs to the domains of both f and g.

(d) $(\alpha f)'(x) = \alpha f'(x)$ whenever x belongs to the domain of f.

(e) $(hf)'(x) = h(x)f'(x) + f(x)^T\, h'(x)$ whenever x belongs to the domains of both f and h. (Note: By $(hf)(x)$ we mean $h(x)\ f(x)$, that is, the scalar $h(x)$ times the vector $f(x)$. We must also think of $f(x)^T\ h'(x)$ as the product of $f(x)$ written as a column vector and $h'(x)$ written as a row vector.)

Theorem 3.2.2 *If $f : \mathbb{R}^M \rightarrowtail \mathbb{R}^N$ is differentiable at x, it is continuous there.*

Proof. We can write

$$f(x+h) - f(x) = f'(x)(h) + R(h)$$

where

$$\lim_{h \to 0} \frac{|R(h)|}{|h|} = 0.$$

We must have

$$\lim_{h \to 0} \left(f'(x)(h) + R(h) \right) = 0,$$

so that

$$\lim_{h \to 0} f(x+h) = f(x).$$

This establishes continuity at x. □

To every linear transformation $f : \mathbb{R}^K \to \mathbb{R}^N$ where $K \leq N$ we previously associated the number $\mathcal{D}(f)$ which is a kind of "distortion" factor for f. At this point it is helpful to associate with every linear transformation $f : \mathbb{R}^M \to \mathbb{R}^N$ another number, $\|f\|$, the *norm* of f. We define it by

$$\|f\| = \max\{|f(u)| : u \in \mathbb{R}^M \quad \text{and} \quad |u| = 1\}.$$

We also write this as

$$\|f\| = \max_{|u|=1} |f(u)|.$$

Let us introduce a special symbol,

$$S^{M-1} = \{u \in \mathbb{R}^M : |u| = 1\}.$$

This is the unit sphere in $\mathbb{R}^M$ with the origin as center. (It is intuitively reasonable to think of this as an $(M-1)$-dimensional object though we have not defined dimension in this sense.) It is a compact set. Since f is continuous, the restriction of f to S^{M-1} must assume its maximum somewhere on S^{M-1}. This guarantees the existence of $\|f\|$.

Note that for a linear transformation f we must have $|f(v/|v|)| \leq \|f\|$ for every nonzero v in the domain of f. This means that for every v in the domain of f we have $|f(v)| \leq \|f\||v|$. Uniform continuity of linear transformations is an immediate consequence of this fact since

$$|f(x) - f(y)| = |f(x-y)| \leq \|f\||x-y|.$$

The curious reader may wonder what sort of relation, if any, exists between $\mathcal{D}(f)$ and $\|f\|$. Both of them in some sense measure how "big" a linear transformation is, but these two tools are adapted to different tasks. $\mathcal{D}(f)$ measures the factor by which a linear transformation distorts volume. It also, in some sense, measures how close the linear transformation is to being one-to-one. It is a very geometric concept. The norm of a linear transformation, however, has absolutely nothing to do with being one-to-one or with volume. It can help establish local bounds for how far apart two points can be spread by a differentiable transformation. Sometimes it gives lower bounds for how close together two points can be shoved by such a transformation. This is very useful in analysis, as can be seen in the proof of the next theorem.

Theorem 3.2.3 (The chain rule) *If $f : \mathbb{R}^M \rightarrowtail \mathbb{R}^N$ is differentiable at x and $g : \mathbb{R}^N \rightarrowtail \mathbb{R}^P$ is differentiable at $f(x)$, then $g \circ f$ is differentiable at x and $(g \circ f)'(x) = \big(g'(f(x))\big) \circ f'(x)$.*

Proof. We can write

$$\begin{aligned}(g \circ f)(x+h) - (g \circ f)(x) &= g(f(x)+k) - g(f(x)) \\ &= \big(g'(f(x))\big)(k) + R_0(k)\end{aligned}$$

(where $k = f(x+h) - f(x)$ and $\lim_{k \to 0} \frac{|R_0(k)|}{|k|} = 0$)

$$\begin{aligned}&= \big(g'(f(x))\big)(f(x+h) - f(x)) + R_0(k) \\ &= \big(g'(f(x))\big)\big(f'(x)(h) + R_1(h)\big) + R_0(k),\end{aligned}$$

(where $\lim_{h \to 0} \frac{|R_1(h)|}{|h|} = 0$)

$$= \big\{\big(g'(f(x))\big) \circ (f'(x))\big\}(h) + \big(g'(f(x))\big)(R_1(h)) + R_0(k).$$

Since the domain of $g \circ f$ is an open set and x belongs to that domain, note that $(g \circ f)(x+h) - (g \circ f)(x)$ is defined provided only that h is sufficiently small. Next note that $(g'(f(x))) \circ f'(x)$ is a linear transformation. It follows from the definition of differentiability that if we can show

$$\lim_{h \to 0} \frac{|\left(g'(f(x))\right)(R_1(h)) + R_0(k)|}{|h|} = 0,$$

we are done.

Choose $\varepsilon > 0$. We may, without loss of generality, assume $\varepsilon \leq 1$. Since $f'(x)$ and $g'(f(x))$ are linear transformations, if we set $M_0 = \|g'(f(x))\|$ and $M_1 = \|f'(x)\|$, then we know that

$$|g'(f(x))(u)| \leq M_0|u| \quad \text{and} \quad |f'(x)(v)| \leq M_1|v|.$$

Because $|R_0(k)|/|k| \to 0$ as $k \to 0$, there must be a $\delta_0 > 0$ with the property that

$$|R_0(k)| \leq \varepsilon|k| \quad \text{whenever} \quad |k| < \delta_0.$$

By virtue of the fact that f is continuous at x, there must exist a $\delta_1 > 0$ such that

$$|k| = |f(x+h) - f(x)| < \delta_0 \quad \text{whenever} \quad |h| < \delta_1.$$

Because $|R_1(h)|/|h| \to 0$ as $h \to 0$, we may also assume δ_1 to be chosen in such a fashion that

$$|R_1(h)| \leq \varepsilon|h| \leq |h| \quad \text{whenever} \quad |h| < \delta_1.$$

It follows that for $|h| < \delta_1$ we have

$$\begin{aligned}
\frac{|\left(g'(f(x))\right)(R_1(h)) + R_0(k)|}{|h|} &\leq \frac{|\left(g'(f(x))\right)(R_1(h))|}{|h|} + \frac{|R_0(k)|}{|h|} \\
&\leq \frac{M_0\varepsilon|h|}{|h|} + \frac{\varepsilon|k|}{|h|} \\
&= M_0\varepsilon + \frac{\varepsilon|f(x+h) - f(x)|}{|h|} \\
&= M_0\varepsilon + \frac{\varepsilon}{|h|}|f'(x)(h) + R_1(h)| \\
&\leq M_0\varepsilon + \frac{\varepsilon|f'(x)(h)|}{|h|} + \frac{\varepsilon|R_1(h)|}{|h|} \\
&\leq M_0\varepsilon + \frac{\varepsilon M_1|h|}{|h|} + \frac{\varepsilon|h|}{|h|} \\
&= (M_0 + M_1 + 1)\varepsilon,
\end{aligned}$$

and we are done. □

Let us write the chain rule in matrix notation. Let $h = g \circ f$. We write $f = (f_1, \ldots, f_N)$ and $g = (g_1, \ldots, g_P)$ and $h = (h_1, \ldots, h_P)$ to indicate the components of the various functions. Then the chain rule amounts to the equation

$$\begin{pmatrix} D_1h_1(x) & \dots & D_Mh_1(x) \\ \dots & & \\ D_1h_P(x) & \dots & D_Mh_P(x) \end{pmatrix}$$
$$= \begin{pmatrix} D_1g_1(f(x)) & \dots & D_Ng_1(f(x)) \\ \dots & & \\ D_1g_P(f(x)) & \dots & D_Ng_P(f(x)) \end{pmatrix} \begin{pmatrix} D_1f_1(x) & \dots & D_Mf_1(x) \\ \dots & & \\ D_1f_N(x) & \dots & D_Mf_N(x) \end{pmatrix}.$$

As a final topic, let us touch on the connection between derivatives and local maxima and minima.

Definition 3.2.1 Suppose $f : A \to \mathbb{R}$ where $A \subseteq \mathbb{R}^M$. We say f has a *local maximum* at $a \in A$ provided there is an open set U containing a with the property that $f(x) \le f(a)$ for all $x \in U \cap A$. Local minimum is defined in a similar way. We say that f has a *maximum* (or a *global maximum*) at $a \in A$ provided $f(x) \le f(a)$ for all $x \in A$. A similar definition can be given for a minimum.

Theorem 3.2.4 *If a differentiable function $f : \mathbb{R}^M \rightarrowtail \mathbb{R}$ has a local maximum or a local minimum at a, then the directional derivative $D_v f(a) = 0$ for all unit vectors v. In particular $D_i f(a) = 0$ for $i = 1, 2, \dots, M$.*

Proof. Suppose f has a local maximum at a. Recall that

$$D_v f(a) = \lim_{\lambda \to 0} \frac{f(a + \lambda v) - f(a)}{\lambda}.$$

For $\lambda > 0$ and sufficiently small we see that

$$\frac{f(a + \lambda v) - f(a)}{\lambda} \le 0,$$

so we must have $D_v f(a) \le 0$. Similarly for $\lambda < 0$ and sufficiently close to 0 we see that

$$\frac{f(a + \lambda v) - f(a)}{\lambda} \ge 0,$$

so that we must have $D_v f(a) \ge 0$. Thus $D_v f(a) = 0$ and we are done. □

Exercises

1. Prove Theorem 3.2.1.

2. Let $f(t) = (\cosh(t), \sinh(t))$ and $g(x, y) = (x^2 - y^2, 2xy)$. Let $h = g \circ f$ and compute $h'(t)$ both directly and using the chain rule.

3. Show that if
$$A = \begin{pmatrix} \cos(\theta) & -\sin(\theta) \\ \sin(\theta) & \cos(\theta), \end{pmatrix}$$
then $\|A\| = 1$.

4. One learns in calculus that if w is a function of x, y, and z and each of x, y, and z is in turn a function of u and v, then

$$\frac{\partial w}{\partial u} = \frac{\partial w}{\partial x}\frac{\partial x}{\partial u} + \frac{\partial w}{\partial y}\frac{\partial y}{\partial u} + \frac{\partial w}{\partial z}\frac{\partial z}{\partial u}$$

and

$$\frac{\partial w}{\partial v} = \frac{\partial w}{\partial x}\frac{\partial x}{\partial v} + \frac{\partial w}{\partial y}\frac{\partial y}{\partial v} + \frac{\partial w}{\partial z}\frac{\partial z}{\partial v}.$$

Show this is merely a particular instance of the chain rule.

5. Suppose $f : \mathbb{R}^M \rightarrowtail \mathbb{R}$ is a differentiable function, x is a point in the domain of f, and v is a unit vector in $\mathbb{R}^M$. Define $g : \mathbb{R} \to \mathbb{R}^M$ by $g(\lambda) = x + \lambda v$ and set $h = f \circ g$. Show that $h'(0) = D_v f(x)$.

6. Suppose $f : \mathbb{R}^M \rightarrowtail \mathbb{R}$ is a differentiable function on an open set containing the line segment L from point a to point b in $\mathbb{R}^M$. Define $g : \mathbb{R} \to \mathbb{R}^M$ by $g(\lambda) = (1 - \lambda)a + \lambda b$ and set $h = f \circ g$. Show that $h'(\lambda) = (b - a) \cdot [(\nabla f)(g(\lambda))]$.

7. If $f : A \to \mathbb{R}$ where $A \subseteq \mathbb{R}^M$, define what it means to say f has a local minimum at $a \in A$. Define what it means to say f has a minimum at $a \in A$.

3.3 Taylor's Theorem, the Mean Value Theorem, and Related Results

In order to derive or even state Taylor's theorem, it is convenient to introduce the differential operator $v \cdot \nabla$ where v is a vector. We define

$$(v \cdot \nabla) f = v \cdot (\nabla f)$$

where f is a real-valued function. If $f : \mathbb{R}^M \rightarrowtail \mathbb{R}$ and $v = (v_1, v_2, \ldots, v_M)$, this amounts to

$$(v \cdot \nabla) f = v_1 D_1 f + v_2 D_2 f + \cdots + v_M D_M f.$$

For $j = 1, 2, 3, \ldots$, we set

$$(v \cdot \nabla)^{j+1} f = (v \cdot \nabla)\left\{[v \cdot \nabla]^j f\right\}.$$

For completeness we also set

$$(v \cdot \nabla)^0 f = f.$$

Example 3.3.1 Suppose $f : \mathbb{R}^2 \rightarrowtail \mathbb{R}$ and $v = (v_1, v_2)$. Then

$$\begin{aligned}(v \cdot \nabla)^2 f &= (v \cdot \nabla)(v_1 D_1 f + v_2 D_2 f)\\ &= v_1 D_1(v_1 D_1 f + v_2 D_2 f) + v_2 D_2(v_1 D_1 f + v_2 D_2 f)\\ &= v_1^2 D_1^2 f + 2 v_1 v_2 D_1 D_2 f + v_2^2 D_2^2 f.\end{aligned}$$

This means that, in effect,

$$(v \cdot \nabla)^2 = (v_1 D_1 + v_2 D_2)^2 = v_1^2 D_1^2 + 2v_1 v_2 D_1 D_2 + v_2^2 D_2^2$$

as though one were squaring an algebraic expression. Similarly, for any natural number k we have

$$(v \cdot \nabla)^k = (v_1 D_1 + v_2 D_2)^k$$

where the expression on the right is expanded according to the pattern of the binomial theorem.

Theorem 3.3.1 (Taylor's theorem) *If $f : \mathbb{R}^M \rightarrowtail \mathbb{R}$ is a C^{r-1} function, $r \geq 1$, at every point of the line segment L joining given points a and b and if all the partial derivatives of f of order $r - 1$ are differentiable at every point of L, then there is some point $c \in L$ distinct from a and b such that*

$$f(b) = \sum_{j=0}^{r-1} \frac{1}{j!} \left\{[(b-a)\cdot\nabla]^j f\right\}(a) + \frac{1}{r!}\left\{[(b-a)\cdot \nabla]^r f\right\}(c).$$

Proof. Set $g(\lambda) = (1-\lambda)a + \lambda b$. As λ ranges from 0 to 1, we see that $g(\lambda)$ ranges over precisely the points of L. Set $h(\lambda) = (f \circ g)(\lambda)$ and define

$$\begin{aligned} H(\lambda) =& h(1) - h(\lambda) - h'(\lambda)(1-\lambda) - \frac{h^{(2)}(\lambda)}{2!}(1-\lambda)^2 - \dots \\ & - \frac{h^{(r-1)}(\lambda)}{(r-1)!}(1-\lambda)^{r-1} - C\,(1-\lambda)^r \end{aligned}$$

where C is a constant chosen so that $H(0) = 0$. We see that

$$\begin{aligned} h(1) &= f(b), \\ h'(\lambda) &= \{[(b-a)\cdot\nabla]\, f\}\,(g(\lambda)), \end{aligned}$$

and

$$h^{(j)}(\lambda) = \left\{[(b-a)\cdot\nabla]^j f\right\}(g(\lambda))$$

for $j = 1, 2, \dots, r$. Notice we already have $H(1) = 0$ and H must be differentiable and continuous over the interval $[0, 1]$. By the mean value theorem for functions of a single variable, there must be some $\lambda_0 \in (0, 1)$ such that

$$H'(\lambda_0) = H(1) - H(0) = 0.$$

Now

$$\begin{aligned} H'(\lambda) &= -h'(\lambda) + h'(\lambda) - h^{(2)}(\lambda)\,(1-\lambda) + h^{(2)}(\lambda)(1-\lambda) - \frac{h^{(3)}(\lambda)}{2!}(1-\lambda)^2 \\ &\quad + \dots + \frac{h^{(r-1)}(\lambda)}{(r-2)!}(1-\lambda)^{r-2} - \frac{h^{(r)}(\lambda)}{(r-1)!}(1-\lambda)^{r-1} + rC(1-\lambda)^{r-1} \\ &= -\frac{h^{(r)}(\lambda)}{(r-1)!}(1-\lambda)^{r-1} + rC(1-\lambda)^{r-1}. \end{aligned}$$

From $H'(\lambda_0) = 0$ we deduce that

$$C = \frac{h^{(r)}(\lambda_0)}{r!}.$$

Set $c = g(\lambda_0)$. This gives us our desired point on L. Since

$$\begin{aligned} h(0) &= f(a), \\ h^{(j)}(0) &= \left\{[(b-a)\cdot\nabla]^j f\right\}(a), \end{aligned}$$

and

$$h^{(r)}(\lambda_0) = \left\{[(b-a)\cdot\nabla]^j f\right\}(c),$$

it follows that the equation $H(0) = 0$ is equivalent to

$$f(b) = \sum_{j=0}^{r-1} \frac{1}{j!} \left\{[(b-a)\cdot\nabla]^j f\right\}(a) + \frac{1}{r!}\left\{[(b-a)\cdot\nabla]^r f\right\}(c),$$

and we are done. □

Definition 3.3.1 By the *Taylor polynomial* of degree k centered at a for the function f we mean

$$p_k(x) = \sum_{j=0}^{k} \frac{1}{j!}\left\{[(x-a)\cdot\nabla]^j f\right\}(a).$$

Example 3.3.2 Let us find the Taylor polynomial of degree 3 centered at $(0, 0)$ for $f(x, y) = \sin(y)/(1-x)$. We must have

$$\begin{aligned} p_3(x, y) = {} & f(0,0) \\ & + xD_1 f(0,0) + yD_2 f(0,0) \\ & + \frac{1}{2}\left(x^2 D_{11} f(0,0) + 2xyD_{12} f(0,0) + y^2 D_{22} f(0,0)\right) \\ & + \frac{1}{6}\left(x^3 D_{111} f(0,0) + 3x^2 y D_{112} f(0,0) + 3xy^2 D_{122} f(0,0)\right. \\ & \left. + y^3 D_{222} f(0,0)\right). \end{aligned}$$

It is straightforward to compute that

$$\begin{aligned} f(0,0) &= D_1 f(0,0) = D_{11} f(0,0) = D_{22} f(0,0) \\ &= D_{111} f(0,0) = D_{122} f(0,0) = 0, \\ D_2 f(0,0) &= D_{12} f(0,0) = 1, \\ D_{112} f(0,0) &= 2, \end{aligned}$$

and

$$D_{222}f(0,0) = -1.$$

Therefore

$$p_3(x, y) = y + xy + x^2y - \frac{1}{6}y^3.$$

Corollary 3.3.1 (Mean value theorem) *If $f : \mathbb{R}^M \rightarrowtail \mathbb{R}$ is differentiable at every point of the line segment L joining a and b, then there is some point $c \in L$ distinct from a and b such that $f(b) - f(a) = \nabla f(c) \cdot (b - a)$.*

Proof. This is simply Taylor's theorem with $r = 1$. □

We will call an open subset U of $\mathbb{R}^M$ *connected* provided that for every $x, y \in U$ we can find a sequence of points $x_0, x_1, \ldots, x_K$ such that

(1) $x = x_0$ and $y = x_K$,

and

(2) for $i = 1, \ldots, K$ the line segment from x_{i-1} to x_i lies in U.

Theorem 3.3.2 *If f is a real-valued, differentiable function such that $f' = 0$ on an open, connected set U, then f is a constant function on U.*

Proof. Choose x and y, distinct points of U. We consider only the case where the line segment L, having x and y as endpoints, lies entirely in U. Set $g(\lambda) = f((1-\lambda)x + \lambda y)$ where $0 \leq \lambda \leq 1$. Then g is a differentiable function and $g'(\lambda) = (\nabla f((1-\lambda)x + \lambda y)) \cdot (y - x)$. Every point of the form $(1-\lambda)x + \lambda y$ lies on L, hence in U, so $g'(\lambda) = 0$. By the mean value theorem, there must be some $\lambda_0 \in (0, 1)$ such that $g(1) - g(0) = g'(\lambda_0) = 0$. This amounts to $f(x) = f(y)$, which means f is constant. □

The sense of the next theorem is that the order in which one carries out differentiations is irrelevant as long as the function is continuously differentiable to a sufficiently high degree.

Theorem 3.3.3 *Suppose $f : \mathbb{R}^M \rightarrowtail \mathbb{R}$ is a C^r function and $i_1, i_2, \ldots, i_r$ are numbers drawn from $\{1, 2, \ldots, M\}$, repetitions being allowed. If $j_1, j_2, \ldots, j_r$ is any permutation of $i_1, i_2, \ldots, i_r$, then*

$$D_{i_1 i_2 \ldots, i_r} f = D_{j_1 j_2 \ldots, j_r} f.$$

Proof. We need consider only the case where $f\ \mathbb{R}^2 \rightarrowtail \mathbb{R}$ is a C^2 function and show $D_{12}f = D_{21}f$. Let (x, y) be a point in the domain of f. We can find a ball B centered at (x, y) which lies in the domain of f. The numbers h and k are nonzero numbers sufficiently small so that the points considered below will all lie in B. We set

$$\Delta = f(x+h, y+k) - f(x+h, y) - f(x, y+k) + f(x, y)$$

and $g(u, v) = f(u, v) - f(u, y)$ so that

$$\Delta = g(x+h, y+k) - g(x, y+k).$$

Then by the mean value theorem we can write

$$\begin{aligned}\frac{\Delta}{hk} &= \frac{D_1 g(u, y+k)}{k} \\ &= \frac{D_1 f(u, y+k) - D_1 f(u, y)}{k} \\ &= D_{12} f(u, v)\end{aligned}$$

for some u between x and $x+h$ and some v between y and $y+k$. Since f is C^2, we can let $h, k \to 0$ and obtain

$$\lim_{h,k \to 0} \frac{\Delta}{hk} = D_{12} f(x, y).$$

However we can use a similar argument to show this same limit is $D_{21} f(x, y)$, and so we are done. □

Exercises

1. Assuming the hypotheses and definitions of Theorem 3.3.1, prove that

$$h^{(j)}(\lambda) = \left\{[(b-a)\cdot \nabla]^j f\right\}(g(\lambda)) \quad \text{for} \quad j = 1, 2, \ldots, r.$$

2. (a) Assume that $M = 2$ and write the third degree Taylor polynomial for f without using the $\sum$ or ∇ notation. (Note that as a mnemonic we may write

$$((x-a)\cdot\nabla)^j = ((x_1 - a_1)D_1 + (x_2 - a_2)D_2)^j$$

and then expand the second expression according to the pattern of the binomial theorem.)

(b) Now assume that $M = 3$ and write out the second degree Taylor polynomial for f without using the $\sum$ or ∇ notation.

3. Find the Taylor polynomial of fourth degree centered at $(0, 0)$ for $f(x, y) = 1/[(1-x)(1-y)]$.

4. Find the Taylor polynomial of second degree centered at $(0, 0)$ for $f(x, y, z) = z^2 e^x \cos(y)$.

5. Suppose the equation

$$f(x, y) = \sum_{i=0}^{\infty}\sum_{j=0}^{\infty} \alpha_{ij}(x - c_1)^i (y - c_2)^j$$

holds in some open set containing the point $c = (c_1, c_2)$. Assume that f can be differentiated to arbitrarily high orders by simply differentiating the series termwise. Show

$$\alpha_{ij} = \frac{\binom{k}{i}}{k!} D_1^i D_2^j f(c_1, c_2),$$

where $k = i + j$ and $\binom{k}{i}$ is a binomial coefficient.

6. Show that an open ball is an open connected set.

7. Show that if U and V are open connected subsets of $\mathbb{R}^M$ and $U \cap V \neq \emptyset$, then $U \cup V$ is also an open connected set.

8. True or false: If U and V are open connected sets with nonempty intersection, then $U \cap V$ is also open and connected.

9. True or false: If U and V are open connected subsets of $\mathbb{R}^M$ and $\mathbb{R}^N$ respectively, then $U \times V$ is an open, connected subset of $\mathbb{R}^{M+N}$.

10. Complete the proof of Theorem 3.3.2 by considering the case where x and y are not endpoints of a line segment lying in U.

11. Give an example of a differentiable function f such that $f' = 0$ on an open set but f is not a constant function.

12. Look at the proof of Theorem 3.3.3 and construct an argument to show that

$$\lim_{h,k \to 0} \frac{\Delta}{hk} = D_{21} f(x, y).$$

3.4 Norm Properties

Before confronting the inverse function theorem, we need to know more about the properties of the norm.

Recall that for a linear transformation $f : \mathbb{R}^M \to \mathbb{R}^N$ we define

$$\|f\| = \max_{|u|=1} |f(u)|.$$

If A is the matrix of the linear transformation f, we shall feel free to write $\|A\|$. For the identity matrix I, trivially, we have $\|I\| = 1$.

More generally consider the linear transformation $f : \mathbb{R}^N \to \mathbb{R}^N$ with matrix

$$A = \begin{pmatrix} \alpha_1 & 0 & \dots & 0 \\ 0 & \alpha_2 & \dots & 0 \\ \dots & & & \\ 0 & 0 & \dots & \alpha_N \end{pmatrix}.$$

Note that $f(e_i) = \alpha_i e_i$ so that $|f(e_i)| \le \max_j |\alpha_j|$. But there must be some k such that $f(e_k) = \max_j |\alpha_j|$. If $x = (x_1, x_2, \dots, x_N)$ is an arbitrary unit vector, then

$$\begin{aligned} |f(x)| &= |(\alpha_1 x_1, \alpha_2 x_2, \dots, \alpha_N x_N)| \\ &= \sqrt{\alpha_1^2 x_1^2 + \alpha_2^2 x_2^2 + \cdots + \alpha_N^2 x_N^2} \\ &\le \sqrt{(x_1^2 + x_2^2 + \cdots + x_N^2)\left(\max_j |\alpha_j|\right)^2} \\ &= \max_j |\alpha_j|. \end{aligned}$$

So by the definition of norm, $\|f\| = \max_j |\alpha_j|$.

There is an alternate and useful characterization of the norm whose proof we leave as an exercise.

Theorem 3.4.1 *If $f : \mathbb{R}^K \to \mathbb{R}^N$ is a linear transformation, then $\|f\|$ is the smallest constant C satisfying $|f(x)| \le C|x|$ for all $x \in \mathbb{R}^K$.*

Although we have called $\|f\|$ the norm of f, we have not shown any connection with the theory of norms on vector spaces (normed spaces). We now remedy that omission. Note that for two given Euclidean spaces the set of linear transformations from one space to another is itself a vector space. The derivation of this fact was an exercise in Chapter 1.

Theorem 3.4.2 *If $\mathcal{L}$ is the set of linear transformations of $\mathbb{R}^K$ to $\mathbb{R}^N$, then $\|\cdot\|$ is a norm on $\mathcal{L}$.*

Theorem 3.4.3 *If the linear transformation f has an inverse, then*

$$\|f\|\|f^{-1}\| \ge 1 \quad \textit{and} \quad |f(x)| \ge \frac{|x|}{\|f^{-1}\|} \quad \textit{for all} \quad x.$$

Proof. Let u be any unit vector. Then

$$1 = |u| = |f(f^{-1}(u))| \le \|f\||f^{-1}(u)| \le \|f\|\|f^{-1}\||u| = \|f\|\|f^{-1}\|.$$

This assures us that $\|f^{-1}\| \ne 0$. For any vector x we must have

$$|x| = |f^{-1}(f(x))| \le \|f^{-1}\||f(x)|$$

and hence

$$|f(x)| \ge \frac{|x|}{\|f^{-1}\|}. \qquad \square$$

We now indicate how the norm can be used to give information about the behavior of differentiable functions.

Theorem 3.4.4 *Suppose that $f : \mathbb{R}^M \rightarrowtail \mathbb{R}^N$ is C^1 and a is in the domain of f. Then for every $M > \|f'(a)\|$ there is an open set U containing a with the property that for all $x, y \in U$ we have $|f(x) - f(y)| \le M|x - y|$.*

Proof. Choose $M > \|f'(a)\|$ and set $\varepsilon = M - \|f'(a)\|$. Define $g(x) = f(x) - [f'(a)](x)$. We must have $g'(x) = f'(x) - f'(a)$ and, in particular, $g'(a) = 0$. We may write $g = (g_1, g_2, \ldots, g_N)$ where each g_i is a real-valued $\mathcal{C}^1$ function. Note that $\nabla g_i(a) = 0$ for all i. We may suppose U is an open ball containing a and is chosen in such a way that $|\nabla g_i(u)| < \varepsilon/N$ for $i = 1, 2, \ldots, N$ and for $u \in U$.

To show we have the right U, choose $x, y \in U$. By the mean value theorem, for $i = 1, 2, \ldots, N$, we can find x_i on the line segment from x to y such that

$$g_i(x) - g_i(y) = \nabla g_i(x_i) \cdot (x - y).$$

Since U is a ball, each x_i must lie in U. Therefore

$$\begin{aligned}
|g(x) - g(y)| &\le \sum_{i=1}^{N} |g_i(x) - g_i(y)| \\
&= \sum_{i=1}^{N} |\nabla g_i(x_i) \cdot (x - y)| \\
&\le \sum_{i=1}^{N} |\nabla g_i(x_i)||x - y| \\
&\le \sum_{i=1}^{N} \frac{\varepsilon}{N}|x - y| \\
&= \varepsilon\, |x - y|.
\end{aligned}$$

However we also know that

$$\begin{aligned}
|g(x) - g(y)| &= |f(x) - [f'(a)](x) - f(y) + [f'(a)](y)| \\
&\ge |f(x) - f(y)| - |[f'(a)](x - y)| \\
&\ge |f(x) - f(y)| - \|f'(a)\||x - y|.
\end{aligned}$$

It follows that

$$|f(x) - f(y)| \le (\varepsilon + \|f'(a)\|)|x - y| = M|x - y|. \qquad \square$$

Theorem 3.4.5 *Suppose* $f : \mathbb{R}^N \rightarrowtail \mathbb{R}^N$ *is* $\mathcal{C}^1$ *and* $\det(f'(a)) \neq 0$. *Then for every* $M > \|(f'(a))^{-1}\|$ *there is an open set* U *containing* a *with the property that for all* $x, y \in U$ *we have*

$$|f(x) - f(y)| \ge \frac{1}{M}|x - y|.$$

Proof. Choose $M > \|(f'(a))^{-1}\|$. Let $g(x) = f(x) - [f'(a)](x)$ and set

$$\varepsilon = \frac{1}{\|(f'(a))^{-1}\|} - \frac{1}{M}.$$

We know from the proof of the last theorem that there is an open set U containing a with the property that for all $x, y \in U$ we have $|g(x) - g(y)| \leq \varepsilon|x - y|$. Choose $x, y \in U$. Then the desired conclusion follows from

$$\begin{aligned} |f(x) - f(y)| &= |g(x) - [f'(a)](x) - g(y) + [f'(a)](y)| \\ &\geq |f'(a)(x - y)| - |g(x) - g(y)| \\ &\geq \frac{1}{\|(f'(a))^{-1}\|}|x - y| - \varepsilon|x - y| \\ &= \frac{1}{M}|x - y|. \end{aligned}$$

□

Exercises

1. Show that if f is orthogonal, then $\|f\| = 1$.
2. Prove Theorem 3.4.1.
3. Prove Theorem 3.4.2.
4. True or false: If $f : \mathbb{R}^K \to \mathbb{R}^N$ where $K \leq N$, then $\mathcal{D}(f) = \|f\|$.
5. Give an example of a linear transformation f for which $\|f\|\|f^{-1}\| > 1$.
6. Give an example of a linear transformation which is not one-to-one but has positive norm.
7. Prove that for any $x = (x_1, x_2, \ldots, x_N) \in \mathbb{R}^N$ we have
$$|x| \leq |x_1| + |x_2| + \cdots + |x_N|.$$
8. Show that if $f : \mathbb{R} \rightarrowtail \mathbb{R}$ is a differentiable function, then $\|f'(x)\| = |f'(x)|$.
9. If $f : \mathbb{R}^2 \to \mathbb{R}^2$ is given by $f(r, \theta) = (r \cos(\theta), r \sin\theta))$, then show that $\|f'(r, \theta)\| = \max\{1, |r|\}$. (Hint: Consider the action of $f'(r, \theta)$ on a vector of the form $(\cos(\alpha), \sin(\alpha))$.)

3.5 The Inverse Function Theorem

We know from introductory calculus that if a real-valued function has a nonzero derivative over an interval, then it must have an inverse function and the inverse function must be differentiable. For instance, $\sin(x)$ has a positive derivative over $-\frac{\pi}{2} < x < \frac{\pi}{2}$, and one can extract from this portion of the sine graph an inverse function known as $\sin^{-1}$ or arcsin. If we examine another portion of the sine graph over which the derivative is nonzero, say, $\frac{\pi}{2} < x < \frac{3\pi}{2}$, we find another inverse function associated with that portion of the graph.

This state of affairs generalizes to higher dimensions, though the form is not quite the same. Consider a function $f : \mathbb{R}^N \rightarrowtail \mathbb{R}^N$. The condition that the derivative not be zero now becomes $\det(f'(x)) \neq 0$. Another difference is that if I and J are intervals and $g : I \to J$ has a nonzero derivative over all of I, then there is a *global* inverse function $g^{-1} : J \to I$. On the other hand, if $f : \mathbb{R}^N \rightarrowtail \mathbb{R}^N$ has the property that $\det(f'(x))$ is never zero, it is still possible that f will not be one-to-one on its domain. (The reader is asked to show the truth of this by considering $f(x, y) = (e^x \cos(y), e^x \sin(y))$ in the exercises.) What is true is that around every point of the domain there exists a "small" open set on which f is one-to-one.

The goal of this section is called the inverse function theorem. This theorem lends itself to being broken into a number of smaller, separate results, each of which is of interest in its own right.

Theorem 3.5.1 *Suppose $f : \mathbb{R}^N \rightarrowtail \mathbb{R}^N$ is C^1 and* $\det(f'(a)) \neq 0$. *Then there is an open set U containing a with the property that f is one-to-one on U.*

Proof. Since

$$0 < \|(f'(a))^{-1}\| < 2\|(f'(a))^{-1}\|,$$

there must be an open set U containing a such that

$$|f(x) - f(y)| \geq \frac{1}{2\|(f'(a))^{-1}\|}|x - y|$$

for all $x, y \in U$. This implies the desired result. □

Example 3.5.1 Let $(x, y) = f(r, \theta) = (r\cos(\theta), r\sin(\theta))$. This is the familiar transformation from polar to Cartesian coordinates. We see that

$$\left[f'(r,\theta)\right] = \begin{pmatrix} \cos\theta & -r\sin\theta \\ \sin\theta & r\cos\theta \end{pmatrix},$$

and hence

$$(f'(r, \theta)) = r$$

which is nonzero everywhere except where $r = 0$. Let

$$U = \left\{(r, \theta) : r > 0 \quad \text{and} \quad -\frac{\pi}{2} < \theta < \frac{\pi}{2}\right\}$$

and

$$V = \{(x, y) : x > 0\}.$$

Then f takes U onto V in a one-to-one fashion. See Figure 3.5.1. The inverse function is

$$g(x, y) = \left(\sqrt{x^2 + y^2}, \tan^{-1}(y/x)\right).$$

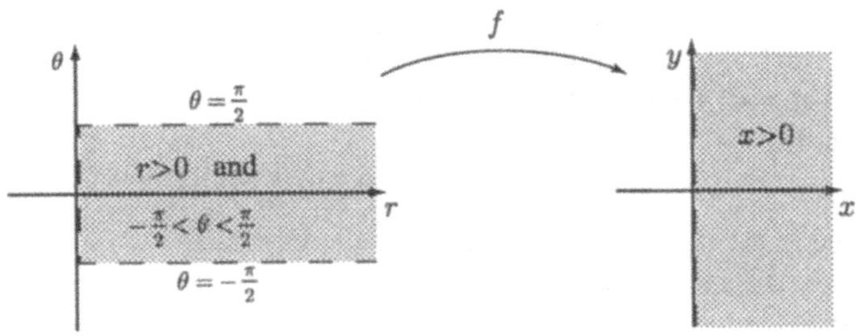

FIGURE 3.5.1.

Example 3.5.2 We take the same $f(r, \theta)$ as in the last example. Now let

$$U = \{(r, \theta) : r > 0 \quad \text{and} \quad 0 < \theta < \pi\}$$

and

$$V = \{(x, y) : y > 0\}.$$

Again f takes U onto V in a one-to-one fashion. See Figure 3.5.2. The inverse function is now

$$g(x, y) = \left(\sqrt{x^2 + y^2},\ \cot^{-1}(x/y)\right).$$

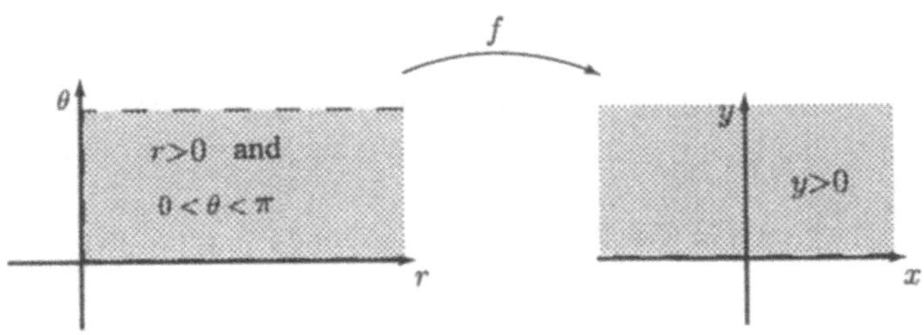

FIGURE 3.5.2.

In the last two examples f and g are both C^∞ functions. In proving the full inverse function theorem we want to show more than the existence of a local inverse function. We would like to show that if f is one-to-one and C^r on U, then the inverse function g must also be C^r. Our goal requires, as an intermediate step, to first consider when a function is *open*.

Theorem 3.5.2 *Suppose* $f : \mathbb{R}^N \rightarrowtail \mathbb{R}^N$ *is* C^1 *and* $\det(f'(a)) \neq 0$ *for all* a *in the domain of* f. *Then* f *is an open map. (That is, for every open subset* U *of* $\mathbb{R}^N$, *the set* $f(U)$ *is open.)*

Proof. Choose a from the domain of f. We will show that $f(a)$ must lie in an open ball which is also a subset of the range of f. There must be an open ball U centered at a such that:

$\overline{U}$, that is, the closure of U, is a subset of the domain of f,

f is one-to-one on some open set containing $\overline{U}$, and

$\det(f'(x)) \neq 0$ for all $x \in \overline{U}$.

Let Bd(U) stand for the boundary of U, that is, $\text{Bd}(U) = \overline{U} - U$. We see that $f(a) \notin f(\text{Bd}(U))$ and $f(\text{Bd}(U))$ is a compact set, so there must exist an open ball B centered at $f(a)$ such that $\overline{B}$ and $f(\text{Bd}(U))$ are disjoint. We shall show there is an $\varepsilon > 0$ such that

$$A = \{y \in B : |y - f(a)| < \varepsilon\}$$

is contained in the range of f. Note that A is an open ball.

Let $\varepsilon =$ the distance from $\overline{B}$ to $f(\text{Bd}(U))$. This is a positive number. Choose $y \in A$ and define a C^1 function $F : \overline{U} \to \mathbb{R}$ by $F(x) = |f(x) - y|^2$. Since $\overline{U}$ is compact, there must be some $b \in \overline{U}$ at which F assumes its minimum. By definition of ε we have $|f(x) - y| \geq \varepsilon$ for all $x \in \text{Bd}(U)$ but $|f(a) - y| < \varepsilon$, hence F cannot assume its minimum on Bd(U). Thus b must be an interior point of U, and it follows that $D_i F(b) = 0$ for $i = 1, 2, \ldots, N$. Let us introduce the notation $f = (f_1, f_2, \ldots, f_N)$ and $y = (y_1, y_2, \ldots, y_N)$. Since $\det(f'(b)) \neq 0$, the row vectors of

$$\begin{pmatrix} D_1 f_1(b) & \ldots & D_N f_1(b) \\ \ldots & & \\ D_1 f_N(b) & \ldots & D_N f_N(b) \end{pmatrix}$$

are linearly independent. Then since the equations $D_i F(b) = 0$ imply

$$\sum_{j=1}^{N} (f_j(b) - y_j) D_i f_j(b) = 0$$

for $i = 1, 2, \ldots, N$, we must have $f_j(b) - y_j = 0$ for $j = 1, 2, \ldots, N$. Thus $f(b) = y$ and we are done. □

Example 3.5.3 The function $f : \mathbb{R} \to \mathbb{R}$ defined by $f(x) = x^3$ is an open map because it takes open intervals to open intervals. However $g : \mathbb{R} \to \mathbb{R}$ defined by $g(x) = x^2$ is not open because g maps the open set $\mathbb{R}$ onto $[0, \infty)$ which is not open.

Theorem 3.5.3 *Suppose* $f : \mathbb{R}^N \rightarrowtail \mathbb{R}^N$ *is a one-to-one* C^r *function* ($r \geq 1$) *with the property that* $\det(f'(x)) \neq 0$ *for all* x *in the domain of* f. *Then* g, *the inverse function of* f, *is also* C^r. *Furthermore* $g'(y) = (f'(g(y)))^{-1}$.

Proof. Choose b in the domain of g and let $a = g(b)$. We will first show that g is differentiable at b. Notice that $\|(f'(a))^{-1}\| > 0$. Choose $\varepsilon > 0$ and $M > \|(f'(a))^{-1}\|$. We can find an open set U that contains a and is a subset of the domain of f and satisfies

$$|f(x) - f(a) - f'(a)(x - a)| \leq \varepsilon |x - a| \quad \text{for all} \quad x \in U$$

and

$$|f(x) - f(t)| \geq \frac{1}{M} |x - t| \quad \text{for all} \quad x, t \in U.$$

Let $V = f(U)$. This is an open set in $\mathbb{R}^N$. Our last condition on U can be rewritten as

$$|g(u) - g(v)| \leq M |u - v| \quad \text{for all} \quad u, v \in V.$$

Now in what follows it is to be understood that $y \in V$ and $x = g(y)$ (or, equivalently, $y = f(x)$). We must have

$$\begin{aligned}
|g(y) - g(b) - (f'(a))^{-1}(y - b)| &= |x - a - (f'(a))^{-1}(f(x) - f(a))| \\
&= \left| \left((f'(a))^{-1} \circ f'(a)\right) \left(x - a - (f'(a))^{-1}(f(x) - f(a))\right) \right| \\
&\leq \|(f'(a))^{-1}\| \, |f'(a)(x - a) - (f(x) - f(a))| \\
&\leq \|(f'(a))^{-1}\| \varepsilon |x - a| \\
&= \|f'(a))^{-1}\| \varepsilon |g(y) - g(b)| \\
&\leq \|(f'(a))^{-1}\| \varepsilon M |y - b|.
\end{aligned}$$

Thus g is differentiable at b with $g'(b) = (f'(a))^{-1}$.

We let $g = (g_1, g_2, \ldots, g_N)$. We know that $g'(y) = (f'(g(y)))^{-1}$ and therefore each $D_i g_j(y)$ is a rational function of functions of the form $D_p f_q(g(y))$. We now construct an induction argument. Since g is differentiable, it must be continuous, and we know f is C^1. It follows that each $D_p f_q\ (g(y))$ is continuous, and hence each $D_i g_j(y)$ is continuous. Thus g is C^1. Now suppose we have shown that g is C^k for some $k < r$. It follows that each $D_p f_q(g(y))$ must be C^k, and hence each $D_i g_j(y)$ must be C^k. Therefore g is a C^{k+1} function. We conclude that g must be a C^r function. □

We will summarize the preceding results in the inverse function theorem. But before stating that result, let us introduce some terms.

Given a function $f : A \to B$, if $C \subseteq A$ we denote the *restriction of f to C* by $f|C$, and we mean by this the function which satisfies $(f|C)(x) = f(x)$ and has C as its domain.

Definition 3.5.1 If U and V are open subsets of $\mathbb{R}^N$ and $f : U \to V$, then we call f a *C^r diffeomorphism of U onto V* ($r \geq 1$) provided that:

(1) f is a one-to-one, C^r function,

(2) $V = f(U)$,

and

(3) $\det(f'(x)) \neq 0$ for all $x \in U$.

By previous results, we know that compositions of C^r diffeomorphisms are also C^r diffeomorphisms and that the inverse of a C^r diffeomorphism is a C^r diffeomorphism.

Theorem 3.5.4 (The inverse function theorem) *If $f : \mathbb{R}^N \rightarrowtail \mathbb{R}^N$ is a C^r function ($r \geq 1$) and $\det(f'(a)) \neq 0$, then there are open sets U and V in $\mathbb{R}^N$ such that $a \in U$ and $f|U$ is a C^r diffeomorphism of U onto V.*

Exercises

1. Let $(x, y) = f(r, \theta) = (r\cos(\theta), r\sin(\theta))$ and set

$$U = \left\{(r, \theta) : r > 0 \quad \text{and} \quad \frac{\pi}{2} < \theta < \frac{3\pi}{2}\right\}.$$

   Find a subset V of $\mathbb{R}^2$ such that f takes U onto V in a one-to-one fashion, and find the inverse function $g : V \to U$.

2. Let $(u, v) = f(x, y) = (x, xy)$. (a) Find the set of (x, y) for which $\det(f'(x, y)) \neq 0$. (b) f takes the set $U = \{(x, y) : x > 0\}$ in a one-to-one fashion onto some open subset V of $\mathbb{R}^2$. Describe V and find the inverse function g on V.

3. Let $(u, v) = f(r, \alpha) = (r\cosh(\alpha), r\sinh(\alpha))$. (a) Find the set of points (r, α) at which $\det(f'(r, \alpha)) \neq 0$. (b) Let $U = \{(r, \alpha) : r > 0\}$. Find a set V such that f takes U onto V in a one-to-one fashion and the inverse function g on V.

4. Suppose (X, d) is a metric space and A is a nonempty, compact subset of X and x is a point of X such that $x \notin A$. Show there is some $a \in A$ such that

$$d(x, a) = \inf\{d(x, y) : y \in A\}.$$

   This number is called *the distance from x to A* and is sometimes denoted by $d(x, A)$.

5. Refer to the last exercise and show there is some open set U containing x such that $\overline{U}$ and A are disjoint.

6. Let A and B be nonempty, disjoint, compact subsets of a metric space X with metric d. Show that there exist $a \in A$ and $b \in B$ such that

$$d(a, b) = \inf\{d(x, y) : x \in A \quad \text{and} \quad y \in B\}.$$

   This number is called *the distance from A to B* and is sometimes denoted by $d(A, B)$. Conclude that $d(A, B) > 0$.

7. Define $f : \mathbb{R}^2 \to \mathbb{R}^2$ by $f(x, y) = (e^x\cos(y), e^x\sin(y))$. Show that $\det(f'(x, y)) \neq 0$ for all (x, y) but that f is not one-to-one.

8. Show that if g is a C^r diffeomorphism of U onto V and f is a C^r diffeomorphism of V onto W, then $f \circ g$ is a C^r diffeomorphism of U onto W.

9. Give an example of a continuous function $f : \mathbb{R}^2 \to \mathbb{R}^2$ that is not an open mapping.

3.6 Some Consequences of the Inverse Function Theorem

One of the main consequences of the inverse function theorem is the *implicit function theorem*. This says that if one has a system of equations (not necessarily linear) in which there are more variables than there are equations, then one can often solve for some of the variables in terms of the others.

Example 3.6.1 An important application of this notion of implicitly defined functions is the practice of implicit differentiation. The reader should be familiar with the mechanics of this from an introductory calculus course. Suppose, for instance, that we have a C^1 function $f : \mathbb{R}^3 \rightarrowtail \mathbb{R}$ and suppose further that it is possible to solve

$$f(x, y, z) = 0$$

for z as a C^1 function in terms of x and y. To find the partial derivatives of z, we think of the equation

$$f(x, y, z(x, y)) = 0$$

and differentiate first with respect to x and then with respect to y, being careful each time to apply the chain rule. This leads to

$$\frac{\partial f}{\partial x} + \frac{\partial f}{\partial z}\frac{\partial z}{\partial x} = 0 \quad \text{and} \quad \frac{\partial f}{\partial y} + \frac{\partial f}{\partial z}\frac{\partial z}{\partial y} = 0.$$

We can now solve for $\partial z/\partial x$ and $\partial z/\partial y$ assuming $\partial f/\partial z \neq 0$.

Two things are noteworthy about this process. One is that we do not have to be able to solve explicitly for z as a function of x and y. As a matter of fact, the method is most useful when we *cannot* solve explicitly for z. The second point is that to solve for these partials we need to know that $\partial f/\partial z \neq 0$.

But how can we be sure our assumption that we can solve for z in terms of x and y is valid? We use a clever trick which brings the inverse function theorem into the picture. Recall that we started with $f : \mathbb{R}^3 \rightarrowtail \mathbb{R}$. Define $h : \mathbb{R}^3 \rightarrowtail \mathbb{R}^3$ by

$$h(x, y, z) = (x, y, f(x, y, z)).$$

Notice that the matrix for $h'(x, y, z)$ is

$$\begin{pmatrix} 1 & 0 & 0 \\ 0 & 1 & 0 \\ \frac{\partial f}{\partial x} & \frac{\partial f}{\partial y} & \frac{\partial f}{\partial z} \end{pmatrix}$$

so that $\det\big(h'(x, y, z)\big) = \partial f/\partial z$. This means that if we can find an open set of points (x, y, z) over which $\partial f/\partial z$ is nonzero, then h will possess an inverse (locally) over such a set. (As above, the condition $\partial f/\partial z \neq 0$ turns out to be important.) Further, if f is C^r, we may assume the same about the inverse of h. Let us write

$$h(x, y, z) = (x, y, u)$$

where it is understood that we are confining our attention to an open set of points over which h is one-to-one. We can equally well write

$$(x, y, z) = h^{-1}(x, y, u)$$

where h^{-1} is also operating on some open subset of $\mathbb{R}^3$. Suppose we consider only points in this open set for which $u = 0$. For such points we have

$$(x, y, z) = h^{-1}(x, y, 0).$$

This means that for some open set of points (x, y) in $\mathbb{R}^2$, it follows that $z = z(x, y)$, a C^r function of x and y. For the points in question we must have

$$(x, y, 0) = h(x, y, z) = (x, y, f(x, y, z))$$

so that $f(x, y, z(x, y)) = 0$. Thus our assumption that we can solve for z in terms of x and y is justified provided we are dealing with points for which $\partial f/\partial z \neq 0$.

Example 3.6.2 Consider a case where we can actually find the implicitly defined functions. Let $f : \mathbb{R}^3 \to \mathbb{R}^2$ be defined by

$$f(x, y, z) = (3xy - z,\ 4y - x^2 y).$$

Suppose we confine our attention to those (x, y, z) which satisfy $f(x, y, z) = (0, 0)$. One readily shows that

$$y = \frac{1}{4 - x^2} \quad \text{and} \quad z = \frac{3x}{4 - x^2}.$$

That is, y and z are implicitly determined as functions of x.

What about the trick we used in the last example to introduce the inverse function theorem? Does that idea play any role here? It does, and we make that role explicit. Define

$$h(x, y, z) = (x, 3xy - z, 4y - x^2 y).$$

This is a slight modification of f and maps from $\mathbb{R}^3$ to $\mathbb{R}^3$. We have introduced as the first coordinate of h the variable in terms of which we solve for y and z. The matrix of $h'(x, y, z)$ is

$$\begin{pmatrix} 1 & 0 & 0 \\ 3y & 3x & -1 \\ -2xy & 4 - x^2 & 0 \end{pmatrix}.$$

We see that h is (locally) one-to-one precisely when $\det\big(h'(x, y, z)\big) = 4 - x^2 \neq 0$. In this case, setting $3xy - z$ and $4y - x^2 y$ equal to zero, we see that there should be open sets on which

$$(x, y, z) = h^{-1}(x, 0, 0),$$

that is, open sets on which y and z are both functions of x. Note that the condition $4 - x^2 \neq 0$ emerges from both the explicit solution for y and z and from our technique of bringing the inverse function theorem into the problem.

Before proving the implicit function theorem, we need a definition.

Definition 3.6.1 Suppose $f : \mathbb{R}^M \rightarrowtail \mathbb{R}^N$ is a differentiable function and $f = (f_1, \ldots, f_N)$. Let $i_1, i_2, \ldots, i_K$ be integers chosen from the set $\{1, 2, \ldots, N\}$ and $j_1, j_2, \ldots, j_K$ be integers chosen from the set $\{1, 2, \ldots, M\}$. We define

$$\frac{\partial(f_{i_1}, \ldots, f_{i_K})}{\partial(x_{j_1}, \ldots, x_{j_K})}(x) = \det \begin{pmatrix} D_{j_1} f_{i_1}(x) & \ldots & D_{j_K} f_{i_1}(x) \\ & \ldots & \\ D_{j_1} f_{i_K}(x) & \ldots & D_{j_K} f_{i_K}(x) \end{pmatrix}.$$

This is a *Jacobian*, the determinant of a square submatrix of the matrix of $f'(x)$.

For instance, in the last example we had $f_1(x, y, z) = 3xy - z$ and $f_2(x, y, z) = 4y - x^2y$, so

$$\frac{\partial(f_1, f_2)}{\partial(y, z)} = 4 - x^2.$$

In the next theorem and subsequent ones, it is sometimes convenient to consider Euclidean spaces of the form $\mathbb{R}^{K+M}$ or, equivalently, $\mathbb{R}^K \times \mathbb{R}^M$. When given points $z \in \mathbb{R}^{K+M}$ we shall often write $z = (x, y)$ where $x \in \mathbb{R}^K$ and $y \in \mathbb{R}^M$.

Theorem 3.6.1 (The implicit function theorem) *Suppose $f : \mathbb{R}^{K+M} \rightarrowtail \mathbb{R}^K$ is a C^r function and $f(a, b) = 0$ where $a \in \mathbb{R}^K$ and $b \in \mathbb{R}^M$. If $f = (f_1, \ldots, f_K)$, let us further suppose that*

$$\frac{\partial(f_1, \ldots, f_K)}{\partial(x_1, \ldots, x_K)}(a, b) \neq 0.$$

Then there exist an open set U in $\mathbb{R}^{K+M}$, an open set W in $\mathbb{R}^M$, and a function $g : W \to \mathbb{R}^K$ such that

(1) g is a C^r function,

(2) $(a, b) \in U$ and $b \in W$,

(3) $g(b) = a$,

(4) $f(g(y), y) = 0$ for all $y \in W$,

and

(5) g is uniquely determined in the sense that if $(x, y) \in U$ and $f(x, y) = 0$, then $x = g(y)$.

Proof. Define $h : \mathbb{R}^{K+M} \rightarrowtail \mathbb{R}^{K+M}$ by $h(x, y) = (f(x, y), y)$ where $x \in \mathbb{R}^K$ and $y \in \mathbb{R}^M$. The matrix of h' is

$$\begin{pmatrix} D_1 f_1 & \ldots & D_K f_1 & D_{K+1} f_1 & \ldots & D_{K+M} f_1 \\ \ldots & & & \ldots & & \\ D_1 f_K & \ldots & D_K f_K & D_{K+1} f_K & \ldots & D_{K+M} f_K \\ 0 & \ldots & 0 & 1 & \ldots & 0 \\ \ldots & & & \ldots & & \\ 0 & \ldots & 0 & 0 & \ldots & 1 \end{pmatrix},$$

so that

$$\det(h') = \frac{\partial(f_1, \ldots, f_K)}{\partial(x_1, \ldots, x_K)}.$$

Because

$$\frac{\partial(f_1, \ldots, f_K)}{\partial(x_1, \ldots, x_K)}(a, b) \neq 0,$$

there must exist open sets U and V of $\mathbb{R}^{K+M}$ such that $h(U) = V$, the point (a, b) lies in U, and h has a C^r inverse on V.

Let us write $h(x, y) = (u, y)$ where $(x, y) \in U$ and $(u, y) \in V$. We define

$$W = \{y \in \mathbb{R}^M : (0, y) \in V\}.$$

Since $h(a, b) = (0, b) \in V$, we know W is nonempty. We want to show that W is an open subset of $\mathbb{R}^M$. Choose $y \in W$. We know that $(0, y) \in V$ and that there must exist $\delta > 0$ with the property that if $|(u, z) - (0, y)| < \delta$, then $(u, z) \in V$. It follows that for every $z \in \mathbb{R}^M$ which satisfies $|z - y| < \delta$, since $|(0, z) - (0, y)| = |z - y|$, we must have $(0, z) \in V$ and hence $z \in W$. Therefore W is open in $\mathbb{R}^M$.

For every $(0, y) \in V$, there is a unique $(x, y) \in U$ such that $h(x, y) = (0, y)$ and hence $(x, y) = h^{-1}(0, y)$. The map

$$y \mapsto (0, y) \mapsto h^{-1}(0, y) = (x, y) \mapsto x$$

is clearly a C^r map, and we define $g : W \to \mathbb{R}^K$ by $g(y) = x$. Our definition of g is constructed in such a way that

$$f(x, y) = 0 \text{ and } (x, y) \in U \text{ if and only if } x = g(y) \text{ and } y \in W.$$

It follows immediately from this that $f(g(y), y) = 0$ for all $y \in W$ and that $b = g(a)$. Thus the theorem is established. □

Our next result says that if a C^r function $f : \mathbb{R}^N \rightarrowtail \mathbb{R}^N$ satisfies $\det(f'(a)) \neq 0$ at some point a, then it is possible to give a local decomposition of f into functions that leave some of the variables of $\mathbb{R}^N$ fixed. We will need this result to prove the change of variables theorem for integrals later.

Example 3.6.3 Consider $f : \mathbb{R}^2 \to \mathbb{R}^2$ defined by $f(x, y) = (x^2 - y^2, 2xy)$. Set

$$h(x, y) = (x, 2xy) \quad \text{and} \quad k(x, y) = (x^2 - \frac{y^2}{4x^2}, y).$$

Note that h keeps the first variable and k keeps the second variable of $\mathbb{R}^2$ fixed and that $f(x, y) = (k \circ h)(x, y)$ whenever $x \neq 0$.

Before proving the result, we must take a detour back to matrix theory.

Lemma 3.6.1 *If A is an $N \times N$ matrix with nonzero determinant, then for every natural number $K < N$ there is a $K \times K$ submatrix of A with nonzero determinant.*

Proof. Let A and $K < N$ be given. Let $a_1^T, \ldots, a_N^T$ be column vectors of A. Choose any K of these, say $a_1^T, \ldots, a_K^T$. Since $\det A \neq 0$, we know that $a_1, \ldots, a_N$ are linearly independent, so $a_1, \ldots, a_K$ must also be linearly independent. Let B be the $N \times K$ matrix with $a_1^T, \ldots, a_K^T$ as column vectors. We must have $\mathcal{D}(B) > 0$. By the Binet–Cauchy theorem, this means there is a $K \times K$ submatrix C of B such that $\det C \neq 0$. But C is also a submatrix of A, and we are done. □

Theorem 3.6.2 *Suppose $f : \mathbb{R}^N \rightarrowtail \mathbb{R}^N$ (where $N \geq 2$) is a C^r function (where $r \geq 1$) and $\det(f'(a)) \neq 0$. Then for all natural numbers K and M such that $K + M = N$, there exist open sets U, V, and W in $\mathbb{R}^N$ and functions G and H such that*

(1) $a \in U$,

(2) G and H are C^r diffeomorphisms of U onto V and V onto W respectively,

(3) $f|U = H \circ G$,

and

(4) up to a permutation of indices,

for all $(x, y) \in U$ where $y \in \mathbb{R}^M$ we have $G(x, y) = (z, y)$ for some $z \in \mathbb{R}^K$,

for all $(x, y) \in V$ where $x \in \mathbb{R}^K$ we have $H(x, y) = (x, w)$ for some $w \in \mathbb{R}^M$.

Proof. Let $f = (f_1, \ldots, f_N)$. We may, without loss of generality, suppose that

$$\frac{\partial(f_1, \ldots, f_K)}{\partial(x_1, \ldots, x_K)}(a) \neq 0.$$

Let $\Pi : \mathbb{R}^N \to \mathbb{R}^K$ be the projection $\Pi(x, y) = x$ where $x \in \mathbb{R}^K$ and $y \in \mathbb{R}^M$. Define $F : \mathbb{R}^N \rightarrowtail \mathbb{R}^N$ by $F(x, y) = ((\Pi \circ f)(x, y), y)$ where $x \in \mathbb{R}^K$ and $y \in \mathbb{R}^M$. Note that F is a C^r function and

$$\det(F'(x, y)) = \frac{\partial(f_1, \ldots, f_K)}{\partial(x_1, \ldots, x_K)}(x, y),$$

so $\det(F'(a)) \neq 0$. By the inverse function theorem there must exist open sets U, V, and W in $\mathbb{R}^N$ such that $f|U$ and $F|U$ are C^r diffeomorphisms of U onto W and U onto V, respectively. Let $G = F|U$ and $H = f \circ (G^{-1})$. Clearly G and H are C^r diffeomorphisms of U onto V and V onto W, respectively, and $f|U = H \circ G$. G is of the desired form, and it is straightforward to show that the same is true of H. This establishes the theorem. □

A diffeomorphism that acts as an identity on one or more of its coordinates is known as a *primitive* diffeomorphism. Consequently the last theorem may be roughly paraphrased as saying that every diffeomorphism, around every point of its domain, may be locally factorized into a composition of primitive diffeomorphisms.

The inverse function theorem gives information about what happens when we have a function $f : \mathbb{R}^N \to \mathbb{R}^N$. What happens when we have $f : \mathbb{R}^M \to \mathbb{R}^N$ and either $M < N$ or $M > N$? Can we find a result analogous to the inverse function theorem?

Example 3.6.4 Consider $f : \mathbb{R} \to \mathbb{R}^2$ defined by $f(\theta) = (\cos(\theta),\ \sin(\theta))$. This function wraps the real line about the unit circle in the plane. We see that

$$[f'(\theta)] = \begin{pmatrix} -\sin(\theta) \\ \cos(\theta) \end{pmatrix}.$$

In the inverse function theorem we talked about $\det(f')$, and here that makes no sense. However we can talk about $\mathcal{D}(f')$. (Recall that for a linear transformation $g : \mathbb{R}^K \to \mathbb{R}^N$ where $K \leq N$ we have $\mathcal{D}(g) = \sqrt{\det(g^\diamond \circ g)}$.) We see that $\mathcal{D}(f'(\theta)) = 1$; it is not zero. This means that if we write $f = (f_1, f_2)$, then for each θ, either $f_1'(\theta) \neq 0$ or $f_2'(\theta) \neq 0$ or both. Notice that $f_1'(\theta)$ and $f_2'(\theta)$ are Jacobians. The fact that they do not both vanish can be used to establish the existence of local "inverses". For example, let

$$U = \left(-\frac{\pi}{2}, \frac{\pi}{2}\right) \quad \text{and} \quad V = \{(x, y) \in \mathbb{R}^2 : x > 0\}.$$

These are open sets in $\mathbb{R}$ and $\mathbb{R}^2$, respectively. We see that $f_2'(\theta) \neq 0$ over U and that f is one-to-one over U. Define $F : V \to U$ by

$$F(x, y) = \arctan(y/x).$$

It is easily seen that $F(f(\theta)) = \theta$ for all $\theta \in U$, so that F is a kind of "local left inverse" to f.

We sketch the general principle involved in the last example. Suppose $f : \mathbb{R}^K \rightarrowtail \mathbb{R}^{K+M}$ is a C^r function ($r \geq 1$) and

$$\frac{\partial(f_1, \ldots, f_K)}{\partial(x_1, \ldots, x_K)}(a) \neq 0,$$

where $f = (f_1, \ldots, f_K, f_{K+1}, \ldots, f_{K+M})$. Let $\Pi : \mathbb{R}^{K+M} \to \mathbb{R}^K$ be the projection $\Pi(x, y) = x$ where $x \in \mathbb{R}^K$ and $y \in \mathbb{R}^M$. It is easily checked that

$$\det\left((\Pi \circ f)'(a)\right) = \frac{\partial(f_1, \ldots, f_K)}{\partial(x_1, \ldots, x_K)}(a) \neq 0.$$

Since $\Pi \circ f : \mathbb{R}^K \rightarrowtail \mathbb{R}^K$, by the inverse function theorem there exist open subsets U and V of $\mathbb{R}^K$ such that $\Pi \circ f$ carries U onto V in a one-to-one fashion and the inverse map $g : V \to U$ is a C^r function. If we set $F = g \circ \Pi$, it is easily seen that $F(f(x)) = x$ for all $x \in U$.

Now what if we have $f : \mathbb{R}^{K+M} \rightarrowtail \mathbb{R}^K$? Is there some result analogous to the inverse function theorem? Suppose that f is C^r and

$$\frac{\partial(f_1, \ldots, f_K)}{\partial(x_1, \ldots, x_K)}(a, b) \neq 0$$

where we think of a typical x in the domain of f as having the form $(x_1, \ldots, x_K, x_{K+1}, \ldots, x_{K+M})$ and $a \in \mathbb{R}^K$ and $b \in \mathbb{R}^M$. It is straightforward to show there is a C^r function $F : \mathbb{R}^{K+M} \rightarrowtail \mathbb{R}^{K+M}$ which contains (a, b) in its range and satisfies $f \circ F = \Pi$ on the domain of F, where $\Pi(x, y) = x$ for $x \in \mathbb{R}^K$ and $y \in \mathbb{R}^M$. (The key to this result is to introduce the map $h(x, y) = (f(x, y), y)$.) Notice that since f maps from a space of higher dimension to one of lower dimension, $f \circ F$ cannot be the identity map. It is instead a projection, which is the closest one can come to the identity in this situation.

Example 3.6.5 Define $f : \mathbb{R}^2 \to \mathbb{R}$ by $f(x, y) = x^2 + y^2$. Clearly f maps each circle $x^2 + y^2 = r^2$ in the plane to the square of its radius, r^2. Since $f'(x, y) = (2x, 2y)$, the dimensions are wrong to talk about $\mathcal{D}(f')$. However we do have that

$$\mathcal{D}\left((f'(x, y))^{\circ}\right) = 2\sqrt{x^2 + y^2},$$

which is nonzero everywhere except at the origin. This tells us that at least one of $\partial f/\partial x$ or $\partial f/\partial y$ must always be nonzero away from the origin. Let

$$U = \{(x, y) \in \mathbb{R}^2 : (x, y) \neq (0, 0)\} \quad \text{and} \quad V = \{(r, \theta) \in \mathbb{R}^2 : r > 0\}.$$

Then the function $F : V \to U$ defined by $F(r, \theta) = (\sqrt{r}\cos(\theta), \sqrt{r}\sin(\theta))$ is easily seen to satisfy

$$f(F(r, \theta)) = r$$

for all $(r, \theta) \in V$. Thus $f \circ F$ operates, at least on V, as a projection.

Exercises

1. For $f(\theta) = (\cos(\theta), \sin(\theta))$, give a similar construction to that of Example 3.6.4 of F where $U = (0, \pi)$ and V is the upper half plane in $\mathbb{R}^2$.

2. Let
$$f(t) = ((1 + t^2)\cos(2\pi t), (1 + t^2)\sin(2\pi t), 2\pi t).$$
Show that $\mathcal{D}(f'(t)) \geq 2\pi$ and find $F : \mathbb{R}^3 \to \mathbb{R}$ such that $F(f(t)) = t$.

3. Let $f(t) = (\cosh(t), \sinh(t))$ and $U = \mathbb{R}$. Find an open subset V of $\mathbb{R}^2$ and F such that $F(V) = U$ and $F(f(t)) = t$ whenever $t \in U$.

4. Count the number of projections Π from $\mathbb{R}^N$ to subspaces of the form $\mathbb{R}^K$, where $K = 1, 2, \ldots, N$.

5. True or false: If $f : \mathbb{R}^M \rightarrowtail \mathbb{R}^N$ where $M < N$ and $\mathcal{D}(f'(x)) \neq 0$ for all x in the domain of f, then f is one-to-one.

6. Show that the projection map $\Pi : \mathbb{R}^{K+M} \to \mathbb{R}^K$ defined by $\Pi(x, y) = x$, where $x \in \mathbb{R}^K$ and $y \in \mathbb{R}^M$, is an open map.

7. Devise a precise definition of what is meant by a projection $\Pi : \mathbb{R}^{K+M} \to \mathbb{R}^K$ which will include $\Pi(x, y, z, w) = (x, z)$ as a special case.

8. Let $f(x, y) = xy$. Find an open set V in $\mathbb{R}^2$ and $G : V \to \mathbb{R}$ such that $f(G(u, v)) = u$.

9. Let $f(x, y) = x^2 - y^2$ and $V = \{(r, \alpha) : r > 0\}$. Find $G : V \to \mathbb{R}$ such that $f(G(r, \alpha)) = r$.

10. Suppose $f = (f_1, f_2) : \mathbb{R}^3 \rightarrowtail \mathbb{R}^2$ is a $\mathcal{C}^1$ function. Suppose further that we can solve $f(x, y, z) = (0, 0)$ for $y = y(x)$ and $z = z(x)$, $\mathcal{C}^1$ functions. Show that

$$\frac{dy}{dx} = -\frac{\left(\frac{\partial(f_1,f_2)}{\partial(x,z)}\right)}{\left(\frac{\partial(f_1,f_2)}{\partial(y,z)}\right)} \quad \text{and} \quad \frac{dz}{dx} = -\frac{\left(\frac{\partial(f_1,f_2)}{\partial(y,x)}\right)}{\left(\frac{\partial(f_1,f_2)}{\partial(y,z)}\right)}.$$

11. (a) Explain the role the implicit function theorem plays in computing dy/dx given $x^2 - y^3 + 12y = 0$. (b) Find open intervals of x-values over which one can find y as a function of x. (c) Sketch the graph of $x^2 - y^3 + 12y = 0$.

12. Suppose $f : \mathbb{R}^K \rightarrowtail \mathbb{R}^{K+M}$ is a $\mathcal{C}^1$ function. Show that $\mathcal{D}(f'(x)) \neq 0$ if and only if there exist $i_1, i_2, \ldots, i_K$ such that

$$\frac{\partial\left(f_{i_1}, f_{i_2}, \ldots, f_{i_K}\right)}{\partial(x_1, x_2, \ldots, x_K)}(x) \neq 0.$$

13. Suppose $f : \mathbb{R}^{K+M} \rightarrowtail \mathbb{R}^K$ is a $\mathcal{C}^1$ function. Show that $\mathcal{D}\left((f'(x)^\diamond\right) \neq 0$ if and only if there exist $i_1, i_2, \ldots, i_K$ such that

$$\frac{\partial(f_1, f_2, \ldots, f_K)}{\partial\left(x_{i_1}, x_{i_2}, \ldots, x_{i_K}\right)}(x) \neq 0.$$

14. Complete the proof of Theorem 3.6.2 by showing that for all $(x, y) \in V$ there is some $w \in \mathbb{R}^M$ such that $H(x, y) = (x, w)$.

15. Define $f : \mathbb{R}^2 \to \mathbb{R}^2$ by $f(x, y) = (x\cos(y), x\sin(y))$. Let $a = (1, \pi/4)$ and find G and H as in Theorem 3.6.2.

16. Let $f(x, y, z) = (x, xy, xyz)$ and $V = \{(x, y, z) : x, y > 0\}$. Show that f is a diffeomorphism of V onto V. Then factor f into a composition of two primitive diffeomorphisms, each of which changes exactly one of the variables and leaves the other two intact.

17. Suppose $f : \mathbb{R}^K \rightarrowtail \mathbb{R}^{K+M}$ is a $\mathcal{C}^r$ function and $\mathcal{D}(f'(a)) \neq 0$. Show there exist an open subset V of $\mathbb{R}^{K+M}$ and a $\mathcal{C}^r$ function $F : V \to \mathbb{R}^K$ such that $f(a) \in V$, and such that for all x in some open set containing a we have $F(f(x)) = x$.

18. Let $f : \mathbb{R}^{K+M} \to \mathbb{R}^K$ satisfy the assumptions of the implicit function theorem. The theorem guarantees that there exists a function g with certain properties. This is an existence result. In general, the function cannot be described explicitly. Nevertheless, the partial derivatives of g can be calculated. The formulas can be obtained easily from the chain rule by differentiating the equation $f(g(y), y) = 0$.

    Let $g = (g_1, \ldots, g_K)$ and $y = (y_1, \ldots, y_M)$. If $1 \leq i \leq K$ and $1 \leq j \leq M$, then show that

$$\frac{\partial g_i}{\partial y_j} = -\frac{\frac{\partial(f_1,\ldots,f_K)}{\partial(x_1,\ldots,x_{i-1},y_j,x_{i+1},\ldots,x_K)}}{\frac{\partial(f_1,\ldots,f_K)}{\partial(x_1,\ldots,x_K)}}$$

    where $\frac{\partial(f_1,\ldots,f_K)}{\partial(x_1,\ldots,x_{i-1},y_j,x_{i+1},\ldots,x_K)}$ is to be interpreted in the obvious way.

3.7 Lagrange Multipliers

In applications we often have to find relative minima or maxima of a real-valued function on a set defined implicitly by an equation $f(x) = 0$. Such a problem is called a *constrained extremum problem.* Now we are going to discuss a method for solving a constrained extremum problem by the so-called Lagrange multipliers. First we give a precise definition of the constrained relative maximum and minimum.

Let $f : \mathbb{R}^{K+M} \rightarrowtail \mathbb{R}^K$. Suppose the set $S = \{x \in \mathbb{R}^{K+M} : f(x) = 0\}$ is nonempty. Let F be a real-valued function on S. We say that F has a *constrained relative maximum* at x_0, or more specifically, F has a *relative maximum at x_0 subject to the constraint $f(x) = 0$*, if there exists a neighborhood U of x_0 in $\mathbb{R}^{K+M}$ such that $F(x) \leq F(x_0)$ for all $x \in S \cap U$. The constrained relative minimum is defined in a similar way.

Theorem 3.7.1 *(Lagrange multiplier rule). Let $f : \mathbb{R}^{K+M} \rightarrowtail \mathbb{R}^K$ be a C^1 function such that $\frac{\partial(f_1,\ldots,f_K)}{\partial(x_1,\ldots,x_K)} \neq 0$ on some open set V and let $F : \mathbb{R}^{K+M} \rightarrowtail \mathbb{R}$ be a C^1 function. If F has a local extremum at $(a, b) \in V$, where $a \in \mathbb{R}^K$ and $b \in \mathbb{R}^M$, subject to the constraint $f(x, y) = 0$, then there exist numbers $\lambda_1, \ldots, \lambda_K$ such that*

$$\nabla F(a, b) = \lambda_1 \nabla f_1(a, b) + \cdots + \lambda_K \nabla f_K(a, b).$$

Proof. By the implicit function theorem, there exists a neighborhood U of (a, b), a neighborhood W of b, and a $\mathcal{C}^1$ function g from W into $\mathbb{R}^K$ such that $g(b) = a$ and $f(g(y), y) = 0$ for all $y \in W$. Applying the chain rule yields a matrix product which must be zero:

$$\begin{pmatrix} \frac{\partial f_1}{\partial x_1}(a,b) & \cdots & \frac{\partial f_1}{\partial x_K}(a,b) & \frac{\partial f_1}{\partial y_1}(a,b) & \cdots & \frac{\partial f_1}{\partial y_M}(a,b) \\ \vdots & & \vdots & \vdots & & \vdots \\ \frac{\partial f_K}{\partial x_1}(a,b) & \cdots & \frac{\partial f_K}{\partial x_K}(a,b) & \frac{\partial f_K}{\partial y_1}(a,b) & \cdots & \frac{\partial f_K}{\partial y_M}(a,b) \end{pmatrix} \begin{pmatrix} \frac{\partial g_1}{\partial y_1}(b) & \cdots & \frac{\partial g_1}{\partial y_M}(b) \\ \vdots & & \vdots \\ \frac{\partial g_K}{\partial y_1}(b) & \cdots & \frac{\partial g_K}{\partial y_M}(b) \\ 1 & & 0 \\ & \ddots & \\ 0 & & 1 \end{pmatrix} = 0.$$

Denote the column vectors of the second matrix by $v_1, \ldots, v_M$. Clearly, $v_1, \ldots, v_M$ are linearly independent. Since $\frac{\partial(f_1,\ldots,f_K)}{\partial(x_1,\ldots,x_K)}(a, b) \neq 0$, the row vectors of the first matrix, $\nabla f_1(a, b), \ldots, \nabla f_K(a, b)$, are linearly independent. Moreover, each of the vectors $\nabla f_1(a, b), \ldots, \nabla f_K(a, b)$ is orthogonal to each of the vectors $v_1, \ldots, v_M$. Consequently, $\nabla f_1(a, b), \ldots, \nabla f_K(a, b), v_1, \ldots, v_M$ span $\mathbb{R}^{K+M}$. Thus, there exist numbers $\lambda_1, \ldots, \lambda_K, \gamma_1, \ldots, \gamma_M$ such that

$$\nabla F(a, b) = \lambda_1 \nabla f_1(a, b) + \cdots + \lambda_K \nabla f_K(a, b) + \gamma_1 v_1 + \cdots + \gamma_M v_M.$$

It suffices to prove that $\gamma_1 = \cdots = \gamma_M = 0$, or that $\nabla F(a, b)$ is orthogonal to each of the vectors $v_1, \ldots, v_M$. Indeed, if F has a local extremum at $(a, b) \in V$ subject to the constraint $f(x, y) = 0$, then the function $H(y) = F(g(y), y)$ has a local extremum at b and thus $\nabla H(b) = 0$. But

$$\nabla H(b) = \nabla F(a, b) \begin{pmatrix} \frac{\partial g_1}{\partial y_1}(b) & \cdots & \frac{\partial g_1}{\partial y_M}(b) \\ \vdots & & \vdots \\ \frac{\partial g_K}{\partial y_1}(b) & \cdots & \frac{\partial g_K}{\partial y_M}(b) \\ 1 & & 0 \\ & \ddots & \\ 0 & & 1 \end{pmatrix}$$

which completes the proof. □

It is interesting to note that the details of this proof tell us that the vector $\nabla F(a, b)$ is perpendicular to the level "surface" $f(x, y) = 0$ at the point (a, b). Let us try to establish this fact on an intuitive level.

In general, the graph of $f(x, y) = 0$, where $f : \mathbb{R}^{K+M} \rightarrowtail \mathbb{R}^K$, can be thought of as a kind of M-dimensional "surface" in $\mathbb{R}^{K+M}$. (For instance, if $f(x_1, x_2, x_3) =$

$4 - x_1^2 - x_2^2 - x_3^2$, then taking $K = 1$ and $M = 2$, we see that $f(x_1, x_2, x_3) = 0$ gives us a 2-dimensional object, a sphere.) It can be shown that the vectors $v_1, \ldots, v_M$ in the proof are tangent vectors to the "surface" $f(x, y) = 0$ and that they span the "tangent plane" to the surface at the point (a, b). (To return to the example of the sphere just given, if we take (a, b) to be the point $\left(2/\sqrt{3}, 2/\sqrt{3}, 2/\sqrt{3}\right)$ in the first octant, we can set $x = x_1$, $y = (x_2, x_3)$, and $g(y) = g(x_2, x_3) = \sqrt{4 - x_2^2 - x_3^2}$. We leave it to the reader to check that

$$v_1 = \begin{pmatrix} -1 \\ 1 \\ 0 \end{pmatrix} \quad \text{and} \quad v_2 = \begin{pmatrix} -1 \\ 0 \\ 1 \end{pmatrix},$$

thought of as vectors that emanate from the point (a, b), span the tangent plane to the sphere at the given point.) Since $\nabla F(a, b)$ is orthogonal to $v_1, \ldots, v_M$, it must be perpendicular to the surface $f(x, y) = 0$ at the point (a, b).

We close with a sample application of the Lagrange multiplier rule.

Example 3.7.1 Find the lengths of axes of the ellipse formed by the intersection of the elipsoid $\frac{x^2}{a^2} + \frac{y^2}{b^2} + \frac{z^2}{c^2} = 1$ $(a > b > c > 0)$ with the plane $Ax + By + Cz = 0$ $(ABC \neq 0)$.

To apply the method of Lagrange multipliers we define

$$f(x, y, z) = (f_1(x, y, z), f_2(x, y, z)) = \left(\frac{x^2}{a^2} + \frac{y^2}{b^2} + \frac{z^2}{c^2} - 1, Ax + By + Cz\right)$$

and

$$F(x, y, z) = x^2 + y^2 + z^2.$$

One can check that the assumptions of Theorem 3.7.1 are satisfied. Thus there exist λ_1 and λ_2 such that the equation

$$\nabla F(x, y, z) = \lambda_1 \nabla f_1(x, y, z) + \lambda_2 \nabla f_2(x, y, z)$$

is satisfied at the extremal points. Hence

$$2x = \lambda_1 \frac{2x}{a^2} + \lambda_2 A, \quad 2y = \lambda_1 \frac{2y}{b^2} + \lambda_2 B, \quad 2z = \lambda_1 \frac{2z}{c^2} + \lambda_2 C. \tag{3.1}$$

Multiplying these equations by x, y, and z, respectively, and then adding them together we get

$$2(x^2 + y^2 + z^2) = 2\lambda_1 \left(\frac{x^2}{a^2} + \frac{y^2}{b^2} + \frac{z^2}{c^2}\right) + \lambda_2(Ax + By + Cz).$$

Using $\frac{x^2}{a^2} + \frac{y^2}{b^2} + \frac{z^2}{c^2} = 1$ and $Ax + By + Cz = 0$, we simplify the above equation to

$$x^2 + y^2 + z^2 = \lambda_1.$$

If we denote an extremal value of F by M, then we have $\lambda_1 = M$. If we assume that $\lambda_2 \neq 0$, then $ABC \neq 0$ and (3.1) imply that none of the expressions $a^2 - M, b^2 - M$, and $c^2 - M$ is zero. Solving (3.1) for x, y, and z, we get

$$x = \frac{\lambda_2 A a^2}{2(a^2 - M)}, \quad y = \frac{\lambda_2 B b^2}{2(b^2 - M)}, \quad x = \frac{\lambda_2 C c^2}{2(c^2 - M)}.$$

Finally, since $Ax + By + Cz = 0$, we have

$$\frac{A^2a^2}{a^2 - M} + \frac{B^2b^2}{b^2 - M} + \frac{C^2c^2}{c^2 - M} = 0.$$

Solving this equation for M we obtain the desired two values.

Exercises

1. Prove (3.1) in Example 3.7.1.

2. Prove that in Example 3.7.1 $\mathcal{D}(f') \neq 0$ on the intersection of the two surfaces. (Hint: Assume $\mathcal{D}(f') = 0$ at some (x, y, z) on the intersection of the two surfaces and show that $\frac{x}{Aa^2} = \frac{y}{Bb^2} = \frac{z}{Cc^2}$.)

3. In Example 3.7.1 we assume that $\lambda_2 \neq 0$. Solve the problem for $\lambda_2 = 0$.

4. Find the minimum and maximum values of the quadratic form

$$f(x_1, \ldots, x_n) = \sum_{i,j=1}^{n} a_{ij} x_i x_j$$

on the unit sphere $x_1^2 + \cdots + x_n^2 = 1$.

5. Find the maximum value of the function $f(x, y, z) = x^2y^2z^2$ subject to the constraint $x^2 + y^2 + z^2 = R^2$. Use the result to prove that

$$(x^2y^2z^2)^{1/3} \leq \frac{x^2 + y^2 + z^2}{3}.$$

Use a similar method to prove that $(x_1 \cdots x_n)^{1/n} \leq \frac{x_1 + \cdots + x_n}{n}$ for any positive numbers $x_1, \ldots, x_n$.

4
THE LEBESGUE INTEGRAL

The standard approach to the Lebesgue integral is based on measure theory. One first develops a substantial amount of measure theory and then defines the Lebesgue integral in terms of measure. This makes the theory of the integral more complicated and unnecessarily increases the level of abstraction. In this book we are going to follow the approach used in *An Introduction to Analysis: From Number to Integral* by Jan Mikusiński and Piotr Mikusiński. In that book the Lebesgue integral in $\mathbb{R}$ is defined directly without mentioning measure theory.

A remarkable fact is that so much of the theory, theorem statements, and even proofs, is the same for both $\mathbb{R}$ and $\mathbb{R}^N$. Because of this we give here a more concise, though still complete, development of the theory. We state the fundamental theorem of calculus but refer the reader to the aforementioned book for its proof. An important new feature of the Lebesgue theory in $\mathbb{R}^N$ is Fubini's theorem. This theorem permits us to reduce the evaluation of an integral to the evaluation of integrals over lower dimensional spaces.

Although measure theory is not needed for the development of the Lebesgue integral, basic facts of the Lebesgue measure are easily derived from properties of the integral.

4.1 A Bird's-Eye View of the Lebesgue Integral

We begin this chapter with an overview, a kind of aerial inspection of the most prominent features of the theory of Lebesgue integration. The reader who understands these features may, if he or she wishes, proceed directly to the change of variables formula and integration over manifolds as set forth in the next chapter. Most of the results of this overview section are illustrated but not verified; proofs are postponed to subsequent sections.

An integral may be regarded as a map which takes functions to real numbers

$$f \mapsto \int f.$$

For instance, if one is considering continuous functions on the interval [0, 1], one might look at the map

$$f \mapsto \int_0^1 f(t)dt$$

where $\int_0^1 f(t)dt$ is the ordinary integral of an introductory calculus course.

There are many different kinds of integrals. One is the Riemann integral of introductory calculus. Another is the Lebesgue integral which is discussed in this chapter. Several properties are commonly accepted as desirable for an integral:

a) All functions on which the integral operates are taken to have a common domain, X. In this chapter we shall be interested in the case where $X = \mathbb{R}^N$ and later in the case where X is a manifold. (If we are only interested in the behavior of a function f on some subset A of X, we can assume f takes on the value 0 at all points x not in A.)

b) The set of functions $f: X \to \mathbb{R}$ on which the integral operates (in other words, the domain of $\int$), is closed under addition and scalar multiplication, $f + g$ and αf.

c) The integral is a linear operator. That is, assuming f, g belong to the domain of $\int$ and α is a real number, then $\int(f+g) = \int f + \int g$ and $\int \alpha f = \alpha \int f$.

What characterizes the Lebesgue integral is that if a function f can be "decomposed" in a certain way, which we shall specify, into an infinite series

$$f \text{ “=” } f_1 + f_2 + f_3 + \dots$$

and each f_i is integrable, then f must also be integrable and must satisfy

$$\int f = \int f_1 + \int f_2 + \int f_3 + \dots.$$

To make this program work, we must specify the integrals of certain simple functions, the building blocks in a sense of all the others, and explain which representations of f as a series are "good" ones. The functions with which we shall be concerned are real-valued functions with domain $\mathbb{R}^N$.

By a *brick* we mean a subset of $\mathbb{R}^N$ of the form

$$[a_1, b_1) \times \cdots \times [a_N, b_N)$$

where $a_k < b_k$ for $k = 1, \dots, N$. It will be convenient to denote a brick by $[a, b)$ where $a = (a_1, \dots, a_N)$ and $b = (b_1, \dots, b_N)$. The "volume" of a brick $[a, b)$ will be denoted by $\mu([a, b))$, that is,

$$\mu([a,b)) = \mu([a_1,b_1) \times \cdots \times [a_N,b_N)) = (b_1 - a_1)\dots(b_N - a_N).$$

This number will be also called the *measure* of $[a, b)$. Note that

$$\mu([a,b)) = |V((b_1 - a_1, 0, \dots, 0), (0, b_2 - a_2, 0, \dots, 0), \dots, (0, \dots, 0, b_N - a_N))|$$

where V is the oriented volume of the N-dimensional parallelepiped discussed in Chapter 1.

A function is called a *brick function* if there exists a brick $[a, b)$ such that $f(x) = 1$ if $x \in [a, b)$ and $f(x) = 0$ otherwise. The set $[a, b)$ is then called the *support* of f. By the *integral* $\int f$ *of a brick function* f whose support is $[a, b)$ we mean the measure of $[a, b)$:

$$\int f = \mu([a,b)).$$

A linear combination of a finite number of brick functions is called a *step function* . Thus a function f is a step function if there exist numbers $\lambda_1, \dots, \lambda_n$ and brick functions $f_1, \dots, f_n$ such that

$$f(x) = \lambda_1 f_1(x) + \cdots + \lambda_n f_n(x) \quad \text{for every} \quad x \in \mathbb{R}^N.$$

By the *integral* $\int f$ of f we mean

$$\int f = \lambda_1 \int f_1 + \cdots + \lambda_n \int f_n.$$

Definition 4.1.1 (Integrable functions) A function f from $\mathbb{R}^N$ into $\mathbb{R}$ is called *Lebesgue integrable* (or just *integrable*) if there exists a sequence of step functions $\{f_n\}$ such that

$$\text{I} \quad \sum_{k=1}^{\infty} \int |f_k| < \infty,$$

and

$$\text{II} \quad f(x) = \sum_{k=1}^{\infty} f_k(x) \text{ at all } x \in \mathbb{R}^N \text{ at which } \sum_{k=1}^{\infty} |f_k(x)| < \infty.$$

If I and II are satisfied, we write $f \simeq f_1 + f_2 + \dots$ or $f \simeq \sum_{k=1}^{\infty} f_k$ and say that f *expands into a series of step functions*.

We then define

$$\int f = \sum_{k=1}^{\infty} \int f_k.$$

There are two important points to note about this definition. First, it is shown in later sections that the value of the integral is independent of the particular expansion into step functions. That is, if $f \simeq \sum_{k=1}^{\infty} f_k$ and $f \simeq \sum_{k=1}^{\infty} g_k$ are both true, we must have

$$\sum_{k=1}^{\infty} \int f_k = \sum_{k=1}^{\infty} \int g_k.$$

Second, if we go back to the definition and require each f_k to be not a step function but simply integrable, then if I and III both hold, it is shown later that we must still have $\int f = \sum_{k=1}^{\infty} \int f_k$. This is useful because sometimes it is convenient to expand a given f in terms of something other than brick functions.

Example 4.1.1 Suppose f is the characteristic function of the singleton point $p \in \mathbb{R}$, that is,

$$f(x) = \begin{cases} 1 & \text{if } x = p, \\ 0 & \text{if } x \neq p. \end{cases}$$

We want to show that f is integrable and compute its integral. Define

$$f_0(x) = \begin{cases} 1 & \text{if } p \leq x < p + 1, \\ 0 & \text{otherwise}, \end{cases}$$

$$f_n(x) = \begin{cases} 1 & \text{if } p + (1/2^n) \leq x < p + (1/2^{n-1}), \\ 0 & \text{otherwise}, \end{cases}$$

for $n = 1, 2, 3, \ldots$. Each f_n is a brick function. It is readily seen that $f(x) = f_0(x) - f_1(x) - f_2(x) - \ldots$ whenever $f_0(x) + f_1(x) + f_2(x) + \ldots$ converges (which is always). Further, $\int f_0 + \int f_1 + \int f_2 + \cdots = 1 + 1/2 + 1/4 + 1/8 + \ldots$ which is a convergent geometric series. Therefore,

$$f \simeq f_0 - f_1 - f_2 - \ldots$$

and

$$\int f = \int f_0 - \int f_1 - \int f_2 - \cdots = 1 - 1/2 - 1/4 - 1/8 - \cdots = 0.$$

Example 4.1.2 Let f be the characteristic function of the rational numbers, that is

$$f(x) = \begin{cases} 1 & \text{if } x \in \mathbb{Q}, \\ 0 & \text{if } x \notin \mathbb{Q}. \end{cases}$$

Since $\mathbb{Q}$ is countable, we can think of it as a sequence of distinct numbers, $p_1, p_2, p_3, \ldots$. Let f_i be the characteristic function of p_i for each i. By the last example, we know $\int f_i = 0$. It is easily seen that $f \simeq f_1 + f_2 + f_3 + \ldots$, so that f must also be integrable and $\int f = \int f_1 + \int f_2 + \int f_3 + \ldots = 0$. (In the theory of Riemann integration, this f is not integrable.)

Example 4.1.3 Let f be the characteristic function of $[a, b]$, a closed interval in $\mathbb{R}$. This differs from a brick function by its value at a single point. Set f_1 = the characteristic function of the brick $[a, b)$ and f_2 = the characteristic function of the single point b. We see that $f \simeq f_1 + f_2$ and hence

$$\int f = \int f_1 + \int f_2 = \int f_1 = b - a.$$

(It is typical of Lebesgue integrals that if the value of the function is changed at a finite or even countable number of points, the value of the integral does not change.)

It is straightforward to see from the definition that the Lebesgue integral has the following properties:

Suppose f and g are Lebesgue integrable functions on $\mathbb{R}^N$ and $\alpha \in \mathbb{R}$.

1. $f + g$ is Lebesgue integrable and $\int (f + g) = \int f + \int g$.

2. αf is Lebesgue integrable and $\int \alpha f = \alpha \int f$.

3. If $f \leq g$, then $\int f \leq \int g$.

A more difficult yet very important property is the following:

4. If f is Lebesgue integrable, then so is $|f|$.

If f is a real-valued function with domain in $\mathbb{R}^N$, it will be very convenient to talk about the integral of f over a subset A of $\mathbb{R}^N$. We write this as $\int_A f$ and we define this to be $\int f \chi_A$, that is the integral over $\mathbb{R}^N$ of f times the characteristic function of the set A. The characteristic function evaluated at x, $\chi_A(x)$, has the value 1 if $x \in A$ and the value 0 otherwise.

It makes sense to integrate over almost any subset of $\mathbb{R}^N$, but not quite all. We need to consider this point.

We say a subset A of $\mathbb{R}^N$ is a set of measure zero (or a null set) provided its characteristic function χ_A is integrable and $\int \chi_A = 0$. From what has been said before, we know that single points and countable sets are sets of measure zero. In particular, the set of rational numbers, $\mathbb{Q}$, considered as a subset of $\mathbb{R}$, is a set of measure zero.

Sets of measure zero enjoy the following two properties:

5. Subsets of sets of measure zero are also sets of measure zero. (If $S \subseteq A$ where A is a set of measure zero, it is easy to see that $\chi_S \simeq \chi_A + \chi_A + \chi_A + \dots$. Then $\int \chi_S = \int \chi_A + \int \chi_A + \cdots = 0$.)

6. A set which is a union of a countable number of sets of measure zero is itself a set of measure zero. (If $S = A_1 \cup A_2 \cup \dots$ where each A_i is a set of measure zero, then we have $\chi_S \simeq \chi_{A_1} + \chi_{A_2} + \chi_{A_1} + \chi_{A_2} + \chi_{A_3} + \chi_{A_1} + \dots$.)

It is a useful fact that any function can be integrated over a set of measure zero and the result is always zero. If f is a real-valued function on $\mathbb{R}^N$ and A is a subset of $\mathbb{R}^N$ of measure zero, then $f\chi_A \simeq \chi_A + \chi_A + \chi_A + \dots$ and hence $\int_A f = \int f\chi_A = 0$.

We introduce a new idea, $f = g$ a.e. ($f = g$ almost everywhere), which means $f(x) = g(x)$ for all x except possibly a set of measure zero, and discover a second useful fact: If $f = g$ a.e. and f is integrable, then so is g and $\int f = \int g$. The relation of being equal almost everywhere is an equivalence relation, and so far as integration goes, whenever $f = g$ a.e., we may treat f and g as the same function. (Note: In a later section we prove that $f = g$ a.e. if and only if $\int |f - g| = 0$.)

We can now define a new type of convergence for a sequence of functions. We say $f_n \to f$ a.e. (almost everywhere) provided $f_n(x) \to f(x)$ for all x except those in some set of measure zero. We also define $f_n \to f$ i.n. (in norm) to mean that $\int |f_n - f| \to 0$. With these two types of convergence, we can state two powerful integration theorems.

Monotone convergence theorem. *If $\{f_n\}$ is a monotone sequence of integrable functions and $\int |f_n| \le M$, for some M and all $n \in \mathbb{N}$, then there exists an integrable function f such that $f_n \to f$ i.n. and $f_n \to f$ a.e. It also follows that $\int f_n \to \int f$.*

Dominated convergence theorem. *If a sequence of integrable functions $\{f_n\}$ converges almost everywhere to f and is bounded by an integrable function g (i.e., $|f_n| \le g$ for every $n \in \mathbb{N}$), then f is integrable and $f_n \to f$ i. n., and hence also $\int f_n \to \int f$.*

Example 4.1.4 We want to show that the horizontal coordinate axis in $\mathbb{R}^2$ is a set of measure zero. Consider the set $A = [0, 1) \times \{0\}$. This is a piece of the horizontal axis. Let f be the characteristic function of A and let f_n be the characteristic function of $A_n = [0, 1) \times [0, 1/n)$. Since each f_n is a brick function, it is integrable and $\int f_n = 1/n$. By the monotone convergence theorem, there must exist an integrable g such that $f_n \to g$ both in norm and almost everywhere. Since $f_n \to f$ pointwise, we have $f = g$ a.e. so that f is also integrable and $\int f = \int g$. Then $\int f = \lim_{n\to\infty} \int f_n = 0$.

In a similar way we can show each set of the form $[n, n + 1) \times \{0\}$, where n is an integer, is a set of measure zero in $\mathbb{R}^2$. Since the horizontal axis is a countable union of such sets, it must be a set of measure zero.

In this example, when we say the horizontal axis is a set of measure zero, it is important that we are thinking of it as a subset of $\mathbb{R}^2$. If we were to consider it as a subset of $\mathbb{R}$, then it would *not* be a set of measure zero.

Example 4.1.5 Let Γ be the graph of a continuous, real-valued function f whose domain is an interval J (possibly infinite) in $\mathbb{R}$. It is relatively easy to see that Γ must be a set of measure zero in $\mathbb{R}^2$. This is because any portion of the graph over which the x-coordinate changes by a finite amount can be enclosed in rectangles whose area can be made arbitrarily small. (See Figure 4.1.1.) Then one can apply the monotone convergence theorem as in the last example.

Example 4.1.6 We turn now to integration over sets which are not of measure zero. Let us set

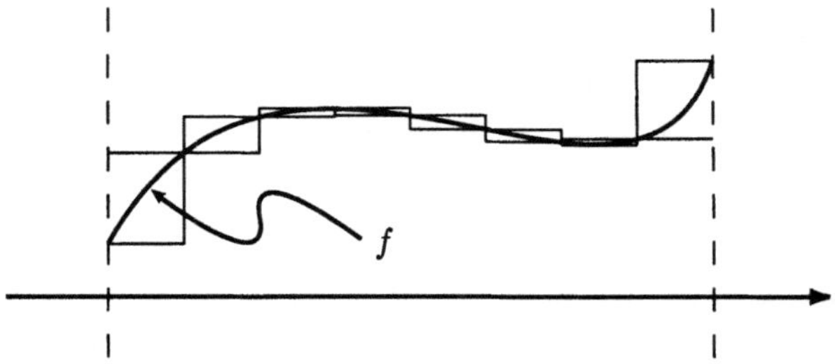

FIGURE 4.1.1.

$$f_n(x) = \begin{cases} \left(\cos\left(\frac{1}{x}\right)\right)^n & \text{if } 0 < x < 1, \\ 0 & \text{otherwise.} \end{cases}$$

Notice $f_n(x) = \pm 1$ only at x of the form $1/m\pi$, where $m \in \mathbb{N}$, and for all other x we have $f_n(x) \to 0$ as $n \to \infty$. Thus $f_n \to 0$ a.e. Assume we know each f_n is integrable. (We touch later on why this true.) Since $|f_n| \leq \chi_{(0,1)}$ and $\chi_{(0,1)}$ is integrable, we can use the dominated convergence theorem to conclude that

$$\lim_{n\to\infty} \int_0^1 \left(\cos\left(\frac{1}{x}\right)\right)^n dx = 0.$$

Example 4.1.7 Let D be the unit disk in $\mathbb{R}^2$ with center at the origin. We show that we can integrate the characteristic function of the disk, χ_D. (The result is presumably the area of the unit disk, π, though we do not show this.)

Note that we can (almost) represent D as the union of a countable number of rectangles $R_1, R_2, R_3, \ldots$, as shown in Figure 4.1.2. Some of the points of the boundary of D might not lie in $R_1 \cup R_2 \cup \ldots$, but the boundary of D is a set of measure zero. Also, any two rectangles R_i overlap in at most a set of measure zero.

Let $f_n = \chi_{R_1} + \chi_{R_2} + \cdots + \chi_{R_n}$. Clearly $\{f_n\}$ is a monotone sequence, and each f_n is a step function, hence integrable. Let R be a rectangle so large that it contains D. For each n, we have $f_n \leq \chi_R$ a.e. (The reason we say almost everywhere is that different rectangles R_i might overlap on their edges so that an f_n function might

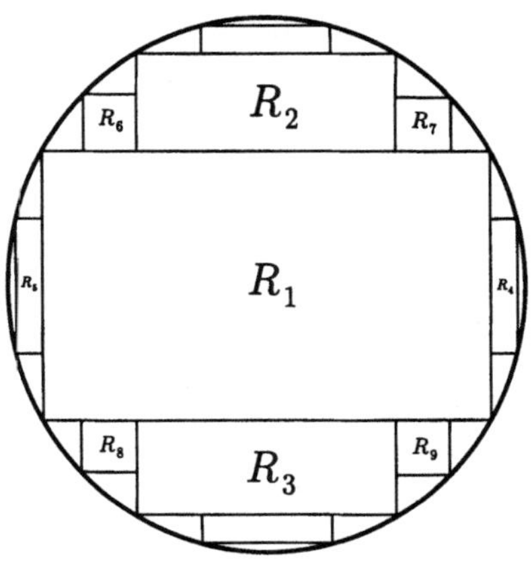

FIGURE 4.1.2.

occasionally take on values of 2 or 3. But this can happen only on a set of measure zero.) Therefore $\int f_n \leq M$ = the area of R. By the monotone convergence theorem, there exists an integrable function f such that $f_n \to f$ a.e. and $\int f_n \to \int f$. Notice that $f_n \to \chi_D$ a.e. We must have $f = \chi_D$ a.e. By virtue of Exercise 5, χ_D must be integrable and we must have $\int \chi_D = \int f = \lim_{n\to\infty} \int f_n$.

There are some sets over which it makes no sense to integrate any function. These are very pathological sets. Over "most" sets one can integrate at least some functions. For example, one can integrate 1 over the unit disk D but not $1/(x^2+y^2)$. The sets over which at least some nontrivial integrations can be performed are called measurable sets.

A set $S \subseteq \mathbb{R}^N$ is called *measurable* if the characteristic function of $S \cap B$ is an integrable function for every brick B.

Measurable sets have the following properties:

7. Bricks are measurable. (Because the intersection of two bricks is either empty or is a brick.)

8. Sets of measure zero are measurable. (Therefore lines in $\mathbb{R}^2$ or planes in $\mathbb{R}^3$ are measurable; they are sets of measure zero in their respective spaces. A closed rectangle in $\mathbb{R}^2$ is measurable because it is a union of a brick and line segments and points which are sets of measure zero.)

9. A countable union or countable intersection of measurable sets is itself a measurable set. (The unit disk D in $\mathbb{R}^2$ must be measurable because we showed D is, up to a set of measure zero, the union of a countable number of rectangles and rectangles are measurable. A set consisting of a single point in $\mathbb{R}^N$, a singleton, must be measurable since it is the intersection of a countable number of bricks.)

10. Complements of measurable sets are measurable. (So the set of points $(x, y) \in \mathbb{R}^2$ defined by $x^2 + y^2 > 1$ is measurable.)

11. Inverse images under continuous maps of measurable sets are again measurable sets. (For example, the vertical line $x = a$ in $\mathbb{R}^2$ is measurable since it is the inverse image of $A = \{a\}$, a measurable set, under the continuous map $f(x, y) = x$.)

12. Open and closed sets in $\mathbb{R}^N$ are measurable. (We see open sets are measurable because it can be shown that we can always write them as a countable union of bricks. Then closed sets must be measurable because they are just the complements of open sets.)

From now on, when we talk about integrating a function over a set A, we assume A is measurable. A given function may or may not be integrable over a measurable set. For instance, the constant function $f(x) = 1$ will not be integrable over $\mathbb{R}^N$ even though $\mathbb{R}^N$ is measurable.

What functions are integrable? Clearly brick functions and step functions. Much more complex integrable functions can be constructed using some sort of limit process such as that contained in the monotone convergence theorem. One fairly general and useful class is the set of continuous functions with compact domains; it can be shown that these are always integrable.

How do we evaluate integrable functions? Limit theorems can be useful here, but the following two results should be more familiar:

The fundamental theorem of calculus. Let f be a real-valued function which is continuously differentiable on the interval $[a, b]$. Then $\int_a^b f' = f(b) - f(a)$.

Fubini's theorem. *If $f : \mathbb{R}^{M+N} \to \mathbb{R}$ is integrable, then using x and y to denote elements of $\mathbb{R}^M$ and $\mathbb{R}^N$ respectively, we have*

$$\int_{\mathbb{R}^M} \left(\int_{\mathbb{R}^N} f(x, y)dy \right) dx = \int_{\mathbb{R}^{M+N}} f = \int_{\mathbb{R}^N} \left(\int_{\mathbb{R}^M} f(x, y)dx \right) dy.$$

We can now integrate simple functions in much of the spirit of an introductory calculus course.

Example 4.1.8 Let

$$f(x, y) = \begin{cases} x^2 \sin y & \text{if } (x, y) \in [0, 1] \times [0, \pi] \\ 0 & \text{otherwise.} \end{cases}$$

f is a continuous function on a compact set, hence integrable. By Fubini's theorem and the definition of f, we can write

$$\int_{\mathbb{R}^2} f = \int_0^\pi \left(\int_0^1 x^2 \sin y dx \right) dy.$$

By the fundamental theorem applied twice, we calculate

$$\int_0^\pi \left(\int_0^1 x^2 \sin y dx \right) dy = \int_0^\pi \frac{1}{3} \sin y dy = \frac{2}{3}.$$

Exercises

1. Show that f is integrable if and only if there exist a sequence of brick functions $\{f_n\}$ and a sequence of real numbers $\{\alpha_n\}$ such that

   I $\sum_{k=1}^{\infty} |\alpha_k| \int f_k < \infty,$

   and

   II $f(x) = \sum_{k=1}^{\infty} \alpha_k f_k(x)$ at every x at which $\sum_{k=1}^{\infty} |\alpha_k| f_k(x) < \infty.$

2. Show that if f is the characteristic function of a singleton point $p \in \mathbb{R}^2$, then f is integrable and $\int f = 0$.

3. Use the definition of Lebesgue integral to show the following:

   (a) If f and g, real-valued functions on $\mathbb{R}^N$, are integrable, then so is $f + g$ and $\int (f + g) = \int f + \int g$.

   (b) If f, a real-valued function on $\mathbb{R}^N$, is integrable and $\alpha \in \mathbb{R}$, then αf is also integrable and $\int \alpha f = \alpha \int f$.

4. Suppose that $f = f_1 + f_2 + f_3 + \ldots$ on $\mathbb{R}^N$, each $f_i \geq 0$, each f_i is integrable, and $\int f_1 + \int f_2 + \int f_3 + \ldots$ is a convergent series. Prove that f must also be integrable.

5. Suppose $f = g$ a.e., that is, $f(x) = g(x)$ for all x except a set of measure zero. Show that if f is integrable, then so is g and $\int f = \int g$.

6. Show that if f and g are integrable functions and $f \leq g$ a.e., then $\int f \leq \int g$.

7. Suppose $f_1 \geq f_2 \geq f_3 \geq \cdots \geq 0$ and each f_i is integrable. If $f(x) = \lim_{i \to \infty} f_i(x)$ for all $x \in \mathbb{R}^N$, show that f is integrable and $\int f_i \to \int f$.

8. Suppose $f_1 \leq f_2 \leq f_3 \leq \cdots \leq M$ on some brick R and each f_i is integrable over R. If $f(x) = \lim_{i \to \infty} f_i(x)$ for $x \in R$, then show that f is integrable over R and that $\int_R f_i \to \int_R f$.

9. Is $\mathbb{R}^N$ a measurable set? How about the empty set?

10. Show that if f is a real-valued, continuous function on $\mathbb{R}$, then the graph of f, considered as a subset of $\mathbb{R}^2$, is measurable.

4.2 Integrable Functions

Now begins the hard work of a rigorous development of Lebesgue integration on $\mathbb{R}^N$.

The definitions of brick, brick function, and step function were given in Section 4.1. We also accept the definition given there of the integral of a step function f. This definition is not fully justified unless we prove that the number $\int f$ is independent of a particular representation of f as a linear combination of brick functions. The proof is sketched in a sequence of exercises at the end of this section.

The integral of step functions has the following simple but important properties:

Lemma 4.2.1 *Let f and g be step functions. Then*

$$\int (f+g) = \int f + \int g, \tag{4.1}$$

$$\int \lambda f = \lambda \int f, \quad \lambda \in \mathbb{R}, \tag{4.2}$$

$$f \le g \quad implies \quad \int f \le \int g, \tag{4.3}$$

$$\left| \int f \right| \le \int |f|, \tag{4.4}$$

If $|f| \le M$ *and the support of* f *is contained in* $[a, b)$, *then*

$$\left| \int f \right| \le M\mu([a, b)). \tag{4.5}$$

We now repeat the definition of an integrable function:

Definition 4.2.1 (Integrable functions) A function f from $\mathbb{R}^N$ into $\mathbb{R}$ is called *Lebesgue integrable* (or just *integrable*) if there exists a sequence of step functions $\{f_n\}$ such that

$$\text{I} \quad \sum_{k=1}^{\infty} \int |f_k| < \infty,$$

and

$$\text{II} \quad f(x) = \sum_{k=1}^{\infty} f_k(x) \text{ at all those points } x \text{ at which } \sum_{k=1}^{\infty} |f_k(x)| < \infty.$$

If I and II are satisfied we write $f \simeq f_1 + f_2 + \dots$ or $f \simeq \sum_{k=1}^{\infty} f_k$ and we say that f *expands into a series of step functions*.

The integral of an integrable function will be defined as $\int f = \int f_1 + \int f_2 + \dots$ As in the case of the integral of step functions, it is necessary to prove that the number $\int f$ is independent of a particular expansion into a series of step functions. This time the problem is much more difficult. The rest of this section will be devoted to the proof of this fact. Once we overcome this initial difficulty, the development of the theory of the Lebesgue integral will be relatively easy.

Following the notation introduced in Chapter 2, by $|x|$ we mean the absolute value of x if x is a real number and the Euclidean norm of x if $x = (x_1, \dots, x_N)$ is a vector in $\mathbb{R}^N$:

$$|x| = \sqrt{x_1^2 + x_2^2 + \dots + x_N^2}.$$

Our first lemma shows that given a nonnegative step function, it is possible to "cut away the edges" of the steps by a small amount in such a way that the integral changes by an arbitrarily small amount. We use this construction only in the proof of the succeeding lemma.

Lemma 4.2.2 *For every nonnegative step function f and $\varepsilon > 0$ there exists a nonnegative step function g and a number $\eta > 0$ such that*

$$g(x) \leq f(y) \text{ for all } x \text{ and } y \text{ such that } |x - y| < \eta, \tag{4.6}$$

and

$$\int (f - g) < \varepsilon. \tag{4.7}$$

Proof. First assume that f is a brick function with the support

$$[a, b) = [a_1, b_1) \times \cdots \times [a_N, b_N).$$

Let η be a positive number such that $2\eta < \min\{b_1 - a_1, \ldots, b_N - a_N\}$ and

$$(b_1 - a_1) \cdots (b_N - a_N) - (b_1 - a_1 - 2\eta) \cdots (b_N - a_N - 2\eta) < \varepsilon.$$

Then the brick function g whose support is $[a_1 + \eta, b_1 - \eta) \times \cdots \times [a_N + \eta, b_N - \eta)$ satisfies (4.6) and (4.7).

Now let $f = \lambda_1 f_1 + \cdots + \lambda_n f_n$ where $f_1, \ldots, f_n$ are brick functions and $\lambda_1, \ldots, \lambda_n > 0$. For every $k = 1, \ldots, n$, let g_k be a brick function and let η_k be a positive number such that

$$g_k(x) \leq f_k(y) \quad \text{for all } x \text{ and } y \text{ such that } \ |x - y| < \eta_k,$$

and

$$\int (f_k - g_k) < \frac{\varepsilon}{n\lambda_k}.$$

Define $g = \lambda_1 g_1 + \cdots + \lambda_n g_n$ and $\eta = \min\{\eta_1, \ldots, \eta_n\}$. Then

$$g(x) = \lambda_1 g_1(x) + \cdots + \lambda_n g_n(x) \leq \lambda_1 f_1(y) + \cdots + \lambda_n f_n(y) = f(y)$$

for all x and y such that $|x - y| < \eta$. Moreover,

$$\begin{aligned}\int (f - g) &= \int (\lambda_1 f_1 + \cdots + \lambda_n f_n - \lambda_1 g_1 - \cdots - \lambda_n g_n) \\ &= \lambda_1 \int (f_1 - g_1) + \cdots + \lambda_n \int (f_n - g_n) < \varepsilon.\end{aligned}$$

Thus g and η satisfy (4.6) and (4.7). The proof is now complete. □

Lemma 4.2.3 *If $\{f_n\}$ is a nonincreasing sequence of step functions which converges to 0 at every point, then $\int f_n \to 0$ as $n \to \infty$.*

Proof. First, since the step functions f_n are non-negative and the sequence $\{f_n\}$ is nonincreasing, the sequence $\left\{\int f_n\right\}$ is nonincreasing and bounded from below by 0. Consequently, the sequence $\left\{\int f_n\right\}$ converges. We have to show that the limit is 0.

Let ε be an arbitrary positive number. By Lemma 4.2.2, for every $n \in \mathbb{N}$ there exist a nonnegative step function g_n and a number $\eta_n > 0$ such that

$$g_n(x) \le f_n(y) \text{ for all } x \text{ and } y \text{ such that } |x-y| < \eta_n, \tag{4.8}$$

and

$$\int f_n < \int g_n + \frac{\varepsilon}{2^n}. \tag{4.9}$$

Obviously, the numbers η_n can be chosen to form a decreasing sequence.

Next define

$$h_1 = g_1$$

and,

$$h_n = g_n - (f_1 - g_1) - (f_2 - g_2) - \cdots - (f_{n-1} - g_{n-1})$$

for $n = 2, 3, \ldots$. Then

$$h_n(x) \le f_n(y) \text{ for all } x \text{ and } y \text{ such that } |x-y| < \eta_n, \tag{4.10}$$

$$\int f_n < \int h_n + \varepsilon, \tag{4.11}$$

$$h_{n+1} \le h_n, \tag{4.12}$$

for all $n \in \mathbb{N}$. Indeed, since $h_n \le g_n$, from (4.8) we obtain

$$h_n(x) \le g_n(x) \le f_n(y) \quad \text{whenever} \quad |x-y| < \eta_n.$$

Moreover, by (4.9), we have

$$\int (f_n - h_n) \le \int (f_n - g_n) + \int (f_1 - g_1) + \cdots + \int (f_{n-1} - g_{n-1}) < \varepsilon,$$

proving (4.11). Finally,

$$\begin{aligned} h_{n+1} &= g_{n+1} - (f_1 - g_1) - (f_2 - g_2) - \cdots - (f_n - g_n) \\ &= g_n - (f_1 - g_1) - (f_2 - g_2) - \cdots - (f_{n-1} - g_{n-1}) + (g_{n+1} - f_n) \\ &= h_n + g_{n+1} - f_n \\ &\le h_n + g_{n+1} - f_{n+1} \le h_n, \end{aligned}$$

because $\{f_n\}$ is nonincreasing and $g_{n+1} - f_{n+1} \le 0$.

We will prove now that for every $\delta > 0$ there exists $n_\delta \in \mathbb{N}$ such that

$$h_n \le \delta \text{ for every } n \ge n_\delta. \tag{4.13}$$

Suppose this is not true for some $\delta > 0$. For $n = 1, 2, \ldots$, let x_n be a point where h_n assumes its maximum. Then there exists a subsequence $\{h_{p_n}\}$ of $\{h_n\}$ such that

$$h_{p_n}(x_{p_n}) > \delta \tag{4.14}$$

for all $n \in \mathbb{N}$. Since the supports of all h_n's are contained in some compact set K there exists a subsequence $\{x_{q_n}\}$ of $\{x_{p_n}\}$ which converges to some $x_0 \in K$. Since $f_n(x_0) \to 0$, there exists an index $m \in \mathbb{N}$ such that

$$f_m(x_0) < \delta. \tag{4.15}$$

On the other hand, since $x_{q_n} \to x_0$, there exists an index $q_k \geq m$ such that

$$|x_{q_k} - x_0| < \eta_m$$

and thus

$$h_m(x_{q_k}) \leq f_m(x_0). \tag{4.16}$$

Since $\{h_n\}$ is a nonincreasing sequence, (4.16) and $q_k \geq m$ imply

$$h_{q_k}(x_{q_k}) \leq h_m(x_{q_k}). \tag{4.17}$$

Combining (4.15), (4.16), and (4.17) we obtain

$$h_{q_k}(x_{q_k}) \leq \delta.$$

But this contradicts (4.14), thus proving (4.13) holds for every δ.

Since $\int h_{n+1} \leq \int h_n$, by (4.12), and $-\varepsilon \leq \int f_n - \varepsilon \leq \int h_n$, by (4.11), the limit $\lim_{n\to\infty} \int h_n$ exists. Moreover, since there is a brick $[a, b)$ such that $\operatorname{supp} h_n \subset [a, b)$ for all $n \in \mathbb{N}$, we have

$$\int h_n \leq \delta \mu([a, b))$$

for $n \geq n_\delta$, by (4.13). Finally, since δ can be an arbitrary small positive number, we conclude that

$$\lim_{n\to\infty} \int h_n \leq 0. \tag{4.18}$$

Now, by (4.11) and (4.18), we have

$$0 \leq \lim_{n\to\infty} \int f_n \leq \lim_{n\to\infty} \int h_n + \varepsilon \leq \varepsilon.$$

Since ε is an arbitrary positive number, we obtain

$$\lim_{n\to\infty} \int f_n = 0,$$

completing the proof. □

Lemma 4.2.4 *If $\{f_n\}$ and $\{g_n\}$ are nondecreasing sequences of step functions and*

$$\lim_{n\to\infty} f_n(x) \leq \lim_{n\to\infty} g_n(x) \quad \textit{for every} \quad x \in \mathbb{R}^N,$$

then

$$\lim_{n\to\infty} \int f_n(x) \leq \lim_{n\to\infty} \int g_n(x).$$

Proof. Let $m \in \mathbb{N}$ be fixed for now. Put

$$h_n = g_n - f_m. \tag{4.19}$$

We decompose h_n into its positive and negative parts

$$h_n = h_n^+ - h_n^-, \tag{4.20}$$

where $h_n^+ = \max(h_n, 0)$ and $h_n^- = \max(-h_n, 0)$. Then the sequence $\{h_n^-\}$ is nonincreasing and $\lim_{n\to\infty} h_n^- = 0$. By Lemma 4.2.3, we thus have

$$\lim_{n\to\infty} \int h_n^- = 0. \tag{4.21}$$

From (4.20) it follows that $\int h_n = \int h_n^+ - \int h_n^-$, and, by (4.21), $\lim_{n\to\infty} \int h_n = \lim_{n\to\infty} \int h_n^+ \geq 0$. But (4.19) implies that $\int h_n = \int (g_n - f_m)$, and thus $\lim_{n\to\infty} \int g_n - \int f_m \geq 0$, i.e., $\int f_m \leq \lim_{n\to\infty} \int g_n$. Now, by letting $m \to \infty$, we obtain

$$\lim_{m\to\infty} \int f_m \leq \lim_{n\to\infty} \int g_n,$$

which is the desired inequality. □

Lemma 4.2.5 *If* $f \simeq f_1 + f_2 + \dots$ *and* $f \geq 0$, *then* $\int f_1 + \int f_2 + \dots \geq 0$.

Proof. Let $\varepsilon > 0$. By $\mathbb{I}$ there exists an $n_0 \in \mathbb{N}$ such that

$$\sum_{k=n_o+1}^{\infty} \int |f_k| < \varepsilon.$$

For $n = 1, 2, \dots$, define

$$g_n = f_1 + \dots + f_{n_o} + |f_{n_o+1}| + \dots + |f_{n_o+n}|$$

and

$$h_n = \max(g_n, 0).$$

Clearly, g_n's and h_n's are step functions and the sequences $\{g_n\}$ and $\{h_n\}$ are nondecreasing. Moreover, since $f \geq 0$, $\mathbb{III}$ implies that the limit $\lim_{n\to\infty} g_n(x)$ is either a nonnegative number or ∞. Therefore $\lim_{n\to\infty} g_n(x) = \lim_{n\to\infty} h_n(x)$ for every $x \in \mathbb{R}$. Thus, by Lemma 4.2.5, we have

$$\lim_{n\to\infty} \int g_n(x) \geq \lim_{n\to\infty} \int h_n(x) \geq 0.$$

Consequently

$$\int f_1 + \dots + \int f_{n_o} + \int |f_{n_o+1}| + \int |f_{n_o+2}| + \dots \geq 0.$$

Since

$$\int f_1 + \cdots + \int \mathrm{f}_{n_o} - \int |f_{n_o+1}| - \int |f_{n_o+2}| + \cdots \leq \sum_{k=1}^{\infty} \int f_k,$$

$$\sum_{k=1}^{\infty} \int f_k \leq \int f_1 + \cdots + \int f_{n_o} + \int |f_{n_o+1}| + \int |f_{n_o+2}| + \ldots,$$

and

$$\sum_{k=n_o+1}^{\infty} \int |f_k| < \varepsilon,$$

we have

$$\sum_{k=1}^{\infty} \int f_k \geq -2\varepsilon.$$

Because ε is an arbitrary positive number, we conclude $\sum_{k=1}^{\infty} \int f_k \geq 0$. □

Lemma 4.2.6 *If $f \simeq f_1 + f_2 + \ldots$ and $f \simeq g_1 + g_2 + \ldots$, then*

$$\int f_1 + \int f_2 + \cdots = \int g_1 + \int g_2 + \ldots.$$

Proof. Since

$$0 \simeq f_1 - g_1 + f_2 - g_2 + \ldots,$$

we have, in view of Lemma 4.2.5,

$$0 \leq \int f_1 - \int g_1 + \int f_2 - \int g_2 + \ldots.$$

Thus

$$\int f_1 + \int f_2 + \cdots \geq \int g_1 + \int g_2 + \ldots$$

Since the roles of both expansions are symmetric, we obtain similarly

$$\int f_1 + \int f_2 + \cdots \leq \int g_1 + \int g_2 + \ldots,$$

which proves the desired equality. □

Now we can give a consistent definition of the integral of an integrable function.

Definition 4.2.2 (Integral) Let $f \simeq f_1 + f_2 + \ldots$. By the *integral* of f we mean the number $\int f = \int f_1 + \int f_2 + \ldots$.

Theorem 4.2.1 *Let f and g be integrable functions. Then*

$$\int (f+g) = \int f + \int g, \tag{4.22}$$

$$\int \lambda f = \lambda \int f, \quad \lambda \in \mathbb{R}, \tag{4.23}$$

$$f \leq g \quad \textit{implies} \quad \int f \leq \int g. \tag{4.24}$$

Note that this theorem shows that the properties (4.1)–(4.3) of the integral of step functions given in Lemma 4.2.1 are not lost in the construction. The same is true for properties (4.4) and (4.5). Proving that (4.4) holds requires some more work. This is done in the next section. Property (4.5) can be proved at this point without great difficulty. We are not going to do that because we prove a more general property of the Lebesgue integral in Section 4.8.

Exercises

1. Show that for any bricks $B_1, \ldots, B_n$ there are disjoint bricks $C_1, \ldots, C_m$ such that each of the bricks $B_1, \ldots, B_n$ can be written as a union of some of the bricks $C_1, \ldots, C_m$.
2. Show that every open set in $\mathbb{R}^N$ is a countable union of disjoint bricks.
3. Show that if $f, f_1, \ldots, f_n$ are brick functions and $f = f_1 + \cdots + f_n$, then $\int f = \int f_1 + \cdots + \int f_n$.
4. Show that if $f_1, \ldots, f_n$ are brick functions and $f = \lambda_1 f_1 + \cdots + \lambda_n f_n$, then there exist brick functions $g_1, \ldots, g_m$ with disjoint supports such that $f = \gamma_1 g_1 + \cdots + \gamma_m g_m$. Then show that $\lambda_1 \int f_1 + \cdots + \lambda_n \int f_n = \gamma_1 \int g_1 + \cdots + \gamma_m \int g_m$.
5. Prove that for a step function f the number $\int f$ is independent of a particular representation of f as a combination of brick functions.
6. Prove Lemma 4.2.1.
7. Show that the use of the same symbol $\int f$ for both the integral of a step function f and the integral of the same function f defined by an expansion into a series of brick functions is justified.
8. Prove Theorem 4.2.1.
9. Show that if f is integrable and g is a step function, then the product fg is an integrable function.
10. Prove that property (4.5) for step functions also holds for general integrable functions.
11. Show that the characteristic function of a singleton is integrable and has integral 0.

4.3 Absolutely Integrable Functions

A function f is called *absolutely integrable* if the function $|f|$ is integrable. A student in an introductory calculus course may learn that the function $f(x) = \frac{\sin x}{x}$ is integrable, but the function $g(x) = \left|\frac{\sin x}{x}\right|$ is not. More precisely, the improper Riemann integral $\int_{-\infty}^{\infty} \frac{\sin x}{x} dx$ converges, while the integral $\int_{-\infty}^{\infty} \left|\frac{\sin x}{x}\right| dx$ diverges. In the theory of Riemann integration, the integrability of f and $|f|$ are not closely linked to one another. In the theory of Lebesgue integration, the situation is simpler: If f is Lebesgue integrable, then so is $|f|$. In other words, every Lebesgue integrable function is absolutely integrable.

Theorem 4.3.1 *If f is integrable, then $|f|$ is integrable. Moreover, if*

$$f \simeq f_1 + f_2 + \dots, \tag{4.25}$$

then

$$\left|\int f_1 + \int f_2 + \dots\right| \le \int |f| \le \int |f_1| + \int |f_2| + \dots \tag{4.26}$$

Proof. The equality

$$f(x) = f_1(x) + f_2(x) + \dots$$

holds at every point at which the series converges absolutely. Let Z denote the set of all those points. Then

$$f(x) = \lim_{n\to\infty} s_n(x) \quad \text{for all} \quad x \in Z,$$

where

$$s_n = f_1 + \dots + f_n.$$

Consequently

$$|f(x)| = \lim_{n\to\infty} |s_n(x)| \quad \text{for all} \quad x \in Z.$$

Denote $g_1 = |s_1| = |f_1|$, $g_2 = |s_2| - |s_1|$, and generally

$$g_n = |s_n| - |s_{n-1}| \quad \text{for} \quad n \ge 2.$$

Note that all g_n's (as well as s_n's) are step functions. Since

$$|g_n| = \left||s_n| - |s_{n-1}|\right| \le |s_n - s_{n-1}| = |f_n|, \tag{4.27}$$

we have

$$\sum_{k=1}^{\infty} \int |g_k| \le \sum_{k=1}^{\infty} \int |f_k| < \infty. \tag{4.28}$$

Although

$$|f(x)| = \sum_{k=1}^{\infty} g_k(x) \quad \text{for all} \quad x \in Z,$$

we cannot claim that $|f| \simeq g_1 + g_2 + \dots$ because the series may converge at some points outside Z to some extraneous values. In order to "spoil" the convergence of the series at those points we modify the series by adding and subtracting terms of the series $\sum_{k=1}^{\infty} f_k$ to get

$$|f| \simeq g_1 + f_1 - f_1 + g_2 + f_2 - f_2 + \dots$$

Indeed, condition $\mathbb{I}$ is still satisfied by (4.28). Moreover, the above series and the series in (4.25) converge absolutely at exactly the same points. Therefore, condition $\mathbb{III}$ is also satisfied, which proves that $|f|$ is integrable, in view of Definition 4.2.1.

To prove the first inequality in (4.26) note that $f \le |f|$ and $-f \le |f|$, and hence $\int f \le \int |f|$ and $-\int f \le \int |f|$, which yields $|\int f| \le \int |f|$. Finally $\int |f| = \int g_1 + \int g_2 + \dots \le \int |f_1| + \int |f_2| + \dots$, by (4.27). The proof is thus completed. □

The following lemma complements nicely the second inequality in (4.26). The property is essential in the proof of a theorem in the next section.

Theorem 4.3.2 *For any integrable function f and any positive number ε there exists an expansion into a series of step functions $f \simeq f_1 + f_2 + \dots$ such that*

$$\int |f_1| + \int |f_2| + \dots < \int |f| + \varepsilon.$$

Proof. Let us first take an arbitrary expansion

$$f \simeq g_1 + g_2 + \dots \tag{4.29}$$

By $\mathbb{I}$, there exists a number $n_0 \in \mathbb{N}$ such that

$$\sum_{k=n_0+1}^{\infty} \int |g_k| < \frac{1}{2}\varepsilon. \tag{4.30}$$

Define

$$f_1 = g_1 + \dots + g_{n_0}, \quad f_2 = g_{n_0+1}, \quad f_3 = g_{n_0+2},$$

and in general

$$f_n = g_{n_0+n-1} \quad \text{for} \quad n \ge 2. \tag{4.31}$$

Then clearly

$$f \simeq f_1 + f_2 + \dots$$

and hence also

$$f - f_1 \simeq f_2 + f_3 + \dots \tag{4.32}$$

Moreover, by (4.30) and (4.31), we have

$$\int |f_2| + \int |f_3| + \dots < \frac{1}{2}\varepsilon. \tag{4.33}$$

The inequality

$$|f_1| \le |f| + |f - f_1|$$

implies

$$\int |f_1| \leq \int |f| + \int |f - f_1|.$$

Now, applying Theorem 4.3.1 to the function $f - f_1$ and using (4.32), we obtain

$$\int |f_1| \leq \int |f| + \int |f_2| + \int |f_3| + \dots$$

Hence, by (4.33),

$$\int |f_1| \leq \int |f| + \frac{1}{2}\varepsilon.$$

By adding this inequality and (4.33) we get the desired result

$$\int |f_1| + \int |f_2| + \dots < \int |f| + \varepsilon. \qquad \square$$

Exercises

1. Let f and g be integrable functions. Show that the functions $\max\{f, g\}$ and $\min\{f, g\}$ are integrable.

2. Show that the relation

$$f \sim g \quad \text{if} \quad \int |f - g| = 0$$

is an equivalence in the space of all integrable functions. Then show that if $f \sim g$, then $\int f = \int g$.

3. Let f be an integrable function. Show that for every $\varepsilon > 0$ there exists a brick function g such that $\int |f - fg| < \varepsilon$.

4. Show that if f is a continuous, nonnegative, integrable function, then f has an expansion $f \simeq f_1 + f_2 + \dots$ such that $|f| \simeq |f_1| + |f_2| + \dots$

4.4 Series of Integrable Functions

In Section 1 we defined an expansion into a series of step functions. The class of functions which expand into series of step functions is essentially larger than the class of step functions. One may expect that if we now consider functions that can be expanded into series of integrable functions we will obtain a new class of functions which is essentially larger than the class of integrable functions. It turns out that this is not so. We do not gain any new functions. This property is closely related to the completeness of the space of all integrable functions, as we will see later. In this section we are going to prove the stated result. Although it should be rather obvious what expanding into a series of integrable functions means, we give a formal definition.

Let f be an arbitrary function. If there are integrable functions $f_1, f_2, \ldots$ such that

$$\text{I} \quad \sum_{k=1}^{\infty} \int |f_k| < \infty,$$

and

$$\text{II} \quad f(x) = \sum_{k=1}^{\infty} f_k(x) \text{ at all those points } x \text{ at which } \sum_{k=1}^{\infty} |f_k(x)| < \infty,$$

then we say that f *expands into a series of integrable functions* and we write

$$f \simeq f_1 + f_2 + \ldots .$$

Theorem 4.4.1 *If a function f expands into a series of integrable functions $f \simeq f_1 + f_2 + \ldots$, then f is integrable and $\int f = \int f_1 + \int f_2 + \ldots$.*

Proof. Let $\sum_{k=1}^{\infty} \varepsilon_k$ be a convergent series of positive numbers. By Theorem 4.3.2 there are expansions

$$f_n \simeq f_{n1} + f_{n2} + \ldots \; (n = 1, 2, \ldots) \tag{4.34}$$

into series of step functions such that

$$\int |f_{n1}| + \int |f_{n2}| + \cdots < \int |f_n| + \varepsilon_n. \tag{4.35}$$

Let

$$\sum_{k=1}^{\infty} g_k \tag{4.36}$$

be a series of step functions arranged from all f_{nk}'s ($n, k \in \mathbb{N}$). Then

$$\int |g_1| + \int |g_2| + \cdots < \int |f_1| + \int |f_2| + \cdots + \varepsilon_1 + \varepsilon_2 + \cdots < \infty. \tag{4.37}$$

Moreover, if series (4.36) converges absolutely at a point x, then all the series in (4.34) as well as the series in $f_1 + f_1 + \ldots$ converge absolutely at x and we have

$$g_1(x) + g_2(x) + \cdots = f_1(x) + f_2(x) + \cdots = f(x).$$

This, together with (4.37), shows that

$$f \simeq g_1 + g_2 + \ldots$$

which proves that f is integrable. Moreover, since

$$\int f_n = \int f_{n1} + \int f_{n2} + \ldots \; (n = 1, 2, \ldots)$$

we have

$$\int f = \int g_1 + \int g_2 + \ldots = \int f_{11} + \int f_{12} + \cdots + \int f_{21} + \int f_{22} + \ldots$$
$$= \int f_1 + \int f_2 + \ldots .$$

□

Corollary 4.4.1 *If $f_1, f_2, \ldots$ are integrable and $\sum_{k=1}^{\infty} \int |f_k| < \infty$, then there exists an integrable function f such that $f \simeq f_1 + f_2 + \ldots$*

In the definition of the expansion $f \simeq f_1 + f_2 + \ldots$ we assume that $f(x) = \sum_{k=1}^{\infty} f_k(x)$ only at those points x at which the series converges absolutely. It is possible to imagine that the series does not converge absolutely at any point. In such a case there would be no connection between the function f and the series. It is not difficult to show that this cannot happen. Actually, condition II implies that the set of points for which the series is not absolutely convergent has to be small, it has to be a set of measure zero. Before we prove this property, we recall the definition of sets of measure zero.

Definition 4.4.1 (Sets of measure zero) A set $S \subseteq \mathbb{R}^N$ is called a *set of measure zero* if the characteristic function of S is integrable and the integral is zero.

In Section 1 we defined the measure of a brick:

$$\mu([a, b)) = \mu([a_1, b_1) \times \cdots \times [a_N, b_N)) = (b_1 - a_1) \cdots (b_N - a_N).$$

We will use the same notation and write $\mu(Z) = 0$ if Z is a set of measure zero. Both are special examples of the Lebesgue measure which is defined for a large class of subsets of $\mathbb{R}^N$.

Theorem 4.4.2 *If $f_1, f_2, \ldots$ are integrable functions and $\sum_{k=1}^{\infty} \int |f_k| < \infty$, then the set of all points x for which the series $\sum_{k=1}^{\infty} f_k(x)$ does not converge absolutely is a set of measure zero.*

Proof. Let S be the set of all points x for which the series $\sum_{k=1}^{\infty} f_k(x)$ does not converge absolutely and let g be the characteristic function of S. Then

$$g \simeq f_1 - f_1 + f_2 - f_2 + f_3 - f_3 + \ldots .$$

Thus g is an integrable function and $\int g = 0$. This means that $\mu(S) = 0$. □

Instead of saying "the set of all points x for which the series $\sum_{k=1}^{\infty} f_k(x)$ does not converge absolutely is a set of measure zero", one can say that "the series $\sum_{k=1}^{\infty} f_k(x)$ converges absolutely everywhere except on a set of measure zero." As we will see, this a very common expression in the theory of the Lebesgue integral. The phrase "everywhere except on a set of measure zero" is usually replaced by the short expression "almost everywhere." In the next section we discuss this concept in more detail.

Exercises

1. Prove that every function that vanishes outside a countable subset of $\mathbb{R}^N$ is integrable and its integral is zero.

2. Give an example of an integrable function which is discontinuous at every point.

3. Give an example of an integrable function which is unbounded on every brick in $\mathbb{R}^N$.

4. Prove Corollary 4.4.1.

5. Let $S \subseteq \mathbb{R}^N$. Prove that $\mu(S) = 0$ if and only if for every $\varepsilon > 0$ there exist bricks $B_1, B_2, B_3, \ldots$ such that $S \subseteq \bigcup_{n=1}^{\infty} B_n$ and $\sum_{n=1}^{\infty} \mu(B_n) < \varepsilon$.

6. Can an open set be of measure zero?

7. True or false? Every set of measure zero is countable.

4.5 Convergence Almost Everywhere

The type of convergence described in Theorem 4.4.2 plays a very important role in the theory of the Lebesgue integral. In that theorem we are talking about absolute convergence of a series, but the same idea can be applied to any sequence of functions.

Definition 4.5.1 (Convergence almost everywhere) We say that a sequence of functions $\{f_n\}$ *converges almost everywhere* to a function f if the set of all points x such that $f_n(x) \nrightarrow f(x)$ is a set of measure zero. If $\{f_n\}$ converges almost everywhere to f, we write $f_n \to f$ a.e.

Convergence almost everywhere has properties similar to pointwise convergence.

Theorem 4.5.1 *If $f_n \to f$ a.e. and $g_n \to g$ a.e., then $f_n + g_n \to f + g$ a.e.,*

$$f_n - g_n \to f - g \quad a.e.,$$
$$f_n g_n \to fg \quad a.e.,$$
$$|f_n| \to |f| \quad a.e.$$

To prove this theorem it is necessary to use some properties of sets of measure zero. Since those properties will be used on several occasions, we give them in the form of a theorem.

Theorem 4.5.2 *A subset of a set of measure zero is of measure zero. A countable union of sets of measure zero is of measure zero.*

Proof. Let $\mu(S) = 0$ and let $Z \subseteq S$. Let f and g be the characteristic functions of S and Z, respectively. Then

$$g \simeq f + f + f + \dots.$$

This proves that g is integrable and the $\int g = 0$. Thus $\mu(Z) = 0$.

Now let $S_1, S_2, \dots$ be sets of measure zero and let $g_1, g_2, \dots$ be the characteristic functions of those sets. If $S = \bigcup_{n=1}^{\infty} S_n$ and g is the characteristic function of S, then

$$g \simeq g_1 + g_2 + g_1 + g_2 + g_3 + g_1 + g_2 + g_3 + g_4 + \dots.$$

The assertion follows easily. □

There is one essential difference between pointwise convergence and convergence almost everywhere. The limit of a sequence convergent at every point is unique, while a sequence convergent almost everywhere has many limits. (For instance, if $f_n \to f$ a.e., if we replace f by a function g which differs from f at a finite number of points, we will still have $f_n \to g$ a.e.) However those limits cannot differ too much: they have to be equal almost everywhere. As one readily expects, two functions f and g are called *equal almost everywhere* if the set of all x for which $f(x) \neq g(x)$ is a set of measure zero. If f equals g almost everywhere, we write $f = g$ a.e. The relation of being equal almost everywhere is an equivalence:

Theorem 4.5.3
$f = f$ a.e.;
If $f = g$ a.e., then $g = f$ a.e.;
If $f = g$ a.e. and $g = h$ a.e., then $f = h$ a.e.

Now we will prove the foretold theorem.

Theorem 4.5.4 *If $f_n \to f$ a.e., then $f_n \to g$ a.e. if and only if $f = g$ a.e.*

Proof. Assume that $f_n \to f$ a.e. and $f_n \to g$ a.e. Let S_f and S_g be the sets of points x such that $f_n(x) \not\to f(x)$ and $f_n(x) \not\to g(x)$, respectively. Then $f(x) = g(x)$ outside of $S_f \cup S_g$. Since $\mu(S_f \cup S_g) = 0$, the proof in this direction is complete. The proof in the other direction is similar. □

The next theorem characterizes equality almost everywhere in terms of the integral.

Theorem 4.5.5 *$f = g$ a.e. if and only if $\int |f - g| = 0$.*

Proof. Assume that $f = g$ a.e. Let S be the set of points x where $f(x) \neq g(x)$ and let h be the characteristic function of S. Then

$$|f - g| \simeq h + h + h + \dots,$$

which proves that $|f-g|$ is an integrable function. Since $\int h = 0$, we have $\int |f-g| = 0$.

Assume now that $\int |f - g| = 0$. Let S and h be as before. Then

$$h \simeq |f - g| + |f - g| + |f - g| + \dots,$$

from which the assertion follows easily. □

Corollary 4.5.1 *$f = 0$ a.e. if and only if $\int |f| = 0$.*

Note that Theorem 4.4.2 implies the following important property of expansions of integrable functions:

Theorem 4.5.6 *If $f \simeq f_1 + f_2 + \dots$, then $f = f_1 + f_2 + \dots$ a.e.*

As one can easily guess, $f = f_1 + f_2 + \dots$ a.e. means that the sequence of partial sums converges to f almost everywhere.

Exercises

1. Prove that if f is an integrable function and $f = g$ a.e., then g is an integrable function and $\int g = \int f$.
2. Prove Theorem 4.5.1.
3. Since $f = 0$ a.e. if and only if $\int |f| = 0$, one may conjecture that $f_n \to 0$ a.e. if and only if $\int |f_n| \to 0$. Is it true?
4. Prove Theorem 4.5.3.
5. Complete the proof of Theorem 4.5.4.
6. Let f be the characteristic function of the set of all rational numbers. Does there exist a sequence of continuous functions convergent to f almost everywhere?
7. Does $f = f_1 + f_2 + \dots$ a.e. imply $f \simeq f_1 + f_2 + \dots$?

4.6 Convergence in Norm

Convergence almost everywhere is a rather intuitive concept. In practice, it is usually not difficult to check convergence almost everywhere. For that reason it is a good form of convergence. Unfortunately this type of convergence lacks a very important and desirable property. If $f_n \to f$ a.e., then it is not necessarily true that $\int f_n \to \int f$. In this section we define a convergence that has that property.

Definition 4.6.1 (Convergence in norm) A sequence of integrable functions f_1, $f_2, \dots$ is said to *converge in norm* to an integrable function f if $\int |f_n - f| \to 0$. To denote convergence in norm we write $f_n \to f$ i.n.

The norm convergence has similar algebraic properties to those of pointwise or uniform convergence:

Theorem 4.6.1 *If $f_n \to f$ i.n. and $g_n \to g$ i.n., then*

$$\begin{aligned} f_n + g_n &\to f + g \text{ i.n.}, \\ f_n - g_n &\to f - g \text{ i.n.}, \\ \lambda f_n &\to \lambda f \text{ i.n. } (\lambda \in \mathbb{R}), \\ |f_n| &\to |f| \text{ i.n.} \end{aligned}$$

It is easy to prove that convergence in norm has the property that convergence almost everywhere lacks.

Theorem 4.6.2 *If $f_n \to f$ i.n., then $\int f_n \to \int f$.*

Proof. Note that $\left|\int f_n - \int f\right| \le \int |f_n - f|$. □

The following theorem is similar to Theorem 4.5.6.

Theorem 4.6.3 *If $f \simeq f_1 + f_2 + \dots$, then $f = f_1 + f_2 + \dots$ i.n., which means that the series converges to f in norm.*

Proof. Given any $\varepsilon > 0$ we choose an index n_0 such that

$$\int |f_{n+1}| + \int |f_{n+2}| + \cdots < \varepsilon \quad \text{for all} \quad n > n_0.$$

Since

$$f - f_1 - \cdots - f_n \simeq f_{n+1} + f_{n+2} + \dots$$

we have

$$\int |f - f_1 - \cdots - f_n| \le \int |f_{n+1}| + \int |f_{n+2}| + \cdots < \varepsilon$$

for all $n > n_0$, which proves the theorem. □

As in the case of convergence almost everywhere, a sequence convergent in norm has many limits, but any two of them are equal almost everywhere.

Theorem 4.6.4 *If $f_n \to f$ i.n. and $f = g$ a.e., then $f_n \to g$ i.n., and conversely, if $f_n \to f$ i.n. and $f_n \to g$ i.n., then $f = g$ a.e.*

Proof. Let $f_n \to f$ i.n. and $f = g$ a.e. Then $g = f + h$ for some function h such that $\int h = 0$ and

$$\int |f_n - g| = \int |f_n - (f + h)| \le \int |f_n - f| + \int |h|.$$

Since $\int |h| = 0$, we have $f_n \to g$ i.n.

Now, assume that $f_n \to f$ i.n. and $f_n \to g$ i.n. Then

$$\int |f-g| \leq \int |f_n - f| + \int |f_n - g|.$$

Since the left hand side does not depend on n and the right hand side tends to 0, we obtain $\int |f-g| = 0$, which proves that $f = g$ a.e. □

Theorem 4.6.5 *If* $\int |f_1| + \int |f_2| + \cdots < \infty$, *then* $f_1 + f_2 + \cdots = f$ *i.n. if and only if* $f_1 + f_2 + \cdots = f$ *a.e.*

Proof. Let g be any function such that $g \simeq f_1 + f_2 + \ldots$. Then $f_1 + f_2 + \cdots = g$ i.n. and $f_1 + f_2 + \cdots = g$ a.e. hold by Theorems 4.5.6 and 4.6.3. If $f_1 + f_2 + \cdots = f$ i.n., then f and g differ by a null function from one another, by 4.6.4. Hence, by 4.5.4, we have $f_1 + f_2 + \cdots = f$ a.e. The proof of the converse is similar. □

The convergence in norm is defined as in any normed space with $\|f\| = \int |f|$. However, this is not a norm in the space of all integrable functions. Clearly there are functions $f \neq 0$ for which $\int |f| = 0$. The same problem was discussed in Section 2.1 in the case of Lebesgue integrable functions on the real line. The solution presented there works in $\mathbb{R}^N$ as well. First we define an equivalence relation

$$f \sim g \quad \text{if} \quad \int |f-g| = 0$$

and then define the space $\mathcal{L}^1(\mathbb{R}^N)$ of all equivalence classes of Lebesgue integrable functions on $\mathbb{R}^N$. Then $\mathcal{L}^1(\mathbb{R}^N)$ with $\|f\| = \int |f|$ is a normed space. In the next section we will prove that it is a Banach space.

Exercises

1. The definition of the norm in $\mathcal{L}^1(\mathbb{R}^N)$ is formally incorrect because an integrable function f is not an element of $\mathcal{L}^1(\mathbb{R}^N)$. To be precise, one should define $\|[f]\| = \int |f|$, where $[f]$ is the equivalence class of f, and then prove that the value of $\int |f|$ is independent of the choice of a particular function representing $[f]$, that is, $\int |f| = \int |g|$ if $f \sim g$. Provide details of this proof. Then show that $\mathcal{L}^1(\mathbb{R}^N)$ with the defined norm is a normed space.

2. Prove Theorem 4.6.1.

3. Give an example of a sequence of integrable functions f_n such that $f_n \to 0$ a.e. and $f_n \not\to 0$ i.n.

4. Give an example of a sequence of integrable functions f_n such that $f_n \to 0$ i.n. and $f_n \not\to 0$ a.e.

5. True or false? If f_n are integrable functions and $f_n \to 0$ uniformly on $\mathbb{R}^N$, then $f_n \to 0$ i.n.

6. Let f be an integrable function and let $v_n \to 0$ in $\mathbb{R}^N$. Define $g_n(x) = f(x - v_n)$. Prove that $g_n \to f$ i.n.

7. True or false? If $f_n \to 0$ i.n. and $g_n \to 0$ i.n., then $f_n g_n \to 0$ i.n.

4.7 Important Convergence Theorems

The main deficiency of the Riemann integral is the lack of good convergence theorems, that is theorems of the form "if $f_n \to f$, then $\int f_n(x)dx \to \int f(x)dx$." Moreover, the space of Riemann integrable functions is not complete with respect to any natural norm. This often causes difficulties in using the Riemann integral. In this section we will prove some theorems which show that the Lebesgue integral has those desired properties. The first theorem is actually the completeness of the space of all Lebesgue integrable functions.

Theorem 4.7.1 (Riesz's first theorem) *Let $\{f_n\}$ be a sequence of integrable functions. If, for every increasing sequence of indices $p_1, p_2, \ldots$, we have*

$$f_{p_{n+1}} - f_{p_n} \to 0 \text{ i.n.} \quad \text{as} \quad n \to \infty,$$

then there is an integrable function f such that $f_n \to f$ i.n.

Proof. Let $\{\varepsilon_n\}$ be a sequence of positive numbers such that $\sum_{k=1}^{\infty} \varepsilon_k < \infty$, and let $\{p_n\}$ be an increasing sequence of natural numbers such that

$$\int |f_k - f_m| < \varepsilon_n \quad \text{for all} \quad k, m > p_n. \tag{4.38}$$

Then

$$\int |f_{p_1}| + \int |f_{p_2} - f_{p_1}| + \int |f_{p_3} - f_{p_2}| + \cdots < \infty.$$

Hence, by Corollary 4.4.1, there exists a function f such that

$$f \simeq f_{p_1} + (f_{p_2} - f_{p_1}) + (f_{p_3} - f_{p_2}) + \cdots.$$

Thus, by Theorem 4.6.3, we have $\int |f_{p_n} - f| \to 0$. Since, by (4.38),

$$\int |f_k - f_{p_m}| < \varepsilon_n \quad \text{for all} \quad k, p_m > p_n,$$

we have

$$\int |f_k - f| \le \int |f_k - f_{p_m}| + \int |f_{p_m} - f|.$$

By letting $m \to \infty$ we obtain

$$\int |f_k - f| < \varepsilon_n \quad \text{for all} \quad k > p_n.$$

Since $\varepsilon_n \to 0$, the above implies that the sequence $\{f_n\}$ converges to f in norm. □

Corollary 4.7.1 *$\mathcal{L}^1(\mathbb{R}^N)$ with the norm $\|f\| = \int |f|$ is a Banach space.*

We leave the proof of the corollary as an exercise.

Convergence in norm does not imply convergence almost everywhere. However, we have the following useful property.

Theorem 4.7.2 (Riesz's second theorem) *If $f_n \to f$ i.n., then there is a subsequence $\{f_{p_n}\}$ of $\{f_n\}$ such that $f_{p_n} \to f$ a.e.*

Proof. Let $\{\varepsilon_n\}$ be a sequence of positive numbers such that $\sum_{k=1}^{\infty} \varepsilon_k < \infty$. Since $\int |f_n - f| \to 0$, there exists an increasing sequence of natural numbers $p_1, p_2, \dots$ such that

$$\int |f_{p_n} - f| < \varepsilon_n \quad \text{for all} \quad n \in \mathbb{N}.$$

Then

$$\int |f_{p_{n+1}} - f_{p_n}| \le \int |f_{p_{n+1}} - f| + \int |f - f_{p_n}| < \varepsilon_{n+1} + \varepsilon_n,$$

and consequently

$$\int |f_{p_1}| + \int |f_{p_2} - f_{p_1}| + \int |f_{p_3} - f_{p_2}| + \cdots < \infty.$$

Hence, there is a function g such that

$$g \simeq f_{p_1} + (f_{p_2} - f_{p_1}) + (f_{p_3} - f_{p_2}) + \dots$$

This, by Theorem 4.5.6, implies $f_{p_n} \to g$ a.e., and also $f_{p_n} \to g$ i.n., by Theorem 4.6.3. In view of Theorem 4.6.4 and the fact that $f_n \to f$ i.n., we have then $f = g$ a.e. Therefore, by Theorem 4.5.4, $f_{p_n} \to f$ a.e., completing the proof. □

We know that if $f_n \to f$ i.n., then $\int f_n \to \int f$. This property of the convergence in norm is not very useful in practice, because it is often difficult to verify convergence in norm. In most cases it is much easier to verify convergence almost everywhere, but this convergence does not guarantee convergence of the integrals. The following two theorems offer a solution to this problem. They give conditions under which convergence almost everywhere implies convergence of the integrals. These theorems belong to the most important theorems in the theory of the Lebesgue integral.

Theorem 4.7.3 (Monotone convergence theorem) *If $\{f_n\}$ is a monotone sequence of integrable functions and $\int |f_n| \le M$, for some M and all $n \in \mathbb{N}$, then there exists an integrable function f such that $f_n \to f$ i.n. and $f_n \to f$ a.e.*

Proof. We may assume that the sequence is nonnegative and non-decreasing, (otherwise we consider $\{f_n - f_1\}$ or $\{f_1 - f_n\}$). Then

$$\int |f_1| + \int |f_2 - f_1| + \cdots + \int |f_n - f_{n-1}| = \int |f_n| \le M.$$

Hence, by letting $n \to \infty$, we obtain

$$\int |f_1| + \int |f_2 - f_1| + \cdots \leq M < \infty.$$

By Corollary 4.4.1, there is an integrable function f such that

$$f \simeq f_1 + (f_2 - f_1) + \ldots$$

Then $f = f_1 + (f_2 - f_1) + \ldots$ i.n. and $f = f_1 + (f_2 - f_1) + \ldots$ a.e., by Theorems 4.6.3 and 4.5.6, which proves the theorem. □

Theorem 4.7.4 (Dominated convergence theorem) *If a sequence of integrable functions $\{f_n\}$ converges almost everywhere to f and is bounded by an integrable function g (i.e., $|f_n| \leq g$ for every $n \in \mathbb{N}$), then f is integrable and $f_n \to f$ i. n., and hence also $\int f_n \to \int f$.*

Proof. For $m, n = 1, 2, \ldots$, define

$$h_{n,m} = \max(|f_n|, \ldots, |f_{n+m}|).$$

Then, for every fixed $n \in \mathbb{N}$, the sequence $\{h_{n,1}, h_{n,2}, \ldots\}$ is nondecreasing and, since

$$\int | h_{n,m} | = \int h_{n,m} \leq \int g < \infty,$$

there is an integrable function h_n such that $h_{n,m} \to h_n$ a.e. as $m \to \infty$. Note that the sequence $\{h_n\}$ is nonincreasing and $0 \leq h_n$ for all $n \in \mathbb{N}$. Thus it converges to a function h at every point and, by the monotone convergence theorem, h is integrable and $h_n \to h$ i.n. Now we will consider two cases.

Case 1: First suppose $f = 0$. Then $f_n \to 0$ a.e., and therefore $h_n \to 0$ a.e. Since the sequence converges in norm, we obtain $h_n \to 0$ i.n. (Why?) Hence

$$\int |f_n| \leq \int h_n \to 0,$$

which proves the theorem in the first case.

Case 2: When f is an arbitrary function, then for every increasing sequence of positive integers $\{p_n\}$ we have

$$g_n = f_{p_{n+1}} - f_{p_n} \to 0 \quad \text{a.e.}$$

and $|g_n| \leq 2g$ for every $n \in \mathbb{N}$. Therefore, the sequence $\{g_n\}$ satisfies the assumptions of Case 1, and consequently $g_n \to 0$ i.n. Hence, by Riesz's first Theorem 4.7.1, the sequence $\{f_n\}$ converges in norm to some integrable function f^*. On the other hand, by Riesz's second Theorem 4.7.2, there exists an increasing sequence of positive integers q_n such that $f_{q_n} \to f^*$ a.e. But $f_{q_n} \to f$ a. e., and thus $f^* = f$ a.e. In view of Theorem 4.6.4, this implies that $f_n \to f$ i.n. □

Exercises

1. In the proof of Theorem 4.7.1 we claim that there exists an increasing sequence of natural numbers $\{p_n\}$ such that $\int |f_k - f_m| < \varepsilon_n$ for all $k, m > p_n$. Justify this claim.

2. Give a proof of Corollary 4.7.1.

3. True or false? If $f_n \to f$ a.e., then there is a subsequence $\{f_{p_n}\}$ of $\{f_n\}$ such that $f_{p_n} \to f$ i.n.

4. Give an example showing that the monotonicity assumption cannot be dropped from the monotone convergence theorem.

5. Give an example showing that the assumption $\int |f_n| \leq M$ cannot be dropped from the monotone convergence theorem.

6. Give an example showing that the assumption $|f_n| \leq g$ cannot be dropped from the dominated convergence theorem.

7. Show that the space of all step functions is dense in $\mathcal{L}^1(\mathbb{R}^N)$.

8. Show that the space of all continuous integrable functions is dense in $\mathcal{L}^1(\mathbb{R}^N)$.

9. Show that the space of all continuous functions with compact support is dense in $\mathcal{L}^1(\mathbb{R}^N)$.

10. Explain why, in the proof of Theorem 4.7.4, case 1, $f_n \to 0$ a.e. implies $h_n \to 0$ a.e. and $h_n \to 0$ i.n.

4.8 Integrals Over a Set

If f is an integrable function, then $\int f$ means the integral over the entire space $\mathbb{R}^N$. In practice we often use integrals over subsets of $\mathbb{R}^N$. To denote the integral over $\Omega \subseteq \mathbb{R}^N$, we write $\int_\Omega f$. As indicated in Section 4.1, we need to be careful with this symbol. There are sets Ω for which $\int_\Omega f$ is meaningless even if f is an integrable function. For that reason we define a class of sets that is called measurable. If Ω is measurable and f is integrable, then $\int_\Omega f$ is always well defined. Actually, in some cases f may be integrable over Ω even though it is not integrable over $\mathbb{R}^N$.

Definition 4.8.1 (Measurable sets.) A set $S \subseteq \mathbb{R}^N$ is called *measurable* if the characteristic function of $S \cap B$ is an integrable function for every brick B.

Theorem 4.8.1

(1) If S is measurable, then the complement of S is measurable.

(2) If $S_1, S_2, \ldots$ are measurable, then $\bigcup_{n=1}^{\infty} S_n$ is measurable.

(3) If $S_1, S_2, \ldots$ are measurable, then $\bigcap_{n=1}^{\infty} S_n$ is measurable.

Definition 4.8.2 (Measure.) Let S be a measurable subset of $\mathbb{R}^N$ and let χ_S denote the characteristic function of S. The *measure* of S is defined as

$$\mu(S) = \int \chi_S \quad \text{if } \chi_S \text{ is an integrable function}$$

and

$$\mu(S) = \infty \quad \text{if } \chi_S \text{ is not an integrable function.}$$

Note that this definition is in agreement with the two instances of the measure already defined: measure of bricks and sets of measure zero.

Now we will prove one of the fundamental properties of the measure, the so-called *σ-additivity* of the measure.

Theorem 4.8.2 *If $S_1, S_2, \ldots$ are disjoint measurable subsets of $\mathbb{R}^N$, then*

$$\mu\left(\bigcup_{n=1}^{\infty} S_n\right) = \sum_{n=1}^{\infty} \mu(S_n).$$

Proof. Let $S = \bigcup_{n=1}^{\infty} S_n$. Suppose first that χ_S is an integrable function. Then $\chi_S \simeq \chi_{S_1} + \chi_{S_2} + \ldots$ and

$$\mu(S) = \int \chi_S = \sum_{n=1}^{\infty} \int \chi_{S_n} = \sum_{n=1}^{\infty} \mu(S_n).$$

Now assume that $\mu(S) = \infty$. We need to show that $\sum_{n=1}^{\infty} \mu(S_n) = \infty$. Suppose this is not true: $\sum_{n=1}^{\infty} \mu(S_n) < \infty$. Then $\sum_{n=1}^{\infty} \int \chi_{S_n} < \infty$ and, by Corollary 4.4.1 and Theorem 4.5.6, there exists an integrable function f such that $f = \chi_{S_1} + \chi_{S_2} + \ldots$ a.e.. Since the sets S_n are disjoint, $\chi_S = \chi_{S_1} + \chi_{S_2} + \ldots$ everywhere and thus $f = \chi_S$ a.e. But this means that χ_S is an integrable function contrary to the assumption that $\mu(S) = \infty$. □

Definition 4.8.3 (Integral over a subset of $\mathbb{R}^N$) Let Ω be a measurable subset of $\mathbb{R}^N$. A function $f\colon \mathbb{R}^N \to \mathbb{R}$ is called *integrable over* Ω if the function $f\chi_\Omega$ is integrable. For a function integrable over Ω, the integral of f over Ω is defined as

$$\int_\Omega f = \int f\chi_\Omega.$$

Note that in the above definition f is a function defined on all of $\mathbb{R}^N$. In practice we often want to integrate a function which is not defined outside of Ω. This technical difficulty can be easily solved by assuming that $f(x) = 0$ for $x \notin \Omega$.

Now we are going to prove some theorems that give conditions under which a function is integrable over a set.

Theorem 4.8.3 *If f is integrable, then f is integrable over every measurable subset of $\mathbb{R}^N$.*

Proof. Let $f \simeq \lambda_1 f_1 + \lambda_2 f_2 + \dots$, where $f_1, f_2, \dots$ are brick functions and $\lambda_1, \lambda_2, \dots$ are real coefficients. If Ω is a measurable set, then

$$f\chi_\Omega \simeq \lambda_1 f_1 \chi_\Omega + \lambda_2 f_2 \chi_\Omega + \dots$$

proving the theorem. □

Definition 4.8.4 (Locally integrable functions.) A function $f : \mathbb{R}^N \to \mathbb{R}$ is called *locally integrable* if the function $f\chi_B$ is integrable for every brick B.

Theorem 4.8.4 *Every continuous function on $\mathbb{R}^N$ is locally integrable.*

Proof. Let f be a continuous function on $\mathbb{R}^N$ and let B be a brick in $\mathbb{R}^N$. For every $n \in \mathbb{N}$ we divide B into n^N disjoint bricks of equal size, say $B_{n,1}, \dots, B_{n,n^N}$, and define

$$f_k(x) = \begin{cases} \sup\limits_{x \in B_{n,k}} f(x) & \text{if } x \in B_{n,k} \text{ for some } k = 1, \dots, n^N, \\ 0 & \text{otherwise,} \end{cases}$$

for $k = 1, 2, \dots$. Using the fact that f is uniformly continuous on B we can prove that $f_k \to f\chi_B$ everywhere on $\mathbb{R}^N$. Since f_k's are integrable (they are step functions), f is bounded on B and $|f_k| \le |f|$ for all $k \in \mathbb{N}$, it follows that $f\chi_B$ is integrable by the dominated convergence theorem. □

Theorem 4.8.5 *If f is a bounded, locally integrable function and Ω is a measurable set such that $\mu(\Omega) < \infty$, then f is integrable over Ω.*

Proof. Let $B_1 \subset B_2 \subset \dots$ be a sequence of bricks such that $\bigcup_{n=1}^{\infty} B_n = \mathbb{R}^N$. For $n = 1, 2, \dots$ define

$$f_n(x) = \begin{cases} f(x) & \text{if } x \in B_n \cap \Omega, \\ 0 & \text{otherwise.} \end{cases}$$

Now the proof can be finished with help from the dominated convergence theorem as in the proof of Theorem 4.8.4. □

Corollary 4.8.1 *If f is a locally integrable function and Ω is a bounded, measurable set, then f is integrable over Ω.*

Corollary 4.8.2 *Let f be a locally integrable function such that $|f| \le M$, for some constant M, and let Ω be a measurable set such that $\mu(\Omega) < \infty$. Then f is integrable over Ω and*

$$\int_\Omega |f| \le M\mu(\Omega).$$

Exercises

1. Prove the dominated convergence theorem for an arbitrary measurable set Ω: If a sequence $\{f_n\}$ of functions integrable over a set Ω converges to f almost everywhere in Ω and is bounded by an integrable function over Ω, then f is integrable over Ω and $\int_\Omega f_n \to \int_\Omega f$.

2. Prove the following

   (a) $\mathbb{R}^N$ is measurable.

   (b) Bricks are measurable.

   (c) Open sets are measurable.

   (d) Closed sets are measurable.

3. Prove Theorem 4.8.1.

4. True or false? If f is a continuous function on some open set U, then f is integrable over every brick $B \subset U$.

5. Prove that the characteristic function of a bounded open set in $\mathbb{R}^N$ is integrable.

6. Prove that the characteristic function of a compact set in $\mathbb{R}^N$ is integrable.

7. Prove that if A and B are measurable sets in $\mathbb{R}^N$ and $A \subseteq B$, then $\mu(A) \leq \mu(B)$.

8. Complete the proof of Theorem 4.8.4 by proving that $f_k \to f\chi_B$.

9. Finish the proof of Theorem 4.8.5.

10. Prove Corollary 4.8.1.

11. Prove Corollary 4.8.2.

12. Prove that if Ω is a bounded measurable set and $f : \overline{\Omega} \to \mathbb{R}$ is continuous, then f is integrable over Ω.

13. Prove that if f is integrable over Ω, then f is integrable over any measurable subset of Ω.

14. A function f is called *measurable* if there exists a sequence of step functions $f_1, f_2, \ldots$ such that $f_n \to f$ a.e. Prove the following:

    (a) The measurable functions form a vector space.

    (b) The product of measurable functions is a measurable function.

    (c) The absolute value of a measurable function is a measurable function.

    (d) If f is a measurable function and $|f| \leq g$ for some locally integrable function g, then f is locally integrable.

15. Prove that f is measurable if and only if the set $\{x : f(x) < \alpha\}$ is measurable for every $\alpha \in \mathbb{R}$.

16. Prove that S is measurable if and only if χ_S is locally integrable.

4.9 Fubini's Theorem

We close this short presentation of the Lebesgue integral in $\mathbb{R}^N$ with one of the most often used properties of the integral. Fubini's theorem concerns iterated integrals. It gives a condition under which the order of integrals can be changed and permits us to reduce the evaluation of an integral to the evaluation of integrals over lower dimensional spaces. We start with a statement about the "size" of vector subspaces of $\mathbb{R}^K$ and an auxiliary lemma.

Consider a line L of the form $y = mx$ in $\mathbb{R}^2$. All 1-dimensional subspaces of $\mathbb{R}^2$ except the y-axis have this form. Considered as a subset of $\mathbb{R}^2$, the line L is a null set, a set of measure zero. To prove this, consider an "interval" J consisting of the points (x, y) of L which satisfy $a \le x < b$ where a and b are given real numbers. To make things simple, we shall consider only the case where $m > 0$. Let $(x_0, y_0), (x_1, y_1), \ldots, (x_n, y_n)$ be $n + 1$ evenly spaced points along L where

$$a = x_0 < x_1 < \cdots < x_n = b \quad \text{and} \quad x_i - x_{i-1} = \Delta x.$$

We construct disjoint rectangles

$$R_i = [x_{i-1}, x_i) \times [y_{i-1}, y_i)$$

which cover J. See Figures 4.9.1 and 4.9.2. Notice that $y_i - y_{i-1} = m\Delta x$ so that the 2-dimensional measure of R_i must be $(x_i - x_{i-1})(y_i - y_{i-1}) = m(\Delta x)^2$. It follows that the 2-dimensional measure of $R_1 \cup R_2 \cup \cdots \cup R_n$ is $mn(\Delta x)^2$ which is the same as $m(b-a)\Delta x$. Since this number can be made arbitrarily small by choosing smaller Δx, it follows that the 2-dimensional measure of J can only be zero. Since the line L is a countable union of intervals of the form of J, we see that L must itself have 2-dimensional measure zero.

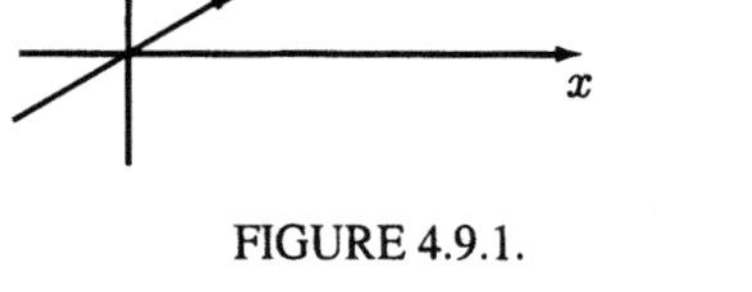

FIGURE 4.9.1.

Similar arguments can be constructed for lines and planes in $\mathbb{R}^3$. More generally, if $M < K$, then the measure of any subset of an M-dimensional vector subspace of $\mathbb{R}^K$ is zero. To be more precise, one should say that the K-dimensional measure of any subset of an M-dimensional subspace of $\mathbb{R}^K$ is zero. The same subset considered as a subset of $\mathbb{R}^M$ may have a positive (or even infinite) M-dimensional measure. In order to specify which measure is applied we will use the notation μ^N for N-dimensional Lebesgue measure.

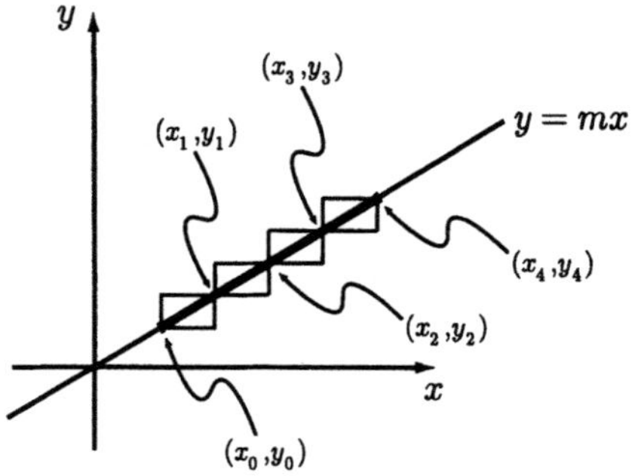

FIGURE 4.9.2.

Lemma 4.9.1 *Suppose that* $f: \mathbb{R}^{M+N} \to \mathbb{R}$ *and*

$$f \simeq \sum_{k=1}^{\infty} \alpha_k \chi_{A_k \times B_k}$$

where A_k *and* B_k *are rectangles in* $\mathbb{R}^M$ *and* $\mathbb{R}^N$, *respectively. For* $x \in \mathbb{R}^M$ *and* $y \in \mathbb{R}^N$ *we adopt the notation* $f_x(y) = f(x, y)$. *Then*

$$f_x \simeq \sum_{k=1}^{\infty} \alpha_k \chi_{A_k}(x) \chi_{B_k}$$

except possibly for a set of x *of* μ^M*-measure zero.*

Note: It is, of course, possible to formulate a similar conclusion for $f_y(x) = f(x, y)$.

Proof. For $p = 1, 2, 3, \ldots$ and $q \in \mathbb{R}$, let

$$Z_{pq} = \left\{ x \in \mathbb{R}^M : \sum_{k=1}^{p} |\alpha_k| \chi_{A_k}(x) \mu^N(B_k) \geq q \right\},$$

$$Z_q = \left\{ x \in \mathbb{R}^M : \sum_{k=1}^{\infty} |\alpha_k| \chi_{A_k}(x) \mu^N(B_k) \geq q \right\},$$

and

$$Z = \left\{ x \in \mathbb{R}^M : \sum_{k=1}^{\infty} |\alpha_k| \chi_{A_k}(x) \mu^N(B_k) = \infty \right\}.$$

We see that

$$Z_q = \bigcap_{r=1}^{\infty} \left(\bigcup_{p=1}^{\infty} Z_{p,q-\frac{1}{r}} \right) \quad \text{and} \quad Z = \bigcap_{q=1}^{\infty} Z_q.$$

Since each Z_{pq} is the set of x for which a step function is greater than or equal to q, it must either be the empty set or the union of a finite number of intervals. So each Z_{pq} is measurable and it follows that the same is true for each Z_q and for Z.

Let

$$h_p(x) = \sum_{k=1}^{p} |\alpha_k| \chi_{A_k}(x) \mu^N(B_k),$$

$$h(x) = \sum_{k=1}^{\infty} |\alpha_k| \chi_{A_k}(x) \mu^N(B_k),$$

and

$$P = \sum_{k=1}^{\infty} |\alpha_k| \mu(A_k \times B_k) = \sum_{k=1}^{\infty} |\alpha_k| \mu^M(A_k) \mu^N(B_k).$$

Since f is integrable, P is a finite number. Note that $\{h_p\}$ is a monotone sequence that converges pointwise to h. Since each $\int h_p$ is bounded by P, we can apply the monotone convergence theorem and conclude that h is integrable over $\mathbb{R}^M$. Note that $\int h = P$. For $q > 0$ since

$$q\chi_{Z_q}(x) = \begin{cases} 0 \text{ if } x \notin Z_q \\ q \text{ if } x \in Z_q \end{cases},$$

we see that $q\chi_{Z_q} \leq h$. Thus $q\mu^M(Z_q) \leq \int h = P$. Since $Z \subseteq Z_q$, we see that $\mu^M(Z) \leq P/q$ for all $q > 0$. Hence Z is a null set.

We now want to show that for all $x \notin Z$ we have

$$f_x \simeq \sum_{k=1}^{\infty} \alpha_k \chi_{A_k}(x) \chi_{B_k}.$$

Choose $x \notin Z$. We see that

$$\sum_{k=1}^{\infty} |\alpha_k| \chi_{A_k}(x) \int \chi_{B_k} = \sum_{k=1}^{\infty} |\alpha_k| \chi_{A_k}(x) \mu^N(B_k) < \infty.$$

Now suppose $y \in \mathbb{R}^N$ is chosen such that

$$\sum_{k=1}^{\infty} |\alpha_k| \chi_{A_k}(x) \chi_{B_k}(y) < \infty.$$

This means that

$$\sum_{k=1}^{\infty} |\alpha_k| \chi_{A_k \times B_k}(x, y) = \sum_{k=1}^{\infty} |\alpha_k| \chi_{A_k}(x) \chi_{B_k}(y) < \infty$$

so that we know

$$\sum_{k=1}^{\infty} \alpha_k \chi_{A_k}(x) \chi_{B_k}(y) = \sum_{k=1}^{\infty} \alpha_k \chi_{A_k \times B_k}(x, y) = f(x, y) = f_x(y),$$

which completes the proof. □

Theorem 4.9.1 (Fubini's theorem) *If $f\colon \mathbb{R}^{M+N} \to \mathbb{R}$ is integrable, then using x and y to denote elements of $\mathbb{R}^M$ and $\mathbb{R}^N$, respectively, we have*

$$\int_{\mathbb{R}^M}\left(\int_{\mathbb{R}^N} f(x,y)dy\right)dx = \int_{\mathbb{R}^{M+N}} f = \int_{\mathbb{R}^N}\left(\int_{\mathbb{R}^M} f(x,y)dx\right)dy.$$

Proof. We prove only the first equality and leave the second one as an exercise. As in the lemma we let $f_x(y) = f(x, y)$. We can write

$$f \simeq \sum_{k=1}^{\infty} \alpha_k \chi_{A_k \times B_k}$$

where A_k and B_k are intervals in $\mathbb{R}^M$ and $\mathbb{R}^N$, respectively. We know that

$$f_x \simeq \sum_{k=1}^{\infty} \alpha_k \chi_{A_k}(x) \chi_{B_k}$$

for all $x \in \mathbb{R}^M$ except possibly a set Z of μ^M-measure zero. Then

$$\int_{\mathbb{R}^N} f_x(y)dy = \sum_{k=1}^{\infty} \alpha_k \chi_{A_k}(x) \mu^N(B_k)$$

whenever $x \notin Z$.

Set

$$h_p = \sum_{k=1}^{p} \alpha_k \mu^N(B_k) \chi_{A_k}$$

and

$$h = \sum_{k=1}^{\infty} \alpha_k \mu^N(B_k) \chi_{A_k}.$$

Each h_p is integrable and the sequence $\{h_p\}$ converges a.e. to h. In the proof of the lemma we showed that

$$\sum_{k=1}^{\infty} |\alpha_k| \mu^N(B_k) \chi_{A_k}$$

is integrable, and clearly

$$|h_p| \le \sum_{k=1}^{\infty} |\alpha_k| \mu^N(B_k) \chi_{A_k}$$

for each p, so the Lebesgue dominated convergence theorem implies that h is integrable over $\mathbb{R}^M$ and that

$$\int_{\mathbb{R}^M} h = \sum_{k=1}^{\infty} \alpha_k \mu^M(A_k) \mu^N(B_k) = \sum_{k=1}^{\infty} \alpha_k \mu(A_k \times B_k).$$

But this amounts to

$$\int_{\mathbb{R}^M}\left(\int_{\mathbb{R}^N} f(x,y)dy\right)dx = \int_{\mathbb{R}^{M+N}} f. \qquad \square$$

Exercises

1. (a) If P is a plane in $\mathbb{R}^3$, show that any subset of P has 3-dimensional measure zero. Use this result to show instantly that any subset of any line in $\mathbb{R}^3$ must have 3-dimensional measure zero.

   (b) Prove that if $M < K$, then the K-dimensional measure of any subset of an M-dimensional vector subspace of $\mathbb{R}^K$ is zero.

2. Show that for $A \subseteq \mathbb{R}^M$ and $B \subseteq \mathbb{R}^N$ and for $x \in \mathbb{R}^M$ and $y \in \mathbb{R}^N$, we have

$$\chi_{A\times B}(x,y) = \chi_A(x)\chi_B(y).$$

3. Show that as stated in the proof of the lemma,

$$Z_q = \bigcap_{r=1}^{\infty}\left(\bigcup_{p=1}^{\infty} Z_{p,q-\frac{1}{r}}\right) \quad \text{and} \quad Z = \bigcap_{q=1}^{\infty} Z_q.$$

4. If $f, g \in \mathcal{L}^1(\mathbb{R}^N)$, then by the convolution $f * g$ of f and g we mean the function defined by

$$(f * g)(x) = \int_{\mathbb{R}^N} f(y)g(x-y)dy.$$

   Prove the following:

   (a) If $f, g \in \mathcal{L}^1(\mathbb{R}^N)$, then $f * g$ is defined almost everywhere and $f * g \in \mathcal{L}^1(\mathbb{R}^N)$,

   (b) $f * g = g * f$,

   (c) $\|f * g\| \leq \|f\| \cdot \|g\|$ where $\|f\| = \int |f|$,

   (d) $(f * g) * h = f * (g * h)$,

   (e) $f * (g + h) = f * g + f * h$.

5. Formulate and prove a version of Lemma 4.9.1 which applies to $f_y(x) = f(x, y)$.

6. Prove the second equality in Fubini's theorem.

5

INTEGRALS ON MANIFOLDS

5.1 Introduction

In mathematics and its applications one may encounter situations in which it is desirable to set up integrals over arcs or surfaces or their higher dimensional generalizations. (These higher dimensional generalizations are called *manifolds*. We shall explain them later in this chapter.) For instance, given a mass distributed along an arc with a known density, find the total mass along the arc. Or given a fluid flow through a surface with a given rate of flow at each point of the surface, find the total flow through the surface at any given instant of time.

But one immediately encounters the fact that an arc or a surface in $\mathbb{R}^3$ is a set of measure zero, and the integral of any function over such a set must be zero. The theory of Lebesgue integration over subsets of $\mathbb{R}^N$, at least in the form which we have so far achieved, is not sufficient to do what we want. We need an extension of the theory.

A vital clue to this extension and an extremely important result in its own right is the change of variables formula for integrals.

Think of a physical space $\mathcal{P}$ which is modelled by $\mathbb{R}^N$. This means we think of $\mathbb{R}^N$ (or open subsets of $\mathbb{R}^N$) as being "pasted" onto $\mathcal{P}$ (or onto subsets of $\mathcal{P}$) by a (usually) one-to-one function $f\colon \mathbb{R}^N \rightarrowtail \mathcal{P}$. If $p = f(x_1, \ldots, x_N)$, then we say p has coordinates $(x_1, \ldots, x_N)$. That is, f amounts to a description of the way in which we assign coordinates to points of $\mathcal{P}$. In general there ought to be many different ways of assigning coordinates to $\mathcal{P}$, but there ought to also be some way of changing from one set of coordinates to any other set. Thus suppose $g\colon \mathbb{R}^N \rightarrowtail \mathcal{P}$ describes a second method of assigning coordinates to points of $\mathcal{P}$. By a change of variables from the coordinates assigned by g to those assigned by f, one ought to mean a function

$h\colon \mathbb{R}^N \rightarrowtail \mathbb{R}^N$ that satisfies $g = f \circ h$. It ought be a one-to-one function so that we also have $f = g \circ (h^{-1})$. In what follows, to simplify the mathematics and produce stronger results, we shall require even more of our change-of-variables functions. We shall think of a *change of variables* as being a C^r diffeomorphism.

Now we turn our attention to considering physical quantities described in terms of different coordinate systems or parameters. Suppose $f, g\colon \mathbb{R}^N \rightarrowtail \mathbb{R}$ both represent the same physical quantity, say the density of a mass distribution, but they look different because each is defined in terms of a different coordinate system. (For instance, we might write down a density function in the plane using Cartesian coordinates thus: $f(x, y) = \sqrt{x^2 + y^2}e^{-2xy}$ where $x > 0$ and $y > 0$. We could describe the same mass distribution by giving a density function in polar coordinates: $g(r, \theta) = r\, e^{-r^2 \sin(2\theta)}$ where $r > 0$ and $0 < \theta < \pi/2$.) Suppose $h\colon \mathbb{R}^N \to \mathbb{R}^N$ is a C^1 diffeomorphism which changes the coordinates on which g operates to those on which f operates. We should *not* expect to have $g = f \circ h$. Rather, we should expect that for any measurable set Ω in $\mathbb{R}^N$ we have

$$\int_{h(\Omega)} f = \int_{\Omega} g.$$

After all, Ω represents a set of physical points as described in one coordinate system, and $h(\Omega)$ represents the same set as described in another coordinate system. In each case the integral gives the total mass distributed over that set.

We want to see how f and g are related to one another; this relation is what we mean by the change-of-variables formula. In the heuristic argument about to follow, the reader may find it helpful to think of f and g as being continuous.

First we should have

$$\frac{\int_{h(\Omega)} f}{\mu(\Omega)} = \frac{\int_{\Omega} g}{\mu(\Omega)}$$

whenever $\mu(\Omega)$, the measure of Ω, is positive. Let Ω be an N-dimensional cube with one vertex at a fixed point x and with sides having the length and direction of $\lambda e_1, \ldots, \lambda e_N$. Then $\mu(\Omega) = \lambda^N$ and

$$\frac{\int_{\Omega} g}{\mu(\Omega)} \approx \frac{g(x)\, \mu(\Omega)}{\mu(\Omega)} = g(x)$$

with the approximation becoming better the closer λ is to zero. For λ having small magnitude, h behaves very much like a linear transformation except that we have $x \mapsto h(x)$ rather than $0 \mapsto 0$. The N-dimensional cube Ω is mapped (approximately) to an N-dimensional parallelepiped with one vertex at $h(x)$ and with sides having the length and direction of $\lambda(h'(x))(e_1), \ldots, \lambda(h'(x))(e_N)$. We know from Chapter 1 that the N-dimensional volume of this parallelepiped is $\lambda^N |\det(h'(x))|$. Then we should have

$$\frac{\int_{h(\Omega)} f}{\mu(\Omega)} \approx f(h(x))|\det(h'(x))|$$

with the approximation growing better the closer λ is to zero. This means that as λ approaches zero we should expect to obtain

$$g(x) = f(h(x))|\det(h'(x))|.$$

Therefore for diffeomorphisms h we expect to have a change of variables formula for integrals that looks like this:

$$\int_{h(\Omega)} f = \int_{\Omega} (f \circ h) \,|\det(h')|.$$

In the next section we shall give a proof of this.

In the last integral $|\det(h')|$ plays the role of a distortion factor for N-dimensional volume. It is not the same everywhere but changes from point and is a sort of *microscopic* distortion factor. Now think of some K-dimensional manifold $\mathcal{M}$ lying in $\mathbb{R}^N$. Any such manifold can be covered by "coordinate patches" which can be defined by C^r diffeomorphisms $h\colon U \to \mathcal{M}$. Recall that for such an h the expression $\mathcal{D}(h')$ plays the role of a distortion factor for K-dimensional volume and reduces to $|\det(h')|$ in the case where $K = N$. This instantly suggests that for $f\colon \mathcal{M} \to \mathbb{R}$ we should define

$$\int_{h(U)} f = \int_{U} (f \circ h)\, \mathcal{D}(h').$$

This is the key to defining integrals over manifolds lying in $\mathbb{R}^N$.

5.2 The Change of Variables Formula

The proof of the change-of-variables formula is the longest and most complex proof to be found in this text. It shall be broken into a number of steps. It is noteworthy for the number of striking and important ideas which must be called upon to attain our goal; Fubini's theorem, the existence of local factorizations of diffeomorphisms into primitive diffeomorphisms, the existence of the Lebesgue number, etc. But before we can begin, we must construct some machinery.

We first describe a particular way in which one can subdivide a brick into smaller and smaller bricks. We shall refer to this particular technique as the *subdivision process*.

Let R be the brick $[\alpha_1, \beta_1) \times \cdots \times [\alpha_N, \beta_N)$ in $\mathbb{R}^N$. For each i let $\gamma_i = (\alpha_i + \beta_i)/2$. Then R may be written as a union of disjoint bricks $R_1,\ R_2, \ldots, R_{2^N}$ where $R_i = I_{i1} \times \cdots \times I_{iN}$ and each I_{ij} has the form $[\alpha_j, \gamma_j)$ or $[\gamma_j, \beta_j)$. This is illustrated for $N = 2$ and 3 in Figures 5.2.1 and 5.2.2. Recall that in a metric space (X, d) the diameter of a bounded set A is $\sigma(A) = \sup\{\, d(x, y)\colon x, y \in A\,\}$. Then $\sigma(R) = \sigma(\overline{R}) = |b - a|$ where $a = (\alpha_1, \ldots, \alpha_N)$ and $b = (\beta_1, \ldots, \beta_N)$. Since we always have $|\gamma_j - \alpha_j| = |\beta_j - \gamma_j| = |\beta_j - \alpha_j|/2$, we must also have

$$\sigma(R_i) = \sqrt{\left(\frac{\beta_1 - \alpha_1}{2}\right)^2 + \cdots + \left(\frac{\beta_N - \alpha_N}{2}\right)^2} = \frac{1}{2}\,|b - a| = \frac{1}{2}\,\sigma(R).$$

If, in a similar fashion to what we have just done, we subdivide the bricks R_i, then subdivide the bricks resulting from *that* subdivision, etc., at each step the diameter of the new bricks is half that of the bricks in the previous step. See Figure 5.2.3 for an illustration of this ongoing subdivision process in the case $N = 2$. The important point is that we may, in this fashion, replace R by bricks all having diameter less than any given $\varepsilon > 0$.

Now we use this subdivision process to establish a couple of useful results.

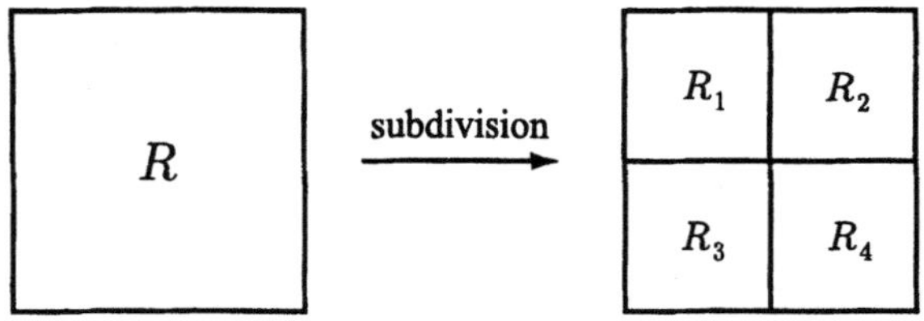

FIGURE 5.2.1.

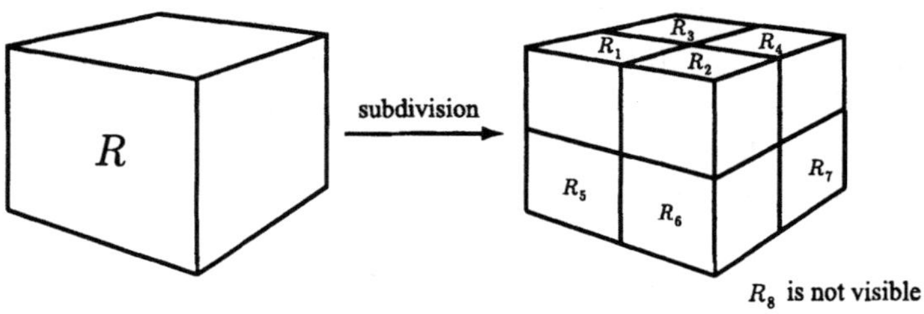

FIGURE 5.2.2.

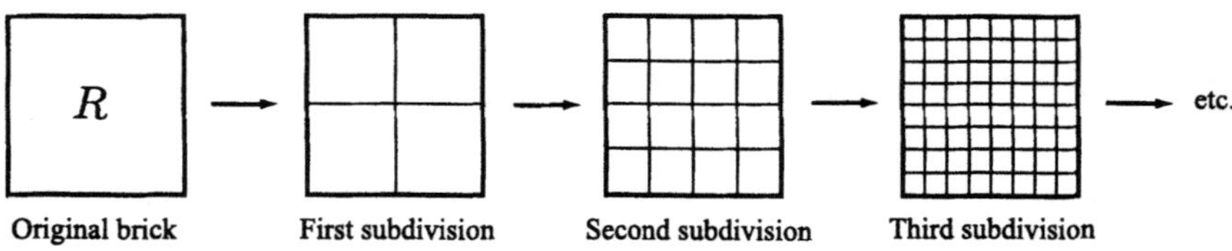

FIGURE 5.2.3.

Lemma 5.2.1 *If C is a closed, nonempty subset of $\mathbb{R}^N$ and U is an open set such that $C \subseteq U$, then there exists a finite or countably infinite set of disjoint bricks $\{R_i\}$ such that $C \subseteq \bigcup_i R_i$ and $\overline{R_i} \subseteq U$ for each i.*

Proof. $\mathbb{R}^N$ is the disjoint union of the bricks $I_1 \times \cdots \times I_N$ where each factor I_j has the form $[k, k+1)$, k an integer. Let $\mathcal{A}$ be the collection of all bricks of this form that intersect C. Clearly $\mathcal{A}$ is countable and $C \subseteq \bigcup \mathcal{A}$. We now construct a second collection of bricks, $\mathcal{B}$. If R is a member of $\mathcal{A}$ such that $\overline{R}$ does not intersect $\mathbb{R}^N - U$, then we put R in $\mathcal{B}$. Suppose R is a member of $\mathcal{A}$ such that $\overline{R}$ *does* intersect $\mathbb{R}^N - U$. Let

$$\rho = \operatorname{dist}(C, \mathbb{R}^N - U).$$

This is a positive number. Apply the subdivision process to R and its subdivisions repeatedly until it is decomposed into a finite set of bricks $A_1, \ldots, A_p$ each having diameter less than $\rho/2$. We put into $\mathcal{B}$ every brick A_i that intersects C. Clearly $R \cap C$ is a subset of the set of bricks A_i which intersects C, and since $\sigma(A_i) = \sigma(\overline{A_i}) < \rho/2$, each $\overline{A_i} \subseteq U$. This completes the construction of $\mathcal{B}$. We see that $\mathcal{B}$ is the promised collection of bricks. □

Lemma 5.2.2 *If U is an open set in $\mathbb{R}^N$ and $f : \mathbb{R}^N \to \mathbb{R}$ is an integrable function such that $\overline{supp(f)} \subseteq U$, then there is an expansion $f \simeq \sum_{i=1}^{\infty} \alpha_i f_i$, where each*

$\alpha_i \in \mathbb{R}$ and each f_i is a brick function, with the property that for each i we have $\overline{supp(f_i)} \subseteq U$.

Proof. We know we can find a countable collection of bricks $R_1, R_2, \ldots$ (or possibly a finite collection) with the property that $\overline{\text{supp}(f)} \subseteq \bigcup_{i=1}^{\infty} R_i$ and $\overline{R_i} \subseteq U$ for each i. We take an expansion of f, say $f \simeq \sum_{i=1}^{\infty} \beta_i g_i$, where each $\beta_i \in \mathbb{R}$ and each g_i is a brick function. Let us set $h_{ij} = g_i \chi_{R_j}$. This is sometimes a brick function and sometimes the zero function. Notice that $\overline{\text{supp}(h_{ij})} \subseteq \overline{R_j} \subseteq U$. It is straightforward to show that $f \simeq \sum_{ij} \beta_i h_{ij}$, and thus we are done. □

It is easy to guess that the next result, a very general one for metric spaces, will eventually be used in conjunction with the subdivision process.

Theorem 5.2.1 *Let C be a compact set in the metric space (X, d) and suppose that $U_1, \ldots, U_n$ are open sets in the same space such that $C \subseteq \bigcup_{i=1}^{n} U_i$. Then there is a positive number λ with the property that for every $A \subseteq C$ such that $\sigma(A) < \lambda$, there is some U_i such that $A \subseteq U_i$.*

(**Note**: λ is called the *Lebesgue number* associated with C and with $U_1, \ldots, U_n$.)

Proof. Suppose the theorem is not true. Then for every natural number m there must be an $A_m \subseteq C$ such that $\sigma(A_m) < 1/m$ and A_m is a subset of no one of the U_i. Choose a point p_m from each A_m. Since C is compact, we may, without loss of generality, suppose that $p_m \to p$ for some $p \in C$. Again, without loss of generality, we may suppose $p \in U_1$. Since U_1 is open, there must be some $\varepsilon > 0$ such that $B(p, \varepsilon) \subseteq U_1$. There must exist an m_0 with the property that for all $m > m_0$ we have $d(p_m, p) < \varepsilon/2$ and $1/m < \varepsilon/2$. Consider any A_m such that $m > m_0$. If $x \in A_m$, we must have

$$d(x, p) \leq d(x, p_m) + d(p_m, p) \leq \sigma(A_m) + (\varepsilon/2) < (\varepsilon/2) + (\varepsilon/2) = \varepsilon.$$

We see from this that $A_m \subseteq B(p, \varepsilon) \subseteq U_1$, which is a contradiction. Thus our theorem is established. □

Example 5.2.1 Let our compact set C be the unit interval, $[0, 1]$. Let our open covering be $U_1 = [0, 3/4)$ and $U_2 = (2/3, 1]$. Then any subset of C of diameter less than 1/12 must be a subset of at least one of U_1, U_2. That is, for this C and this open covering, we have $\lambda = 1/12$.

We now come to our main result.

Theorem 5.2.2 (The change-of-variables theorem) *If g is a C^1 diffeomorphism of U onto V, where U and V are open sets in $\mathbb{R}^N$, and if f is a real-valued function that is integrable over $g(\Omega)$, where Ω is a measurable set and $\overline{\Omega} \subseteq U$, then $(f \circ g) \mid \det(g') \mid$ is integrable over Ω and*

$$\int_{g(\Omega)} f = \int_{\Omega} (f \circ g) \, | \det(g')|.$$

Proof. Notice that since $\chi_\Omega = \chi_{g(\Omega)} \circ g$, the change-of-variables formula amounts to

$$\int f \, \chi_{g(\Omega)} = \int (f \circ g) \, (\chi_\Omega) \, | \det(g')| = \int ((f \, \chi_{g(\Omega)}) \circ g) \, | \det(g')|.$$

Because of this, we restrict our attention to functions f having the property that $\overline{\text{supp}(f)} \subseteq V$ and prove only that

$$\int f = \int (f \circ g) \, | \det(g')|.$$

Step 1. *It is sufficient to prove the change-of-variables theorem when the integrand is a brick function.*

Suppose we know the change-of-variables theorem to hold whenever the integrand is a brick function. Let g be a C^1 diffeomorphism of U onto V and let f be an integrable function such that $\overline{\text{supp}(f)} \subseteq V$. We know that we can write $f \simeq \sum_{i=1}^{\infty} \alpha_i \, f_i$, each $\alpha_i \in \mathbb{R}$ and each f_i a brick function, with the property that for each i we have $\overline{\text{supp}(f_i)} \subseteq V$. By assumption we know that $(f_i \circ g) \, | \det(g')|$ is integrable and

$$\int f_i = \int (f_i \circ g) \, | \det(g')|$$

for each i.

We need to show that $(f \circ g) \, | \det(g')|$ is integrable. Note that

$$\sum_{i=1}^{\infty} |\alpha_i| \int (f_i \circ g) \, | \det(g')| = \sum_{i=1}^{\infty} |\alpha_i| \int f_i < \infty.$$

Next suppose we have an x for which

$$\sum_{i=1}^{\infty} |\alpha_i| (f_i(g(x))) \, | \det(g'(x))| < \infty.$$

Since $| \det(g'(x))|$ is a factor common to each term, we see that $\sum_{i=1}^{\infty} \alpha_i \, f_i(g(x))$ must be absolutely convergent, hence convergent to $f(g(x))$. Therefore

$$\sum_{i=1}^{\infty} \alpha_i \, f_i(g(x)) \, | \det(g'(x))| = f(g(x)) \, | \det(g'(x))|.$$

Thus

$$(f \circ g) \, | \det(g')| \simeq \sum_{i=1}^{\infty} \alpha_i \, (f_i \circ g) \, | \det(g')|.$$

From this last deduction we instantly obtain

$$\int (f \circ g)\,|\det(g')| = \sum_{i=1}^{\infty} \alpha_i \int (f_i \circ g)\,|\det(g')| = \sum_{i=1}^{\infty} \alpha_i \int f_i = \int f,$$

and we are done.

Step 2. *If the change-of-variables theorem holds for two diffeomorphisms, it holds for their composition.*

Suppose g and h are C^1 diffeomorphisms of V onto W and U onto V, respectively, for which the change-of-variables theorem holds. That is, whenever f_1 and f_2 are integrable functions with $\overline{\text{supp}(f_1)} \subseteq V$ and $\overline{\text{supp}(f_2)} \subseteq W$, then

$$\int f_1 = \int (f_1 \circ h)\,|\det(h')| \quad \text{and} \quad \int f_2 = \int (f_2 \circ g)\,|\det(g')|.$$

Then $g \circ h$ is a C^1 diffeomorphism of U onto W, and if f is an integrable function with $\overline{\text{supp}(f)} \subseteq W$, we see that

$$\begin{aligned} \int f &= \int (f \circ g)\,|\det(g')| \\ &= \int (f \circ g \circ h)\,|\det(g' \circ h)|\,|\det(h')| \\ &= \int (f \circ g \circ h)\,|\det(g \circ h)'|. \end{aligned}$$

This is the desired result.

Step 3. *When the domain of the functions has dimension at least 2, it is sufficient to prove the change-of-variables theorem for primitive diffeomorphisms.*

Suppose we know the change-of-variables theorem to be true for primitive C^1 diffeomorphisms with domain of dimension at least 2. Let g be a C^1 diffeomorphism of U onto W where U and W are open subsets of $\mathbb{R}^N$ and $N \geq 2$. Let R be a brick such that $\overline{R} \subseteq W$ and let f be the associated brick function. We need only show that $(f \circ g)\,|\det(g')|$ is integrable and

$$\int f = \int (f \circ g)\,|\det(g')|.$$

By Theorem 3.6.2, for each $p \in \overline{R}$ we can find open sets U_p, V_p, and W_p in $\mathbb{R}^N$ and primitive C^1 diffeomorphisms h_p and k_p of U_p onto V_p and V_p onto W_p, respectively, such that $g|U_p = k_p \circ h_p$. Since $\overline{R}$ is compact, there is a finite number of the W_p sets, say $W_1, W_2, \ldots, W_m$, such that $\overline{R} \subseteq \bigcup_{i=1}^{m} W_i$. For $i = 1, \ldots, m$, let us call the associated U_p, U_i, the associated h_p, h_i, and so forth. Let λ be the Lebesgue number associated with $\overline{R}$ and $W_1, \ldots, W_m$. Apply the subdivision process repeatedly to R until it is decomposed into bricks $R_1, \ldots, R_n$, each having diameter less than λ. Set f_j equal to the restriction of f to R_j so that $f = f_1 + \cdots + f_n$. Let us consider a particular R_j. By the definition of Lebesgue number, $\overline{R_j}$ must be a subset of some

W_i . Since k_i is a primitive diffeomorphism, we see that $(f_j \circ k_i)\,|\det(k_i')|$ must be integrable and

$$\int f_j = \int (f_j \circ k_i)\,|\det(k_i')|.$$

Note that

$$\overline{\operatorname{supp}\left((f_j \circ k_i)\,|\det(k_i')|\right)} \subseteq k_i^{-1}(\overline{R_j}) \subseteq V_i.$$

Since h_i is a primitive diffeomorphism, by our assumption,

$$\int (f_j \circ k_i)\,|\det(k_i')| = \int (f_j \circ k_i \circ h_i)\,|\det(k_i' \circ h_i)|\,|\det(h_i')|.$$

But this last integral reduces to

$$\int (f_j \circ g)\,|\det(g')|.$$

Therefore

$$\int f = \int f_1 + \cdots + \int f_n = \sum_{j=1}^{n} \int (f_j \circ g)\,|\det(g')| = \int (f \circ g)\,|\det(g')|,$$

and we are done.

Step 4. *The change-of-variables theorem for dimension one.*

It is likely the reader has already seen this proof, and indeed seen it in a more general form than we present, in a treatment of analysis for single variable functions. Nevertheless, we include this proof for the sake of completeness.

Let g be a C^1 diffeomorphism of U onto V where U and V are open subsets of $\mathbb{R}$. Let f be the brick function of the brick $[\alpha, \beta)$ where $[\alpha, \beta] \subseteq V$. Then

$$\int f = \beta - \alpha.$$

Either $g'(x) > 0$ for all x or $g'(x) < 0$ for all x. Suppose $g' > 0$. Then

$$\int (f \circ g)\,|\det(g')| = \int_{g^{-1}(\alpha)}^{g^{-1}(\beta)} g' = \beta - \alpha.$$

The proof when $g' < 0$ is similar.

Step 5. *Induction step on the dimension of the domain.*

Suppose we have the change-of-variables theorem for all natural numbers less than N where $N \geq 2$. Let K and M be natural numbers such that $K + M = N$. Let g be a C^1 primitive diffeomorphism of U onto V, where U and V are open subsets of $\mathbb{R}^N$, one that has the form $g(x, y) = (z, y)$, where $x \in \mathbb{R}^K$ and $y \in \mathbb{R}^M$. Finally, let f be a brick function such that $\overline{\operatorname{supp}(f)} \subseteq V$. To complete the proof of the change-of-variables theorem, it is sufficient to show that $(f \circ g)\,|\det(g')|$ is integrable and

$$\int f = \int (f \circ g)\,|\det(g')|.$$

If R is the brick associated with f, then $f \circ g$ is the characteristic function of $g^{-1}(R)$. Note that $g^{-1}(R)$ is a bounded, measurable set and $|\det(g')|$ is a continuous function on the open set U that contains $\overline{g^{-1}(R)} = g^{-1}(\overline{R})$. We know that $|\det(g')|$ is integrable over $g^{-1}(R)$, but this amounts to saying that $(f \circ g)\,|\det(g')|$ is integrable.

We can write $g(x, y) = (h(x, y), y)$ where $h\colon U \to \mathbb{R}^K$ is a C^1 function. Let us define

$$\begin{aligned} h_y(x) &= h(x, y),\\ f_y(x) &= f(x, y),\\ U_y &= \{x \in \mathbb{R}^K :\ (x, y) \in U\}, \end{aligned}$$

and

$$V_y = h_y(U_y).$$

U_y is an open set of $\mathbb{R}^K$ for all y. The following argument shows each h_y is a one-to-one function: Suppose $h_y(x) = h_y(z)$. Then $h(x, y) = h(z, y)$, and hence

$$g(x, y) = (h(x, y), y) = (h(z, y), y) = g(z, y).$$

Since g is one-to-one, we see that $x = z$. Next it is straightforward to see that $\det(h'_y(x)) = \det(g'(x, y)) \neq 0$ for all $x \in U_y$. We thus see that each h_y is a C^1 diffeomorphism and that V_y must be open in $\mathbb{R}^K$. Then by Fubini's theorem and the induction hypothesis we must have

$$\begin{aligned} \int f &= \int_{\mathbb{R}^M} \int_{\mathbb{R}^K} f_y(x)dx\, dy = \int_{\mathbb{R}^M} \int_{\mathbb{R}^K} (f_y \circ h_y)(x)\,|\det(h'_y(x)|\, dx\, dy\\ &= \int (f \circ g)\,|\det(g')|. \end{aligned}$$

This completes the proof of the change-of-variables theorem. □

Example 5.2.2 Let $f(x, y) = e^{-x^2-y^2}$. We wish to evaluate $\int_{\mathbb{R}^2} f$. Let

$$U = (0, \infty) \times (0, 2\pi) \quad \text{and} \quad V = \mathbb{R}^2 - \{(x, y)\colon x \geq 0 \text{ and } y = 0\}.$$

Note that $\{(x, y)\colon x \geq 0 \text{ and } y = 0\}$ is a null set, so that

$$\int_V f = \int_{\mathbb{R}^2} f,$$

assuming f is integrable, which we have not yet shown. It is easily seen that the function $g\colon U \to V$ defined by $g(r, \theta) = (r\cos(\theta), r\sin(\theta))$ is a C^1 diffeomorphism of V onto U and $\det|g'(r, \theta)| = r$. By the change-of-variables theorem and Fubini's theorem, again relying on the as-yet-unproven integrability of f, we have

$$\int_V f = \int_U (f \circ g)\,|\det(g')| = \int_0^{2\pi} \int_0^{\infty} r\, e^{-r^2} dr\, d\theta.$$

For each natural number n let

$$h_n(r,\theta) = \begin{cases} re^{-r^2} & \text{if } 0 < r < n \text{ and } 0 < \theta < 2\pi. \\ 0 & \text{otherwise.} \end{cases}$$

Each h_n is integrable and has integral $\pi(1 - e^{-n^2})$. By the monotone convergence theorem we see that $(f \circ g)\,|\det(g')|$ is integrable and that we must have

$$\int_0^{2\pi} \int_0^{\infty} r\, e^{-r^2} dr\, d\theta = \pi.$$

Using the change-of-variables theorem with g^{-1} as our diffeomorphism, we see that f must be integrable over V, and hence over $\mathbb{R}^2$. We conclude that

$$\int_V f = \pi.$$

Exercises

1. Let R be the brick $[\alpha_1, \beta_1) \times \cdots \times [\alpha_N, \beta_N)$ in $\mathbb{R}^N$. Show that $\sigma(R) = \sigma(\overline{R}) = |b - a|$ where $a = (\alpha_1, \ldots, \alpha_N)$ and $b = (\beta_1, \ldots, \beta_N)$.

2. Complete the proof of Lemma 5.2.2 by showing that $f \simeq \sum_{ij} \beta_i h_{ij}$.

3. What is the Lebesgue number associated with the open covering

$$U_1 = [0, 1/3), \quad U_2 = (1/5,\ 2/3), \quad U_3 = (3/5, 1]$$

of the compact set $[0, 1]$?

4. What is the Lebesgue number of the open covering $U_1,\ U_2$ of the closed, unit disk in $\mathbb{R}^2$ where U_1 and U_2 are the open disks of radius $3/2$ centered at $(-1, 0)$ and $(1, 0)$ respectively?

5. For each following g and Ω, show that g is a diffeomorphism on Ω and rewrite $\int_{g(\Omega)} f$ as $\int_\Omega (f \circ g)\,|\det(g')|$ using the change-of-variables formula:

   (a) $g(x_1, x_2, \ldots, x_N) = (-x_1, x_2, \ldots, x_N)$ and $\Omega = \mathbb{R}^N$.

   (b) $g(x_1, x_2, \ldots, x_N) = (\alpha_1 x_1, \alpha_2 x_2, \ldots, \alpha_N x_N)$, where $\alpha_1, \ldots, \alpha_N$ are given, nonzero constants, and $\Omega = \mathbb{R}^N$.

   (c) $\Omega = \mathbb{R}^N$ and g is the map of $\mathbb{R}^N$ onto itself which interchanges the ith and jth variables for given, distinct indices i and j.

   (d) g is an orthogonal transformation and $\Omega = \mathbb{R}^N$.

   (e) $g(x, y, z) = (x, xy, xyz)$ and $\Omega = \{(x, y, z) \colon x, y, z > 0\}$.

   (f) $g(x, y) = (x^2 - y^2, 2xy)$ and $\Omega = \{(x, y) \colon x, y > 0\}$.

(g) $g(r, \alpha, \beta) = (r\ \cos(\alpha),\ r\ \sin(\alpha)\ \cos(\beta),\ r\ \sin(\alpha)\ \sin(\beta))$ and $\Omega = \{(r, \alpha, \beta) : r > 0 \text{ and } 0 < \alpha, \beta < \pi/2\}$.

6. If g is a $\mathcal{C}^1$ diffeomorphism of U onto V, where U and V are open subsets of $\mathbb{R}^N$ and $\overline{A} \subseteq U$, show that $g(\overline{A}) = \overline{g(A)}$. Can this be proven with weaker hypotheses?

7. Prove: If g is a $\mathcal{C}^1$ diffeomorphism of U onto V, where U and V are open subsets of $\mathbb{R}^N$, and if Ω is a set of measure zero in $\mathbb{R}^N$ such that $\overline{\Omega} \subseteq U$, then $g(\Omega)$ is also a set of measure zero in $\mathbb{R}^N$.

8. Prove: If g is a $\mathcal{C}^1$ diffeomorphism of U onto V, where U and V are open subsets of $\mathbb{R}^N$, and if Ω is a brick in $\mathbb{R}^N$ such that $\overline{\Omega} \subseteq U$, then $g(\Omega)$ is a bounded measurable set in $\mathbb{R}^N$.

9. Prove: If g is a $\mathcal{C}^1$ diffeomorphism of U onto V, where U and V are open subsets of $\mathbb{R}^N$, and if Ω is a measurable set in $\mathbb{R}^N$ such that $\overline{\Omega} \subseteq U$, then $g(\Omega)$ is also a measurable set in $\mathbb{R}^N$.

10. Use Example 5.2.2 to prove that $\int_{-\infty}^{\infty} e^{-x^2} dx = \sqrt{\pi}$.

5.3 Manifolds

Consider the function $f(x, y, z) = (x^2 + y^2 + z^2 - 9)^2 - 36(1 - z^2)$. Then $\nabla f(x, y, z) = (4x(x^2+y^2+z^2-9), 4y(x^2+y^2+z^2-9), 4z(x^2+y^2+z^2-9)+72z)$.

Denote $\mathcal{T} = \{(x, y, z) \in \mathbb{R}^3 : f(x, y, z) = 0\}$. This is a torus. It is easy to check that if $(x, y, z) \in \mathcal{T}$, then $\nabla f(x, y, z) \neq 0$. Therefore the implicit function theorem can be applied at every point of $\mathcal{T}$ (with respect to different variables). This means that locally $\mathcal{T}$ is the graph of a real-valued $\mathcal{C}^\infty$ function of two variables. One can say that $\mathcal{T}$ is a surface that in a neighborhood of any point is like the unit disk in $\mathbb{R}^2$. Sets which can be described in a similar way are very important in analysis. Some people describe them as being "locally" like $\mathbb{R}^K$. They are called manifolds.

Manifolds come in two varieties, with and without boundary. We begin by defining the simplest case, manifolds-without-boundary.

Definition 5.3.1 (First formulation; manifold-without-boundary) For $K \geq 1$, a nonempty subset $\mathcal{M}$ of $\mathbb{R}^N$ is called a $\mathcal{C}^r$ *manifold* (*-without-boundary*) of dimension K if for every point $x_0 \in \mathcal{M}$ there exists a set U which is open in $\mathbb{R}^K$ and a function $g: U \to \mathcal{M}$ such that

(1) $x_0 \in g(U)$,

(2) g is one-to-one and $\mathcal{C}^r$,

(3) $\mathcal{D}(g')$ is never zero,

(4) $g(U)$ is open in $\mathcal{M}$,

and

(5) g^{-1} is continuous.

We call the function g a C^r *coordinate patch* on $\mathcal{M}$ and say that x_0 lies in the coordinate patch or that the coordinate patch contains x_0.

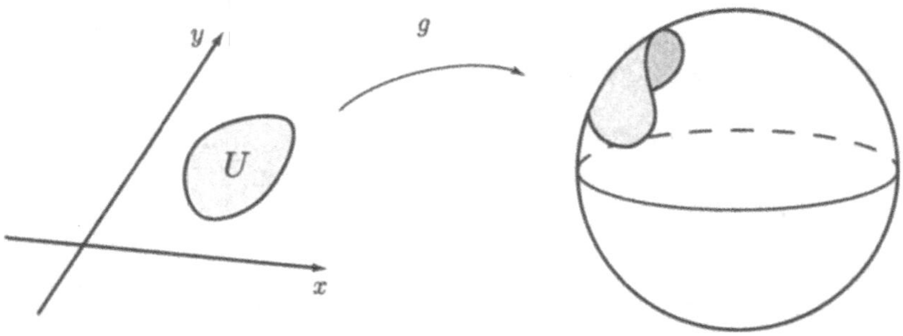

FIGURE 5.3.1.

Examples are the sphere $x^2+y^2+z^2=1$ (Figure 5.3.1) and a torus (Figure 5.3.2). In the figures we indicate the workings of a coordinate patch $g\colon U \to \mathcal{M}$. We can "paste" a coordinate patch over every point in these two manifolds. (Notice also in these examples that although our manifolds are not open in $\mathbb{R}^3$, nevertheless it makes sense to talk about a subset S of $\mathcal{M}$ being open in $\mathcal{M}$. Recall that in accord with the discussion at the end of Section 2.2, if $Y \subseteq X$ where X is a metric space, then a subset S of Y is open in Y provided there is an open subset O of X such that $S = O \cap X$.)

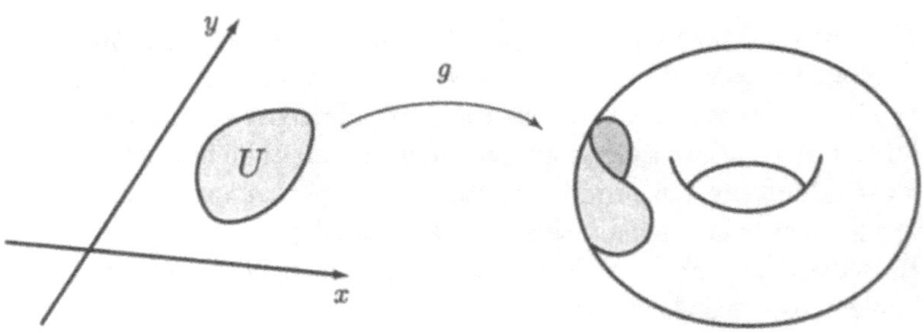

FIGURE 5.3.2.

However it turns out to be convenient to talk about manifolds which have "edges" or "boundaries". An example would be the closed unit disk in $\mathbb{R}^2$. In this case we want our coordinate patches to map not from $\mathbb{R}^K$ into the manifold but from a "Euclidean half-space" into the manifold. The procedure is indicated graphically in Figure 5.3.3 for the closed unit disk. We now proceed to reformulate Definition 5.3.1 to include both manifolds-without-boundary and manifolds-with-boundary.

Definition 5.3.2 (Second formulation; manifold) Let

$$\mathbb{H}^K = \{(x_1, x_2, \ldots, x_K) \in \mathbb{R}^K : x_1 \geq 0\}.$$

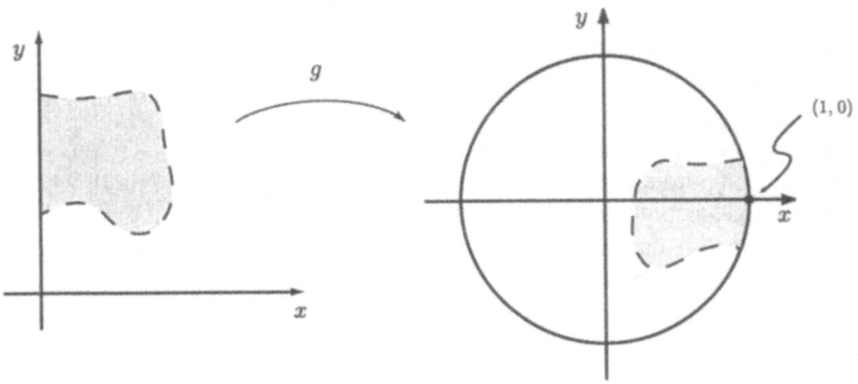

FIGURE 5.3.3.

For $K \geq 1$, a nonempty subset $\mathcal{M}$ of $\mathbb{R}^N$ is called a C^r *manifold of dimension* K if for every point $x_0 \in \mathcal{M}$ there exists a set U which is open either in $\mathbb{R}^K$ or $\mathbb{H}^K$ and a function $g: U \to \mathcal{M}$ such that

(1) $x_0 \in g(U)$,

(2) g is one-to-one and C^r,

(3) $\mathcal{D}(g')$ is never zero,

(4) $g(U)$ is open in $\mathcal{M}$,

and

(5) g^{-1} is continuous.

We call the function g a C^r *coordinate patch* on $\mathcal{M}$ and say that x_0 lies in the coordinate patch or that the coordinate patch contains x_0. (It can be shown that (4) and (5) are superfluous; they are implied by (1), (2), and (3). We leave the proof of this as an exercise.)

If x_0 lies in a coordinate patch $g: U \to \mathcal{M}$ such that U is open in $\mathbb{R}^K$, then we say x_0 is an *interior point* of $\mathcal{M}$. Otherwise, x_0 is a *boundary point* of $\mathcal{M}$. The set of boundary points is called the boundary of $\mathcal{M}$ and is denoted by $\partial\mathcal{M}$.

If g and h are two coordinate patches on $\mathcal{M}$, we say they *overlap* provided we can find x, y, and z such that $g(y) = x = h(z)$, and we say x is *common* to both coordinate patches.

Of course $\mathbb{R}^K$ and $\mathbb{H}^K$ or any nonempty open subset of these spaces is trivially a C^∞ K-manifold.

It should be kept in mind that in everything that follows, we consider only manifolds that lie in some $\mathbb{R}^N$. We do this for convenience. In a more sophisticated treatment of the subject, manifolds are defined without reference to any $\mathbb{R}^N$ which contains them. For example, the physical space-time in which we live is often modeled as a "curved" 4-dimensional manifold, but one does not think of it as lying in some larger $\mathbb{R}^N$.

Example 5.3.1 One way in which $(N-1)$-dimensional manifolds often arise is as the solution sets of equations of the form $f(x)=0$ where $x \in \mathbb{R}^N$. For instance if $f_1(x_1, x_2, x_3) = x_1^2 + x_2^2 + x_3^2 - 4$, then the solution set of $f_1(x) = 0$ is a sphere of radius 2 centered at the origin. If $f_2(x_1, x_2, x_3) = x_1^2 + x_2^2 - 1$, then $f_2(x) = 0$ is the equation of a cylinder centered on the x_3-axis and having radius 1.

We can complicate this construction by taking $f(x) = (f_1(x), f_2(x))$, where f_1 and f_2 are the functions just defined, and asking for the solution set of $f(x) = 0$. This turns out to be a 1-dimensional manifold consisting of two circles, the intersection of the sphere and the cylinder. This works because the sphere and the cylinder are in some sense correctly positioned with respect to one another. The condition that guarantees "correct positioning" is $\mathcal{D}\left((f')^\diamond\right) \neq 0$. At any point where this condition holds, the tangent planes to the surfaces $f_1(x) = 0$ and $f_2(x) = 0$ will not coincide but will instead intersect one another in a line. This in turn forces the two surfaces $f_1(x) = 0$ and $f_2(x) = 0$ to intersect one another locally in a nice, 1-dimensional curve, a manifold. We have not, to this point, defined tangent space or tangent plane, so what we have said has the force of a heuristic discussion rather than a proof.

If $\mathcal{D}\left((f')^\diamond\right) = 0$ at some points, then it is possible that the set $\mathcal{M}$ defined by $f(x) = 0$ will fail to be a manifold or it may fail to be a "smooth" manifold. We illustrate these possibilities in the exercises.

Theorem 5.3.1 *Let $f : \mathbb{R}^{K+M} \rightarrowtail \mathbb{R}^K$ be a C^r function and let*

$$\mathcal{M} = \{x \in \mathbb{R}^{K+M} : f(x) = 0\}$$

be a nonempty set. Then assuming that $\mathcal{D}\left((f')^\diamond\right)$ is never zero on $\mathcal{M}$, it follows that $\mathcal{M}$ is a C^r M-dimensional manifold without boundary.

Proof. Let $c \in \mathcal{M}$ and let us write $c = (a, b)$ where $a \in \mathbb{R}^K$ and $b \in \mathbb{R}^M$. If $f = (f_1, \ldots, f_K)$, we may suppose without loss of generality that

$$\frac{\partial(f_1, \ldots, f_K)}{\partial(x_1, \ldots, x_K)}(a, b) \neq 0.$$

By the implicit function theorem there exist an open set U in $\mathbb{R}^{K+M}$, an open set W in $\mathbb{R}^M$, and a function $g : W \to \mathbb{R}^K$ such that

(1) g is a C^r function,

(2) $(a, b) \in U$ and $b \in W$,

(3) $g(b) = a$,

(4) $f(g(y), y) = 0$ for all $y \in W$,

and

(5) g is uniquely determined in the sense that if $(x, y) \in U$ and $f(x, y) = 0$, then $x = g(y)$.

Let us define $h\colon W \to \mathcal{M}$ by $h(y) = (g(y), y)$. Note that $h(b) = (a, b) = c$. We need only show that h is a C^r coordinate patch in $\mathcal{M}$. It follows that $\mathcal{M}$ is a C^r M-dimensional manifold and c is an interior point of $\mathcal{M}$.

Clearly h is C^r and one-to-one. Notice that $\Pi|h(W) = h^{-1}$ where $\Pi(x, y) = y$ for $x \in \mathbb{R}^K$ and $y \in \mathbb{R}^M$; thus h^{-1} is continuous. Since h^{-1} is continuous and W is open in $\mathbb{R}^M$, it follows that $h(W)$ must be open in $\mathcal{M}$. That is, there must exist an open set O in $\mathbb{R}^{K+M}$ such that $h(W) = \mathcal{M} \cap O$. Finally if $g = (g_1, \ldots, g_K)$, then the matrix for $h'(y)$ must have the form

$$\begin{pmatrix} D_1 g_1(y) & \ldots & D_M g_1(y) \\ \ldots & & \\ D_1 g_K(y) & \ldots & D_M g_K(y) \\ 1 & \ldots & 0 \\ \ldots & & \\ 0 & \ldots & 1 \end{pmatrix}.$$

Since this contains an $M \times M$ identity matrix, $\mathcal{D}(h')$ never equals zero on W. This completes the proof. □

Example 5.3.2 To see that for any $K \in \mathbb{N}$ the K-dimensional unit sphere S^K is a manifold in $\mathbb{R}^{K+1}$ it suffices to define $f\colon \mathbb{R}^{K+1} \to \mathbb{R}$ as

$$f(x_1, \ldots, x_K, x_{K+1}) = x_1^2 + \ldots + x_{K+1}^2 - 1.$$

It is only necessary to compute that

$$\mathcal{D}(f'(x_1, \ldots, x_{K+1})) = 2\sqrt{x_1^2 + \cdots + x_{K+1}^2} = 2$$

for all points on S^K and apply the last theorem.

The following is a trivial consequence of Definition 5.3.1.

Theorem 5.3.2 *If $\mathcal{M}$ is a C^r K-dimensional manifold in $\mathbb{R}^N$, then $K \leq N$.*

The next result says that if two coordinate patches overlap in a C^r manifold, then the transformation at any given point from one set of coordinates to the other set is a one-to-one C^r function. This is such an important property that in a more abstract development of manifold theory it is usually taken as part of the definition of a manifold.

Theorem 5.3.3 *If $g\colon U \to \mathcal{M}$ and $h\colon V \to \mathcal{M}$ are overlapping C^r coordinate patches on the C^r K-dimensional manifold $\mathcal{M} \subseteq \mathbb{R}^N$, then*

$$h^{-1} \circ g\colon U \cap (g^{-1} \circ h)(V) \quad \to \quad (h^{-1} \circ g)(U) \cap V$$

is a one-to-one, onto, C^r function and det($(h^{-1} \circ g)'$) is never zero.

Proof. It is clear that $h^{-1} \circ g$ must be one-to-one and onto. If we can also show it is C^r, it must equally be true that $g^{-1} \circ h$ is C^r, and from the fact that $(g^{-1} \circ h) \circ (h^{-1} \circ g) =$ the identity map, it follows immediately that det($(h^{-1} \circ g)'$) is never zero.

We will show that $h^{-1} \circ g$ is C^r at every point of $U \cap (g^{-1} \circ h)(V)$. Choose $u \in U \cap (g^{-1} \circ h)(V)$. There exist a unique $x \in \mathcal{M}$ and $v \in (h^{-1} \circ g)(U) \cap V$ such that $g(u) = x = h(v)$. Notice that $\mathcal{D}(h'(y)) \neq 0$ in some neighborhood of v.

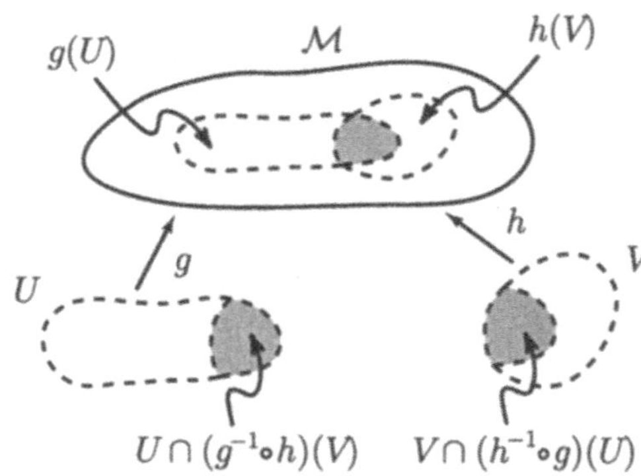

FIGURE 5.3.4.

Suppose $\mathcal{M} \subseteq \mathbb{R}^N$. If $K = N$, then $\det(h'(v)) \neq 0$. It follows from the inverse function theorem that h^{-1} is C^r in a neighborhood of x, hence $h^{-1} \circ g$ is C^r.

Now suppose $K < N$. We may assume without loss of generality that

$$\frac{\partial(h_1, \ldots, h_K)}{\partial(x_1, \ldots, x_K)}(v) \neq 0.$$

Define $H : V \times \mathbb{R}^{N-K} \to \mathbb{R}^N$ by $H(y, w) = (h(y), w)$ where $y \in V$ and $w \in \mathbb{R}^{N-K}$. It is easily shown that

$$\det(H'(y, w)) = \frac{\partial(h_1, \ldots, h_K)}{\partial(x_1, \ldots, x_K)}(y).$$

By the inverse function theorem, H and H^{-1} are C^r in neighborhoods of $(v, 0)$ and $x = (x_1, \ldots, x_N)$ respectively. Now $H^{-1}(r, w) = (h^{-1}(r, w), w)$ where $r \in \mathbb{R}^K$ and $w \in \mathbb{R}^{N-K}$. It follows that h^{-1} is C^r in a neighborhood of x. Thus $h^{-1} \circ g$ is C^r.

□

Theorem 5.3.4 *Let $\mathcal{M}$ be a C^r K-dimensional manifold in $\mathbb{R}^N$. Then x is a boundary point of $\mathcal{M}$ if and only if there is a C^r coordinate patch $g\colon U \to \mathcal{M}$ containing x such that U is open in $\mathbb{H}^K$ and x is the image under g of a point of the form $(0, x_2, x_3, \ldots, x_K)$.*

Proof. Suppose x is a boundary point of $\mathcal{M}$. It follows from the definition of manifold that there must be C^r coordinate patch $g\colon U \to \mathcal{M}$ containing x such that U is open in $\mathbb{H}^K$. We know that x is the image under g of a point $(x_1, x_2, \ldots, x_K)$ of $\mathbb{H}^K$ where $x_1 \geq 0$. If $x_1 > 0$, then x must be an interior point of $\mathcal{M}$, not a boundary point. Hence $x_1 = 0$.

Suppose there is a C^r coordinate patch $g\colon U \to \mathcal{M}$ containing x such that U is open in $\mathbb{H}^K$ and $x = g(0, x_2, x_3, \ldots, x_K)$. We must show that x is not an interior point of $\mathcal{M}$. Assume it is. There must be a C^r coordinate patch $h\colon V \to \mathcal{M}$ containing x such that V is open in $\mathbb{R}^K$. Let

$$C = (h^{-1} \circ g)(U) \cap V \quad \text{and} \quad D = U \cap (g^{-1} \circ h)(V).$$

Then $g^{-1} \circ h$ is a one-to-one $\mathcal{C}^r$ map of C onto D such that $\det((g^{-1} \circ h)')$ is never zero on C. Note $g(U)$ is open in $\mathcal{M}$ by the definition of a coordinate patch, thus $(h^{-1} \circ g)(U)$ is open in V since h is a continuous function. This means that $(h^{-1} \circ g)(U)$ is open in $\mathbb{R}^K$ since V is open in $\mathbb{R}^K$. Therefore C is open in $\mathbb{R}^K$. Since $g^{-1} \circ h$ must be an open map by Theorem 3.5.2, we see that D must be open in $\mathbb{R}^K$. Because $(0, x_2, x_3, \ldots, x_K) \in D$, there must be some open ball E containing $(0, x_2, x_3, \ldots, x_K)$ such that $E \subseteq D$. But $D \subseteq U$, and there is no open ball E containing $(0, x_2, x_3, \ldots, x_K)$ such that $E \subseteq U$. This contradiction establishes the theorem. □

Theorem 5.3.5 *If $K \geq 2$ and $\mathcal{M}$ is a $\mathcal{C}^r$ K-dimensional manifold in $\mathbb{R}^N$ for which $\partial\mathcal{M} \neq \emptyset$, then $\partial\mathcal{M}$ is a $\mathcal{C}^r$ $(K-1)$-dimensional manifold without boundary.*

Proof. Choose $x \in \partial\mathcal{M}$. There is a $\mathcal{C}^r$ coordinate patch $g\colon U \to \mathcal{M}$ containing x such that U is open in $\mathbb{H}^K$ and x is the image under g of a point $(0, x_2, \ldots, x_K)$. Note that $g(y) \in \partial\mathcal{M}$ if and only if y has the form $(0, y_2, \ldots, y_K)$. Let

$$V = \{(y_2, \ldots, y_K)\colon (0, y_2, \ldots, y_K) \in U\}.$$

This is an open subset of $\mathbb{R}^{K-1}$. Define $h\colon V \to \partial\mathcal{M}$ by

$$h(y_2, \ldots, y_K) = g(0, y_2, \ldots, y_K).$$

This is a one-to-one $\mathcal{C}^r$ function and satisfies $h(x_2, \ldots, x_K) = x$.

There must be an open set O in $\mathbb{R}^N$ such that $g(U) = \mathcal{M} \cap O$. Then $h(V) = g(\{0\} \times V) = \partial\mathcal{M} \cap O$; that is, $h(V)$ is open in $\partial\mathcal{M}$.

If $C = h(U)$, we have that $h^{-1} = \Pi \circ (g^{-1}|C)$, where $\Pi(y_1, y_2, \ldots, y_K) = (y_2, \ldots, y_K)$. Thus h^{-1} is continuous.

Finally, let us write $g = (g_1, g_2, \ldots, g_N)$ and note that the matrix of $h'(y_2, \ldots, y_K)$ is

$$\begin{pmatrix} D_2g_1(0, y_2, \ldots, y_K) & \ldots & D_Kg_1(0, y_2, \ldots, y_K) \\ & \ldots & \\ D_2g_N(0, y_2, \ldots, y_K) & \ldots & D_Kg_N(0, y_2, \ldots, y_K) \end{pmatrix}.$$

Since $\mathcal{D}(g') \neq 0$, we know from Theorem 1.9.3 that the column vectors

$$\begin{pmatrix} D_1g_1 \\ \ldots \\ D_1g_N \end{pmatrix}, \begin{pmatrix} D_2g_1 \\ \ldots \\ D_2g_N \end{pmatrix}, \ldots, \begin{pmatrix} D_Kg_1 \\ \ldots \\ D_Kg_N \end{pmatrix}$$

are linearly independent. It follows that

$$\begin{pmatrix} D_2g_1 \\ \ldots \\ D_2g_N \end{pmatrix}, \ldots, \begin{pmatrix} D_Kg_1 \\ \ldots \\ D_Kg_N \end{pmatrix}$$

are linearly independent, and hence again by Theorem 1.9.3 that $\mathcal{D}(h') \neq 0$. Therefore h is a $\mathcal{C}^r$ coordinate patch in $\partial\mathcal{M}$ containing x, and x must be an interior point of $\partial\mathcal{M}$. This establishes the desired result. □

Example 5.3.3 . This allows us to show in a second way that the unit sphere S^{N-1} $(N \geq 2)$ is a manifold because S^{N-1} is the boundary of the unit ball B^N. However we have yet to prove that B^N is itself a manifold. We understand that

$$B^N = \{(x_1, \dots, x_N) : x_1^2 + \dots + x_N^2 \leq 1\}.$$

It is obvious how to find coordinate patches for interior points. It is not so obvious how to define coordinate patches for the boundary points.

For simplicity we consider the point $(1, 0)$ in B^2. (A formula for any other boundary point can be obtained by rotations.) First we flip $\mathbb{H}^2$ over and translate it one unit to the right using the map $h(x_1, x_2) = (1 - x_1, x_2)$. See Figure 5.3.5. Then we map the infinite strip $(0, 1] \times \mathbb{R}$ into the right-hand half of B^2 by taking radial segments with the origin as their initial point and compressing them until they just fit within B^2. More precisely, if $x = (x_1, x_2)$ is a point in the infinite strip, we will map it to the point y on the same radial segment as x where y is the unique point such that $|x|/|p| = |y|/|q|$ where p and q are as indicated in Figure 5.3.6. The transformation $x \mapsto y$ is defined by

$$g(x_1, x_2) = \frac{x_1}{\sqrt{x_1^2 + x_2^2}} (x_1, x_2).$$

We finally define a coordinate patch $f : U \to B^2$ for $(1, 0)$ by taking U to be $\{(x_1, x_2) \in \mathbb{R}^2 : 0 \leq x_1 < 1\}$ (this is an open set in $\mathbb{H}^2$) and defining

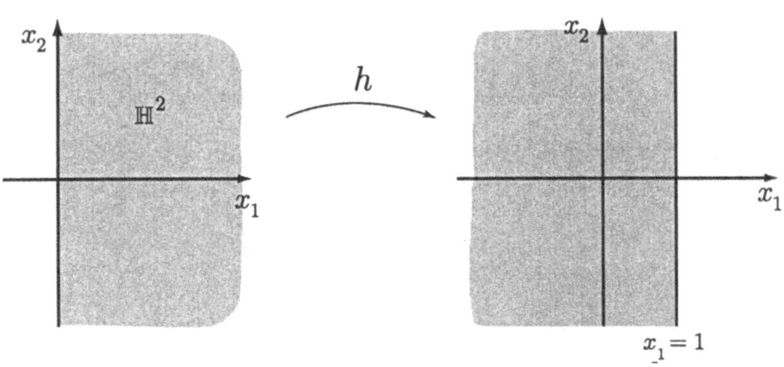

FIGURE 5.3.5.

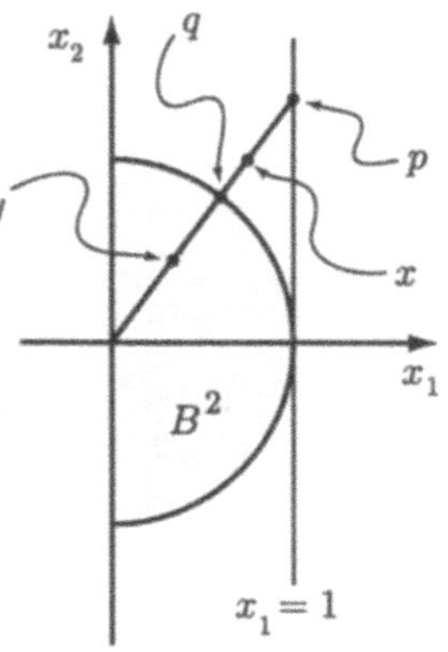

FIGURE 5.3.6.

$$f(x_1, x_2) = (g \circ h)(x_1, x_2) = \frac{1 - x_1}{\sqrt{(1 - x_1)^2 + x_2^2}} (1 - x_1, x_2).$$

More generally we may define a coordinate patch $f: U \to B^N$ for $(1, 0, \ldots, 0)$ by taking

$$f(x_1, \ldots, x_N) = \frac{1 - x_1}{\sqrt{(1 - x_1)^2 + x_2^2 + \ldots + x_N^2}} (1 - x_1, x_2, \ldots, x_N)$$

where $U = \{(x_1, \ldots, x_N) \in \mathbb{R}^N : 0 \leq x_1 < 1\}$.

Theorem 5.3.6 *Let $\mathcal{M}_1$ be an M_1-dimensional manifold in $\mathbb{R}^{N_1}$ and $\mathcal{M}_2$ be an M_2-dimensional manifold in $\mathbb{R}^{N_2}$. If either $\mathcal{M}_1$ or $\mathcal{M}_2$ (or both) has no boundary points, then $\mathcal{M}_1 \times \mathcal{M}_2$ is an $M_1 + M_2$-dimensional manifold in $\mathbb{R}^{N_1+N_2}$.*

Proof. Let $z_0 \in \mathcal{M}_1 \times \mathcal{M}_2$. If $z_0 = (x_0, y_0)$, where $x_0 \in \mathcal{M}_1$ and $y_0 \in \mathcal{M}_2$, respectively, then there exist coordinate patches $g: U \to \mathcal{M}_1$ and $h: V \to \mathcal{M}_2$ containing x_0 and y_0, respectively, such that either U and V are open in $\mathbb{R}^{M_1}$ and $\mathbb{R}^{M_2}$, respectively, or one of them is open in the appropriate $\mathbb{H}^{M_i}$ and the other is open in the appropriate $\mathbb{R}^{M_i}$. Let us suppose, for the sake of simplicity, that U and V are open in $\mathbb{R}^{M_1}$ and $\mathbb{R}^{M_2}$, respectively. The argument we are about to give extends trivially to the other cases.

Define $f(x, y) = (g(x), h(y))$. Then $f: U \times V \to \mathcal{M}_1 \times \mathcal{M}_2$. Since $U \times V$ is open in $\mathbb{R}^{M_1+M_2}$ and $\mathcal{D}(f'(x, y)) \neq 0$ for all $(x, y) \in U \times V$, the function f is easily seen to be a coordinate patch in $\mathcal{M}_1 \times \mathcal{M}_2$ containing z_0. Thus we see that $\mathcal{M}_1 \times \mathcal{M}_2$ is a manifold as promised. □

Example 5.3.4 Since S^1 has no boundary points, it is now clear that the torus $S^1 \times \ldots \times S^1$ is a manifold. In this case it is also possible to describe the coordinate patches explicitly. Consider the function $f: \mathbb{R}^K \to \mathbb{R}^{2K}$ defined by

$$f(t_1, \ldots, t_K) = (\cos t_1, \sin t_1, \cos t_2, \sin t_2, \ldots, \cos t_K, \sin t_K).$$

This function is not a one-to-one function, but for every x on the torus one can easily find an open set U in $\mathbb{R}^K$ such that f restricted to U is a coordinate patch containing x.

Exercises

1. Let (X, d) be a metric space and let Y be a nonempty subset of X. To say that U is open in Y means that for every $p \in Y$ there exists some $\varepsilon > 0$ such that

$$\{y \in Y : d(y, p) < \varepsilon\} \subseteq U.$$

Show that U is open in Y if and only if there is an open set V in X such that $U = Y \cap V$.

2. Suppose that $\mathcal{M}$ is a nonempty subset of $\mathbb{R}^N$, that U is an open set in $\mathbb{R}^K$ (where $1 \leq K \leq N$), and that g is a C^r map of U onto $\mathcal{M}$ with the property that $\mathcal{D}(g')$ is never zero. Prove that $\mathcal{M}$ must be a C^r K-manifold. (Hint: Look at Theorem 3.6.1.)

3. Let $\mathcal{M}$ be the subset of $\mathbb{R}^3$ defined by $x^2 + y^2 - z^2 = 1$.

   (a) Show that $\mathcal{M}$ is a 2-manifold.

   (b) Consider the map $g : \mathbb{R}^2 \to \mathcal{M}$ defined by

$$g(\alpha, \beta) = (\cosh(\alpha)\cos(\beta), \cosh(\alpha)\sin(\beta), \sinh(\alpha)).$$

Show that for every point x in $\mathcal{M}$ we can use g to construct a coordinate patch in $\mathcal{M}$ which contains x.

4. Consider the map $g : \mathbb{R}^3 \to S^3$ defined by

$$g(\alpha, \beta, \gamma) = (\cos(\alpha), \sin(\alpha)\cos(\beta), \sin(\alpha)\sin(\beta)\cos(\gamma), \sin(\alpha)\sin(\beta)\sin(\gamma)).$$

Show that for every point x in S^3 other than those of the form $(x_1, x_2, 0, 0)$ we can use g to construct a coordinate patch in S^3 which contains x. Show how we can modify g to construct coordinate patches for points of the form $(x_1, x_2, 0, 0)$.

5. Let B^3 be the closed, unit ball in $\mathbb{R}^3$. Define $g : (0, 1] \times \mathbb{R}^2 \to \mathbb{R}^3$ by

$$g(r, \alpha, \beta) = (r\cos(\alpha), r\sin(\alpha)\cos(\beta), r\sin(\alpha)\sin(\beta)).$$

   (a) Show that $g((0, 1] \times \mathbb{R}^2) \subseteq B^3$.

(b) Show that we can use g to define coordinate patches in B^3 about every point on the boundary of B^3 except $(\pm 1, 0, 0)$.

(c) Show that we can use $h\colon (0, 1] \times \mathbb{R}^2 \to \mathbb{R}^3$ defined by

$$h(r, \alpha, \beta) = (r \sin(\alpha)\cos(\beta), r\cos(\alpha), r\sin(\alpha)\sin(\beta))$$

to construct coordinate patches in B^3 about $(\pm 1, 0, 0)$.

6. Verify that if $f_1(x_1, x_2, x_3) = x_1^2+x_2^2+x_3^2-4$ and $f_2(x_1, x_2, x_3) = x_1^2+x_2^2-1$, then $\mathcal{D}\Big((f'(x))^{\diamond}\Big) \neq 0$ provided $x = (x_1, x_2, x_3)$ lies on neither the axis $x_1 = x_2 = 0$ nor on the plane $x_3 = 0$. Show this condition is satisfied for all x for which $f_1(x) = f_2(x) = 0$.

7. Prove that if $\mathcal{M}$ is a manifold in $\mathbb{R}^N$, then it can be considered as a manifold in $\mathbb{R}^{N+M}$ for any $M \in \mathbb{N}$.

8. Define $f : \mathbb{R}^2 \to \mathbb{R}$ by $f(x, y) = x^3 - y^2$. Sketch a graph to satisfy yourself that $\mathcal{M} = \{(x, y) \in \mathbb{R}^2 : f(x, y) = 0\}$ is not a "smooth" manifold at the point at which $\mathcal{D}\left((f')^{\diamond}\right) = 0$.

9. Define $f : \mathbb{R}^3 \to \mathbb{R}$ by $f(x, y, z) = x^2 + y^2 - z^2$ and let $\mathcal{M} = \{(x, y, z) \in \mathbb{R}^3 : f(x, y, z) = 0\}$. Show that $\mathcal{M}$ fails to be a manifold at the point at which $\mathcal{D}\left((f')^{\diamond}\right) = 0$.

10. Refer to Example 5.3.3 and verify that $\mathcal{D}(f'(0, 0)) \neq 0$; this is necessary to show f is a coordinate patch.

11. Show that if $\mathcal{M}_1$ is a manifold that has a nonempty boundary and $\mathcal{M}_2$ is a manifold without a boundary, then $\partial(\mathcal{M}_1 \times \mathcal{M}_2) = (\partial\mathcal{M}_1) \times \mathcal{M}_2$.

12. Show that the following is an easy consequence of Definition 5.3.1: If $\mathcal{M}$ is a C^r K-dimensional manifold in $\mathbb{R}^N$, then $K \leq N$.

13. In the definition of coordinate patch, show that conditions (4) and (5) are implied by (1)–(3). (Hint: Given $g : U \to \mathcal{M}$ satisfying (1)–(3), where $\mathcal{M}$ is a K-manifold and $\mathcal{M} \subseteq \mathbb{R}^N$, we may assume

$$\frac{\partial(g_1, \ldots, g_K)}{\partial(x_1, \ldots, x_K)} \neq 0.$$

Define $G : U \times \mathbb{R}^{N-K} \to \mathbb{R}^N$ by $G(x, y) = g(x) + y$, where $x \in U$ and $y \in \mathbb{R}^{N-K}$, and apply the inverse function theorem.)

14. Show that if U and V are open sets in $\mathbb{R}^K$ and g is a C^r diffeomorphism of U onto V and $h\colon V \to \mathcal{M}$ is a C^r coordinate patch in the manifold $\mathcal{M}$, then $h \circ g\colon U \to \mathcal{M}$ is a C^r coordinate patch in $\mathcal{M}$.

15. Find all constants a and b for which the set $\mathcal{M}$ in $\mathbb{R}^3$ defined by $x^2 + y^2 - z^2 + az + b = 0$ is a 2-manifold in $\mathbb{R}^3$.

5.4 Integrals of Real-valued Functions over Manifolds

What makes it hard to define integrals over a manifold is the fact that *globally* a manifold may be very different from $\mathbb{R}^K$. Think of a large torus. A near-sighted ant crawling over the torus might be deceived into believing he was crawling on a plane or on some surface that could be mapped to $\mathbb{R}^2$ by a diffeomorphism. But globally a torus is nothing like a plane, and even for dimension two, one can find an infinite number of surfaces that cannot be mapped to $\mathbb{R}^2$ or to one another by diffeomorphisms. We need to find a way to take our integration theory, which, so to speak, grew up in $\mathbb{R}^K$, and transplant it to all kinds of different manifolds.

What permits us to define integrals on manifolds is the fact that *locally* they look like pieces of $\mathbb{R}^K$ and we know how to integrate over $\mathbb{R}^K$. The way out of our difficulty is to "chop" or partition the manifolds up into "smaller" pieces, each of which can be identified with a piece of $\mathbb{R}^K$. The important thing is to see that the numbers we get, the integrals over manifolds, are independent of the way we partition the manifold. We first consider the kind of pieces we want to work with.

Definition 5.4.1 We say that $(A_1, g_1), (A_2, g_2), (A_3, g_3), \ldots$ is a *measurable decomposition* of the C^r K-dimensional manifold $\mathcal{M}$ provided that

$$\mathcal{M} = \bigcup_{i=1}^{\infty} A_i,$$

$A_1,\ A_2,\ A_3,\ \ldots$ are pairwise disjoint,

$g_i : U_i \to \mathcal{M}$ is a C^r coordinate patch for each i,

$A_i \subseteq g_i(U_i)$ for each i,

and

$g_i^{-1}(A_i)$ is a measurable subset of $\mathbb{R}^K$ for each i.

It is possible to have measurable decompositions with a finite number of members, $(A_1, g_1), (A_2, g_2), \ldots, (A_p, g_p)$, but one can consider this a special case of the definition with each $A_i = \emptyset$ for i sufficiently large.

Example 5.4.1 Consider the map $g: \mathbb{R} \to S^1$ defined by $g(\theta) = (\cos(\theta), \sin(\theta))$. Let

$$U_1 = \left(-\frac{\pi}{2}, \frac{\pi}{2}\right),$$
$$U_2 = (0, \pi),$$
$$U_3 = \left(\frac{\pi}{2}, \frac{3\pi}{2}\right),$$
$$U_4 = (\pi, 2\pi).$$

Let g_i be the restriction of g to each U_i. Then each g_i is a coordinate patch. Let us set

$$A_1 = g_1\left(\left[0, \frac{\pi}{2}\right)\right),$$
$$A_2 = g_2\left(\left[\frac{\pi}{2}, \pi\right)\right),$$
$$A_3 = g_3\left(\left[\pi, \frac{3\pi}{2}\right)\right),$$
$$A_4 = g_4\left(\left[\frac{3\pi}{2}, 2\pi\right)\right).$$

Note that S^1 is the union of the disjoint sets $A_1,\ A_2,\ A_3,\ A_4$. We see that $\{(A_i, g_i)\}_{i=1}^4$ is a measurable decomposition of S^1. See Figure 5.4.1.

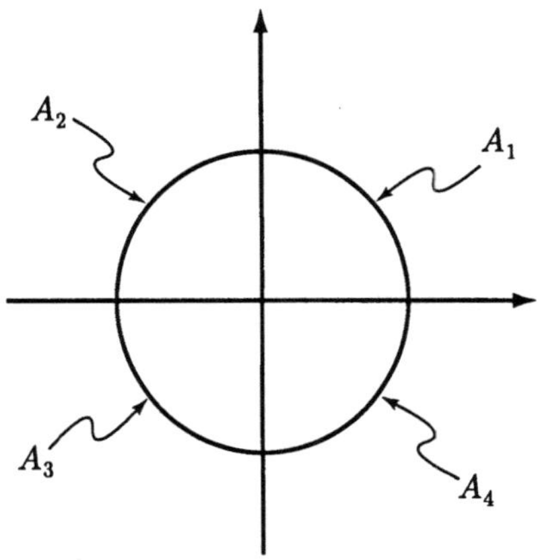

FIGURE 5.4.1.

Theorem 5.4.1 *If S is a subset of $\mathbb{R}^N$ and $\{U_\alpha\}_{\alpha\in A}$ is a collection of open sets in $\mathbb{R}^N$ such that $S \subseteq \bigcup_{\alpha\in A} U_\alpha$, then there is a countable (possibly finite) subcollection $\{U_{\alpha_i}\}_{i=1}^\infty$ of $\{U_\alpha\}_{\alpha\in A}$ such that $S \subseteq U_{\alpha_1} \cup U_{\alpha_2} \cup \ldots$.*

Proof. Let $\mathbb{Q}^N$ be the set of points in $\mathbb{R}^N$ having all rational coordinates. This is a countable set. Let $\mathcal{B}$ be the set of all open balls in $\mathbb{R}^N$ having a point of $\mathbb{Q}^N$ as a center and a rational radius. This is another countable set. For every $x \in S$ there is some $B_x \in \mathcal{B}$ such that $x \in B_x$ and $B_x \subseteq U_\alpha$ for some $\alpha \in A$. The collection of sets of the form B_x is countable, so let us relabel them $B_1, B_2, B_3, \ldots$. Note that $S \subseteq \bigcup_{i=1}^\infty B_i$. For $i = 1, 2, 3, \ldots$, let us choose $\alpha_i \in A$ such that $B_i \subseteq U_{\alpha_i}$. Then $\{U_{\alpha_i}\}_{i=1}^\infty$ is the desired sequence of open sets. □

Theorem 5.4.2 *C^r manifolds always have measurable decompositions.*

Proof. Let $\mathcal{M}$ be a C^r K-dimensional manifold in $\mathbb{R}^N$. For every $x \in \mathcal{M}$ there is a C^r coordinate patch $g_x : U_x \to \mathcal{M}$ such that $x \in g_x(U_x)$. For every U_x there is an open set V_x in $\mathbb{R}^N$ such that $g(U_x) = \mathcal{M} \cap V_x$. Since $\mathcal{M} \subseteq \bigcup_{x\in\mathcal{M}} V_x$, it follows from Theorem 5.4.1 that there must be a countable (or possibly even finite) number of C^r coordinate patches $g_i : U_i \to \mathcal{M}$ such that $\mathcal{M} = \bigcup_{i=1}^{\infty} g_i(U_i)$.

Set $A_1 = g_1(U_1)$. Since U_1 is an open subset of $\mathbb{R}^K$, it is measurable. Now let us suppose we have found $A_1, A_2, \ldots, A_m$ such that

$A_1, A_2, \ldots, A_m$ are pairwise disjoint,

$A_i \subseteq g_i(U_i)$ for $i = 1, \ldots, m$,

$g_i^{-1}(A_i)$ is a measurable subset of $\mathbb{R}^K$ for $i = 1, \ldots, m$,

and

$\bigcup_{i=1}^{m} A_i = \bigcup_{i=1}^{m} g_i(U_i)$.

Let us set

$$A_{m+1} = g_{m+1}(U_{m+1}) - \left(\bigcup_{i=1}^{m} A_i \right).$$

We see that we must have

$A_1, A_2, \ldots, A_{m+1}$ are pairwise disjoint,

$A_i \subseteq g_i(U_i)$ for $i = 1, \ldots, m+1$,

and

$\bigcup_{i=1}^{m+1} A_i = \bigcup_{i=1}^{m+1} g_i(U_i)$.

We need only check the measurability condition. But it is straightforward (exercise) to show that

$$g_{m+1}^{-1}(A_{m+1})$$
$$= U_{m+1} - \left(\bigcup_{1 \le i \le m} \{ (g_{m+1}^{-1} \circ g_i)(g_i^{-1}(A_i)) : g_i(U_i) \cap g_{m+1}(U_{m+1}) \neq \emptyset \} \right),$$

and since $g_{m+1}^{-1} \circ g_i$ is a diffeomorphism, this is a measurable set. Thus by induction we can manufacture a sequence $\{A_i\}_{i=1}^{\infty}$ having the desired properties. In particular

$$\bigcup_{i=1}^{\infty} A_i = \bigcup_{i=1}^{\infty} g_i(U_i) = \mathcal{M}.$$

This completes the proof. □

We return to the problem of integrating a function over a manifold. Suppose f is a real-valued function on the C^r K-dimensional manifold $\mathcal{M}$ and $\{(A_i, g_i)\}_{i=1}^{\infty}$ is a measurable decomposition of $\mathcal{M}$. We would like define the integral of f over $\mathcal{M}$ to be

$$\int_{\mathcal{M}} f = \sum_{i=1}^{\infty} \int (f \circ g_i)\, \mathcal{D}(g_i')\, \phi_i$$

where ϕ_i is the characteristic function of $g_i^{-1}(A_i)$. Notice that $f \circ g_i$ simply gives us the values of f inside the set A_i in the manifold, $\mathcal{D}(g_i')$ is a "distortion" factor associated with the map g_i (it plays a role analogous to $|\det(g')|$ in the change-of-variables formula), and ϕ_i serves to restrict the integration of $(f \circ g_i)\, \mathcal{D}(g_i')$ to the set A_i.

However there is a difficulty here. Given two different measurable decompositions of $\mathcal{M}$ and two sums of the form $\sum_{i=1}^{\infty} \int (f \circ g_i)\, \mathcal{D}(g_i')\, \phi_i$, how do we know these sums give us the same value? In other words, how do we know $\int_{\mathcal{M}} f$ is well defined?

We will, of course, show that this value is independent of which particular measurable decomposition we use, but first let us introduce integrability with respect to a given measurable decomposition.

Definition 5.4.2 Let f be a real-valued function on the C^r K-dimensional manifold $\mathcal{M}$. We say that f is *integrable with respect to a measurable decomposition* $\{(A_i, g_i)\}_{i=1}^{\infty}$ provided that

(1) for each i the function $(f \circ g_i)\mathcal{D}(g_i')\, \phi_i$ is integrable, where ϕ_i is the characteristic function of $g_i^{-1}(A_i)$ in $\mathbb{R}^K$, and

(2) $\sum_{i=1}^{\infty} \int |f \circ g_i|\, \mathcal{D}(g_i')\, \phi_i < \infty$.

Notice that (2) ensures that $\sum_{i=1}^{\infty} \int (f \circ g_i)\, \mathcal{D}(g_i')\, \phi_i$ is a convergent series.

Before showing this sum is independent of the choice of measurable decomposition, let us give an example of evaluating an integral over a manifold.

Example 5.4.2 Suppose f is integrable over the manifold $\mathcal{M}$ described by $x^2 + y^2 - z^2 = 1$. How would we evaluate the integral?

Notice that $\mathcal{M}$ is a hyperboloid of one sheet centered about the z-axis. Define $g\colon \mathbb{R}^2 \to \mathcal{M}$ by

$$g(\alpha, \beta) = (\cosh(\alpha)\cos(\beta), \cosh(\alpha)\sin(\beta), \sinh(\alpha)).$$

Let

$$A_1 = g\Big(\mathbb{R} \times [0, \pi)\Big) \quad \text{and} \quad A_2 = g\Big(\mathbb{R} \times [\pi, 2\pi)\Big).$$

A_1 and A_2 are disjoint and their union is $\mathcal{M}$. A_1 is (essentially) the set of points in $\mathcal{M}$ with positive y-coordinates while A_2 is (essentially) those with negative y-coordinates. If we let g_1 and g_2 be the restriction of g to "small" open sets containing respectively $\mathbb{R} \times [0, \pi)$ and $\mathbb{R} \times [\pi, 2\pi)$, then (A_1, g_1) and (A_2, g_2) constitute a measurable decomposition of $\mathcal{M}$. We compute

$$\mathcal{D}(g'(\alpha, \beta)) = \cosh(\alpha)\, \sqrt{\cosh(2\alpha)}.$$

Then

$$\begin{aligned}\int_{\mathcal{M}} f &= \int_{\mathbb{R}\times[0,\pi)} (f\circ g)\,\mathcal{D}(g') + \int_{\mathbb{R}\times[\pi,2\pi)} (f\circ g)\,\mathcal{D}(g') \\ &= \int_{-\infty}^{+\infty}\int_0^{2\pi} f(\cosh(\alpha)\cos(\beta), \cosh(\alpha)\sin(\beta), \\ &\qquad \sinh(\alpha))\cosh(\alpha)\sqrt{\cosh(2\alpha)}\,d\alpha d\beta.\end{aligned}$$

Theorem 5.4.3 *Let $f\colon \mathcal{M} \to \mathbb{R}$ where $\mathcal{M}$ is a C^r K-dimensional manifold in $\mathbb{R}^N$. If f is integrable with respect to any measurable decomposition of $\mathcal{M}$, then it is integrable with respect to all such measurable decompositions. Furthermore, if $\{(A_i, g_i)\}_{i=1}^{\infty}$ and $\{(B_j, h_j)\}_{j=1}^{\infty}$ are any two measurable decompositions of $\mathcal{M}$, we must have*

$$\sum_{i=1}^{\infty}\int (f\circ g_i)\,\mathcal{D}(g_i')\,\phi_i = \sum_{j=1}^{\infty}\int (f\circ h_j)\,\mathcal{D}(h_j')\,\psi_j$$

where ϕ_i is the characteristic function of $g_i^{-1}(A_i)$ and ψ_j is the characteristic function of $h_j^{-1}(B_j)$.

Proof. Assume f is integrable with respect to $\{(A_i, g_i)\}_{i=1}^{\infty}$. Let us define ϕ_{ij} to be the characteristic function of $g_i^{-1}(A_i \cap B_j)$ and ψ_{ij} to be the characteristic function of $h_j^{-1}(A_i \cap B_j)$. If A_i and B_j do not intersect, then these functions are the zero function. Note that $g_i^{-1}(A_i \cap B_j)$ must be measurable since we can write

$$g_i^{-1}(A_i \cap B_j) = g_i^{-1}(A_i) \cap (g_i^{-1}\circ h_j)\left(h_j^{-1}(B_j)\right)$$

and $g_i^{-1}\circ h_j$ is a diffeomorphism. Similarly, $h_j^{-1}(A_i \cap B_j)$ must be measurable.

By the monotone convergence theorem we must have

$$\sum_{j=1}^{\infty}\int |f\circ g_i|\,\mathcal{D}(g_i')\,\phi_{ij} = \int |f\circ g_i|\,\mathcal{D}(g_i')\,\phi_i.$$

It follows that

$$\sum_{i,j}\int |f\circ g_i|\,\mathcal{D}(g_i')\,\phi_{ij} = \sum_{i=1}^{\infty}\int |f\circ g_i|\,\mathcal{D}(g_i')\,\phi_i < \infty,$$

so that

$$\sum_{i,j}\int (f\circ g_i)\,\mathcal{D}(g_i')\,\phi_{ij}$$

is an absolutely convergent series and can be rearranged any way we wish. Next by the change-of-variables theorem and Theorem 1.9.2 we can write

$$\begin{aligned}\int (f\circ g_i)\,\mathcal{D}(g_i')\,\phi_{ij} &= \int (f\circ g_i\circ(g_i^{-1}\circ h_j))\;\mathcal{D}(g_i'\circ(g_i^{-1}\circ h_j))\\ &\qquad |\det(g_i^{-1}\circ h_j)|\,\phi_{ij}\circ(g_i^{-1}\circ h_j)\\ &= \int (f\circ h_j)\,\mathcal{D}(h_j')\,\psi_{ij},\end{aligned}$$

which tells us in passing that $(f\circ h_j)\,\mathcal{D}(h_j')\,\psi_{ij}$ is integrable. Similarly, we must have

$$\int |f\circ g_i|\,\mathcal{D}(g_i')\,\phi_{ij} = \int |f\circ h_j|\,\mathcal{D}(h_j')\,\psi_{ij}.$$

If follows from this that

$$\sum_{i,j}\int (f\circ h_j)\,\mathcal{D}(h_j')\,\psi_{ij}$$

is an absolutely convergent series which can be rearranged in any fashion. Because

$$(f\circ h_j)\,\mathcal{D}(h_j')\,\psi_j = \sum_{i=1}^{\infty}(f\circ h_j)\,\mathcal{D}(h_j')\,\psi_{ij}$$

and

$$\begin{aligned}\sum_{i=1}^{\infty}\int |f\circ h_j|\,\mathcal{D}(h_j')\,\psi_{ij} &= \sum_{i=1}^{\infty}\int |f\circ g_i|\,\mathcal{D}(g_i')\,\phi_{ij}\\ &\le \sum_{i,j}\int |f\circ g_i|\,\mathcal{D}(g_i')\,\phi_{ij} < \infty,\end{aligned}$$

we see that

$$(f\circ h_j)\,\mathcal{D}(h_j')\,\psi_j \simeq \sum_{i=1}^{\infty}(f\circ h_j)\,\mathcal{D}(h_j')\,\psi_{ij}.$$

Therefore $(f\circ h_j)\,\mathcal{D}(h_j')\,\psi_j$ is integrable and

$$\int (f\circ h_j)\,\mathcal{D}(h_j')\,\psi_j = \sum_{i=1}^{\infty}\int (f\circ h_j)\,\mathcal{D}(h_j')\,\psi_{ij}.$$

It follows that

$$\begin{aligned}\sum_{i=1}^{\infty}\int (f\circ g_i)\,\mathcal{D}(g_i')\,\phi_i &= \sum_{i=1}^{\infty}\sum_{j=1}^{\infty}\int (f\circ g_i)\,\mathcal{D}(g_i')\,\phi_{ij}\\ &= \sum_{j=1}^{\infty}\sum_{i=1}^{\infty}\int (f\circ h_j)\,\mathcal{D}(h_j')\,\psi_{ij}\\ &= \sum_{j=1}^{\infty}\int (f\circ h_j)\,\mathcal{D}(h_j')\,\psi_j,\end{aligned}$$

and we are done. □

Definition 5.4.3 If f is a real-valued function on a C^r manifold $\mathcal{M}$, we say it is *integrable* if it is integrable with respect to any measurable decomposition $\{(A_i, g_i)\}_{i=1}^{\infty}$ of $\mathcal{M}$. If f is integrable, we define

$$\int_{\mathcal{M}} f = \sum_{i=1}^{\infty} \int (f \circ g_i)\, \mathcal{D}(g_i')\, \phi_i$$

where ϕ_i is the characteristic function of $g_i^{-1}(A_i)$.

Example 5.4.3 Sometimes it is desirable to parameterize a manifold using something other than a coordinate patch. Consider, for instance, the unit sphere in $\mathbb{R}^3$ centered at the origin,

$$S^2 = \{\, x \in \mathbb{R}^3 \colon \ |x| = 1 \,\}.$$

We might try to parameterize this with spherical coordinates: Define

$$g\colon [0, \pi] \times [0, 2\pi] \to S^2$$

by

$$g(\phi, \theta) = (\sin(\phi)\cos(\theta), \sin(\phi)\sin(\theta), \cos(\phi)).$$

It is easily calculated that $\mathcal{D}(g'(\phi, \theta)) = \sin(\phi)$ and one would expect that for any integrable function $f\colon S^2 \to \mathbb{R}$ we would have

$$\int_{S^2} f = \int_0^{\pi} \int_0^{2\pi} \Big\{ f(\sin(\phi)\cos(\theta), \sin(\phi)\sin(\theta), \cos(\phi)) \Big\} \sin(\phi)\, d\phi\, d\theta.$$

This is indeed true, but g is not a coordinate patch. It fails to be one-to-one on the boundary of the rectangle $[0, \pi] \times [0, 2\pi]$. If we let B stand for the boundary of this rectangle, we see that $g(B)$ consists of the set of points $(x_1, x_2, x_3) \in S^2$ for which $x_2 = 0$ and $x_1 \geq 0$; that is, $g(B)$ is half of the circle obtained by intersecting the sphere S^2 with the plane $x_2 = 0$. But this is a "null set" in S^2. An integral over such a set must always be zero. We see that $g\,|(0, \pi) \times (0, 2\pi)$ is a coordinate patch and it covers all of S^2 except for the "null set" $g(B)$. This is why our formula for the integral works.

Definition 5.4.4 We define a subset A of the C^r manifold $\mathcal{M}$ to be a *null set* of $\mathcal{M}$ if and only if the characteristic function $\chi_A\colon \ \mathcal{M} \to \mathbb{R}$ is integrable and $\int_{\mathcal{M}} \chi_A = 0$.

The next theorem tells us that when we integrate over a manifold, we may neglect null sets.

Theorem 5.4.4 *If A is a null set of the C^r manifold $\mathcal{M}$ and $f\colon \mathcal{M} \to \mathbb{R}$ is a function such that $f\chi_A$ is integrable, then $\int_{\mathcal{M}} f\chi_A = 0$.*

Proof. Let $\{(A_i, g_i)\}_{i=1}^{\infty}$ be a measurable decomposition of $\mathcal{M}$ with $g_i : U_i \to \mathcal{M}$, and let ϕ_i be the characteristic function of $g_i^{-1}(A_i)$. By the definition of a null set we must have

$$\int (\chi_A \circ g_i)\, \mathcal{D}(g_i')\, \phi_i = 0$$

for all i. Then $(\chi_A \circ g_i)\,\mathcal{D}(g_i')\,\phi_i$ is a null function. Therefore

$$|f \circ g_i|\,(\chi_A \circ g_i)\,\mathcal{D}(g_i')\,\phi_i$$

must also be a null function. Note that $|f \circ g_i|\,(\chi_A \circ g_i) = |f\chi_A| \circ g_i$ so that we now know that $(|f\chi_A| \circ g_i)\,\mathcal{D}(g_i')\,\phi_i$ is a null function. This implies

$$\int ((f\chi_A) \circ g_i)\,\mathcal{D}(g_i')\,\phi_i = 0$$

for all i. Therefore

$$\int_{\mathcal{M}} f\chi_A = 0,$$

and we are done. □

Exercises

1. Show as claimed in the proof of Theorem 5.4.1, for every $x \in S$ there is some $B_x \in \mathcal{B}$ such that $x \in B_x$ and $B_x \subseteq U_\alpha$ for some $\alpha \in A$.

2. Prove that a compact C^r manifold has a finite measurable decomposition.

3. Show as claimed in the proof of Theorem 5.4.2,

$$g_{m+1}^{-1}(A_{m+1})$$
$$= U_{m+1} - \left(\bigcup_{1 \le i \le m} \left\{ (g_{m+1}^{-1} \circ g_i)(g_i^{-1}(A_i)) : g_i(U_i) \bigcap g_{m+1}(U_{m+1}) \neq \emptyset \right\} \right).$$

4. Show that if $g(\phi, \theta) = (\sin(\phi)\cos(\theta),\ \sin(\phi)\sin(\theta),\ \cos(\phi))$, then we have that $\mathcal{D}(g'(\phi, \theta)) = \sin(\phi)$.

5. Show that A is a null set of the C^r manifold $\mathcal{M}$ if and only if for every C^r coordinate patch $g \colon U \to \mathcal{M}$ such that A intersects $g(U)$ we have

$$\int \chi_{g^{-1}(A)}\mathcal{D}(g') = 0.$$

6. Prove that if $\mathcal{M}$ is a C^r K-dimensional manifold with $K \ge 1$, then any singleton set of $\mathcal{M}$ is a null set of $\mathcal{M}$.

7. Prove that if $\mathcal{M}$ is a C^r manifold, then $\partial\mathcal{M}$ is a null set of $\mathcal{M}$.

8. Let $\mathcal{M}$ be a M-dimensional manifold in $\mathbb{R}^N$. A set $S \subseteq \mathcal{M}$ is called *measurable* if for every coordinate patch $g \colon U \to \mathcal{M}$ and every compact subset A of U, the function $\chi_{S \cap g(A)}$ is integrable. Let S be a measurable subset of $\mathcal{M}$ and let

χ_S denote the characteristic function of S. The *measure* (with respect to $\mathcal{M}$) of a measurable set $S \subseteq \mathcal{M}$ is defined as

$$\mu(S) = \int_{\mathcal{M}} \chi_S \quad \text{if } \chi_S \text{ is an integrable function}$$

and

$$\mu(S) = \infty \quad \text{if } \chi_S \text{ is not an integrable function.}$$

Prove the following:

(a) $\mathcal{M}$ is measurable.

(b) Open subsets of $\mathcal{M}$ are measurable.

(c) Closed subsets of $\mathcal{M}$ are measurable.

(d) Countable unions and intersections of measurable subsets of $\mathcal{M}$ are measurable.

(e) Prove that if A and B are measurable subsets of $\mathcal{M}$ and $A \subseteq B$, then $\mu(A) \leq \mu(B)$.

(f) If $S_1, S_2, \ldots$ are disjoint measurable subsets of $\mathcal{M}$, then $\mu\left(\bigcup_{n=1}^{\infty} S_n\right) = \sum_{n=1}^{\infty} \mu(S_n)$.

9. Define the integral over a subset of a manifold $\mathcal{M}$ and locally integrable functions on $\mathcal{M}$ and then check which theorems of Section 4.7 remain true in this setting.

10. State and prove a version of the dominated convergence theorem for integrals on manifolds.

5.5 Volumes in $\mathbb{R}^N$

We illustrate the concept of integration over manifolds (or subsets of manifolds) by deriving formulas for "volumes" of simplexes, arcs, surfaces, balls, spheres, and tori.

By the volume of some $\Omega \subseteq \mathbb{R}^N$ we always mean the integral of the characteristic function of Ω if the integral exists. It is important to remember that the integral is not necessarily the integral over $\mathbb{R}^N$, but the integral over Ω. The use of the word "volume" is rather an abuse of language, because it can mean the arc length, the area, the ordinary volume, as well as a particular number assigned to a K-dimensional manifold in $\mathbb{R}^N$.

We consider simplexes. Let

$$\begin{aligned}
e_0 &= (0, 0, \ldots, 0), \\
e_1 &= (1, 0, \ldots, 0), \\
e_2 &= (0, 1, 0, \ldots, 0), \\
&\ldots \\
e_N &= (0, 0, \ldots, 0, 1).
\end{aligned}$$

The simplex determined by $e_0, e_1, \ldots, e_N$ will be denoted by Δ^N, that is,

$$\Delta^N = \{x \in \mathbb{R}^N : x = t_0e_0 + t_1e_1 + \cdots + t_Ne_N,\ \ 0 \le t_i \le 1,\ \ t_0 + \cdots + t_N = 1\}.$$

The volume of Δ^N is defined as $\int_{\Delta^N} 1$. We desire evaluating this integral. Let $p \in \Delta^N$. Then $p = (t_1, t_2, \ldots, t_N)$ where each t_i satisfies $0 \le t_i \le 1$ and $t_1 + t_2 + \cdots + t_N \le 1$. We see that

$$\begin{aligned}
\int_{\Delta^N} 1 &= \int_0^1 \int_0^{1-t_N} \int_0^{1-(t_{N-1}+t_N)} \cdots \int_0^{1-(t_2+\cdots+t_N)} dt_1\, dt_2 \cdots dt_N \\
&= \int_0^1 \cdots \int_0^{1-(t_3+\cdots+t_N)} [1-(t_2+\cdots+t_N)]\, dt_2 \ldots dt_N \\
&= \int_0^1 \cdots \int_0^{1-(t_4+\cdots+t_N)} \frac{1}{2!}[1-(t_3+\cdots+t_N)]^2\, dt_3 \ldots dt_N \\
&= \int_0^1 \cdots \int_0^{1-(t_5+\cdots+t_N)} \frac{1}{3!}[1-(t_4+\cdots+t_N)]^3\, dt_4 \ldots dt_N \\
&= \cdots = \frac{1}{N!}
\end{aligned}$$

This result can be easily extended to any N-simplex A with vertices $p_0, \ldots, p_N$ in $\mathbb{R}^N$. Define $\varphi\colon \Delta^N \to A$ by

$$\varphi(t_0e_0 + t_1e_1 + \cdots + t_Ne_N) = t_0p_0 + t_1p_1 + \cdots + t_Np_N.$$

That is,

$$\varphi(t_1, \cdots, t_N) = [1-(t_1 + \cdots + t_N)]p_0 + t_1p_1 + \cdots + t_Np_N$$

where $0 \le t_1, t_2, \cdots, t_N$ and $t_1 + t_2 + \cdots + t_N \le 1$. Since

$$\det(\varphi') = \det[p_1 - p_0, \ldots, p_N - p_0],$$

we have

$$\int_A 1 = \int_{\Delta^N} |\det(\varphi')| = \frac{1}{N!}\det[p_1 - p_0, \cdots, p_N - p_0],$$

which is the volume of A.

A C^r manifold C of dimension 1 in $\mathbb{R}^N$ is locally like a line segment. In other words, it is an arc in $\mathbb{R}^N$. The number r describes the smoothness of C. Without loss of generality, we can assume that there exists a one-to-one C^r function g from an interval (a, b) into $\mathbb{R}^N$ such that $\mathcal{D}(g'(t)) \neq 0$ for all $t \in (a, b)$, g^{-1} is continuous, and $C = g(I)$ where I is a subinterval (open, closed, or half open) of (a, b). Let $I = [\alpha, \beta]$ and let f be the density of a mass distributed along C. Then the total mass can be found as the integral

$$\int_C f = \int_\alpha^\beta f(g(t))\mathcal{D}(g'(t))\, dt.$$

If $g = (g_1, \cdots, g_N)$, then

$$\mathcal{D}(g'(t)) = \sqrt{\det\left(\begin{bmatrix} g_1'(t) \\ \vdots \\ g_N'(t) \end{bmatrix} [g_1'(t), \cdots, g_N'(t)]\right)}$$
$$= \sqrt{(g_1'(t))^2 + \cdots + (g_N'(t))^2}.$$

Consequently,

$$\int_C f = \int_\alpha^\beta f(g(t))\sqrt{(g_1'(t))^2 + \cdots + (g_N'(t))^2}\, dt.$$

If the interval (α, β) is bounded and $f \equiv 1$, then the above integral can be interpreted as the arc length. This gives us the familiar formula:

$$\text{Arc length of } \; C = \int_\alpha^\beta \sqrt{(g_1'(t))^2 + \cdots + (g_N'(t))^2}\, dt.$$

A $\mathcal{C}^r$ manifold S of dimension 2 in $\mathbb{R}^N$ is a surface in $\mathbb{R}^N$. If $g = (g_1, \cdots, g_N)$ is a one-to-one $\mathcal{C}^r$ mapping from $(\alpha, \beta) \times (\gamma, \delta)$ into $\mathbb{R}^N$ such that $\mathcal{D}(g'(x, y)) \neq 0$ for all $x \in (\alpha, \beta)$, $y \in (\gamma, \delta)$, g^{-1} is continuous, and $S = g((\alpha, \beta) \times (\gamma, \delta))$, then

$$\int_S f = \int_\alpha^\beta \int_\gamma^\delta f(g(x, y))\mathcal{D}(g'(x, y))\, dt$$
$$= \int_\alpha^\beta \int_\gamma^\delta f(g(x, y)) \sqrt{\sum_{i<j} \left(\frac{\partial(g_i, g_j)}{\partial(x, y)}(x, y)\right)^2}\, dy\, dx.$$

If $f \equiv 1$ and $(\alpha, \beta) \times (\gamma, \delta)$ is bounded, then the above becomes the familiar formula for the surface area:

$$\text{Surface area of } \; S = \int_\alpha^\beta \int_\gamma^\delta \sqrt{\sum_{i<j} \left(\frac{\partial(g_i, g_j)}{\partial(x, y)}(x, y)\right)^2}\, dy\, dx.$$

Now we are going to find the volume of the unit ball in $\mathbb{R}^N$. Let

$$V_N = \int_{\mathbb{R}^N} f(x)\, dx, \quad N = 1, 2, \ldots$$

where f is the characteristic function of the ball

$$B_N = \left\{x = (x_1, \cdots, x_N) : x_1^2 + \cdots + x_N^2 \leq 1\right\}.$$

Then

$$V_N = \int_{\mathbb{R}} \left(\int_{\mathbb{R}^{N-1}} f(x_1, y)\, dy\right) dx_1 = \int_{-1}^1 \left(\int_{\mathbb{R}^{N-1}} f(x_1, y)\, dy\right) dx_1$$

where $y = (x_2, \dots, x_N)$. The change of variables formula implies that for every $x_1 \in [-1, 1]$ we have

$$\int_{\mathbb{R}^{N-1}} f(x_1, y)\, dy = (1 - x_1^2)^{\frac{N-1}{2}} V_{N-1}.$$

Hence

$$V_N = \int_{-1}^{1} (1 - x_1^2)^{\frac{N-1}{2}} V_{N-1}\, dx_1 = \int_{-1}^{1} (1 - x_1^2)^{\frac{N-1}{2}}\, dx_1\, V_{N-1}.$$

Denote

$$I_N = \int_{-1}^{1} (1 - x_1^2)^{\frac{N-1}{2}}\, dx_1.$$

By integrating by parts we obtain a recursion relation

$$I_N = (N-1)(-I_N + I_{N-2}) \quad \text{or} \quad I_N = \frac{N-1}{N} I_{N-2}.$$

Since $V_1 = 2$, $I_1 = 2$, and $I_2 = \frac{\pi}{2}$, we find that

$$V_2 = \pi, \quad V_3 = \frac{4\pi}{3}, \quad V_4 = \frac{\pi^2}{2}, V_5 = \frac{8\pi^2}{15}, V_6 = \frac{\pi^3}{6}.$$

The obtained recursion relations can be used to prove that in general

$$V_N = \frac{\pi^{N/2}}{\Gamma\left(\frac{N}{2} + 1\right)}. \tag{5.1}$$

Note that we were able to find the volume of the unit ball in $\mathbb{R}^N$ without using a parametrization. The situation is more complicated in the case of the "surface area" of the unit sphere in $\mathbb{R}^N$. Denote by A_{N-1} the surface area of the sphere S^{N-1}, that is,

$$A_{N-1} = \int_{S^{N-1}} 1.$$

Let us first consider A_1, the circumference of the unit circle. The parametrization

$$P_1(\theta) = (\cos(\theta), \sin(\theta)) \quad \text{with} \quad 0 < \theta < 2\pi$$

does not cover the whole circle: the point $(1, 0)$ is not included. From Theorem 5.4.3 it follows that

$$A_1 = \int_{S^1} 1 = \int_0^{2\pi} \mathcal{D}(P_1'(\theta))\, d\theta = 2\pi$$

since $\mathcal{D}(P_1'(\theta)) = 1$. Similarly,

$$P_2(\theta, \alpha_1) = (\cos(\theta), \sin(\theta)\, P_1(\alpha_1)) = (\cos(\theta), \sin(\theta)\cos(\alpha_1), \sin(\theta)\sin(\alpha_1))$$

with $0 < \theta < \pi$ and $0 < \alpha_1 < 2\pi$ parametrizes the sphere in $\mathbb{R}^3$ without the circle

$$(\cos(\theta), \sin(\theta), 0), \quad 0 < \theta < \pi.$$

Again, the parametrization P_2 leads to simple integration. It turns out that this case is not special and the method can be extended to any dimension.

We define a parametrization P_N in terms of P_{N-1}:

$$P_N(\theta, \alpha_1, \dots, \alpha_{N-1}) = (\cos(\theta), \sin(\theta)\, P_{N-1}(\alpha_1, \dots, \alpha_{N-1}))$$

with $0 < \theta, \alpha_1, \dots, \alpha_{N-2} < \pi$ and $0 < \alpha_{N-1} < 2\pi$. Then

$$P_N'(\theta) = \begin{pmatrix} -\sin(\theta) & 0 & \dots & 0 \\ \cos(\theta)\, P_{N-1}^1(\alpha) & \sin(\theta)\, \frac{\partial P_{N-1}^1}{\partial \alpha_1}(\alpha) & \dots & \sin(\theta)\, \frac{\partial P_{N-1}^1}{\partial \alpha_{N-1}}(\alpha) \\ \vdots & \vdots & & \vdots \\ \cos(\theta)\, P_{N-1}^{N-1}(\alpha) & \sin(\theta)\, \frac{\partial P_{N-1}^{N-1}}{\partial \alpha_1}(\alpha) & \dots & \sin(\theta)\, \frac{\partial P_{N-1}^{N-1}}{\partial \alpha_{N-1}}(\alpha) \end{pmatrix}$$

where $P_{N-1} = (P_{N-1}^1, \cdots, P_{N-1}^{N-1})$ and $\alpha = (\alpha_1, \cdots, \alpha_{N-1})$. Since

$$\left(P_{N-1}^1(\alpha)\right)^2 + \dots + \left(P_{N-1}^{N-1}(\alpha)\right)^2 = 1$$

and

$$P_{N-1}^1(\alpha)\, \frac{\partial P_{N-1}^1}{\partial \alpha_k}(\alpha) + \dots + P_{N-1}^{N-1}(\alpha)\, \frac{\partial P_{N-1}^{N-1}}{\partial \alpha_k}(\alpha) = 0$$

for $k = 1, \dots, N-1$, we have

$$\mathcal{D}(P_N'(\theta)) = \sqrt{\det \begin{pmatrix} 1 & 0 & \dots & 0 \\ 0 & & & \\ \vdots & & A & \\ 0 & & & \end{pmatrix}} \tag{5.2}$$

where

$$A = \sin^{2N-2}(\theta)\, \left(P_{N-1}'\right)^{\mathrm{T}} \left(P_{N-1}'\right).$$

Consequently,

$$\mathcal{D}(P_N'(\theta)) = \sin^{N-1}(\theta)\, \mathcal{D}(P_{N-1}')$$

and finally

$$A_N = \int_{S^N} 1 = \underbrace{\int_0^{\pi} \cdots \int_0^{\pi}}_{N-1 \text{ times}} \int_0^{2\pi} \sin^{N-1}(\theta)\mathcal{D}(P'_{N-1})(\alpha_1, \ldots, \alpha_{N-1})$$
$$d\theta\, d\alpha_1 \ldots d\alpha_{N-1}$$
$$= \int_0^{\pi} \sin^{N-1}(\theta)d\theta \underbrace{\int_0^{\pi} \cdots \int_0^{\pi}}_{N-2 \text{ times}} \int_0^{2\pi} \mathcal{D}(P'_{N-1})(\alpha_1, \ldots, \alpha_{N-1})$$
$$d\alpha_1, \ldots d\alpha_{N-1}$$
$$= \int_0^{\pi} \sin^{N-1}(\theta)d\theta A_{N-1}.$$

Since $A_1 = 2\pi$, we get

$$A_2 = 4\pi, \quad A_3 = 2\pi^2, \quad A_4 = \frac{8\pi^2}{3}, \quad A_5 = \pi^3.$$

Now we turn our attention to tori. The N-dimensional torus is defined as the Cartesian product of N circles. This gives us a natural parametrization in $\mathbb{R}^{2N}$:

$$\phi(\alpha_1, \ldots, \alpha_N) = (r_1 \cos\alpha_1, r_1 \sin\alpha_1, r_2 \cos\alpha_2, r_2 \sin\alpha_2, \ldots, r_N \cos\alpha_N, r_N \sin\alpha_N)$$

where $r_1, \ldots, r_N > 0$ are fixed and for each i we have $0 < \alpha_i \leq 2\pi$. Calculating the surface area is relatively simple. Indeed, we have

$$\phi'(\alpha_1, \ldots, \alpha_N) = \begin{pmatrix} -r_1 \sin\alpha_1 & 0 & 0 & \ldots & 0 \\ r_1 \cos\alpha_1 & 0 & 0 & \ldots & 0 \\ 0 & -r_2 \sin\alpha_2 & 0 & \ldots & 0 \\ 0 & r_2 \cos\alpha_2 & 0 & \ldots & 0 \\ \vdots & & \ddots & & \vdots \\ 0 & 0 & \ldots & 0 & -r_N \sin\alpha_N \\ 0 & 0 & \ldots & 0 & r_N \cos\alpha_N \end{pmatrix}$$

and

$$\mathcal{D}(\phi'(\alpha_1, \ldots, \alpha_N B)) = \sqrt{\det \begin{pmatrix} r_1^2 & 0 & \ldots & 0 \\ 0 & r_2^2 & & 0 \\ \vdots & & \ddots & \vdots \\ 0 & \ldots & 0 & r_N^2 \end{pmatrix}} = r_1 r_2 \cdots r_N.$$

Consequently, if we denote by S_N the surface area of the N-dimensional torus, then

$$S_N = \underbrace{\int_0^{2\pi} \cdots \int_0^{2\pi}}_{N \text{ times}} r_1 r_2 \cdots r_N d\alpha_1 d\alpha_2 \ldots d\alpha_N = (2\pi)^N r_1 r_2 \cdots r_N.$$

Exercises

In Exercises 1–4, V_N denotes the volume of the unit N-ball.

1. Justify the claim that for every $x_1 \in [-1, 1]$ we have
$$\int_{\mathbb{R}^{N-1}} f(x_1, y)dy = (1 - x_1^2)V_{N-1}.$$

2. Derive simplified formulas for V_N for even N and odd N.

3. Prove (5.1).

4. Find $\lim_{n\to\infty} V_N$. Note that the volume of the unit cube in any dimension is 1.

5. Find the volume of the ellipsoid
$$E = \left\{ x = (x_1, \dots, x_N) \in \mathbb{R}^N : \frac{x_1^2}{a_1^2} + \cdots + \frac{x_N^2}{a_N^2} \leq 1 \right\}$$
where $a_1, \dots, a_N > 0$.

6. Prove (5.2).

7. Consider a torus T parametrized by
$$\phi(t, \alpha_1, \dots, \alpha_N) = (t \cos\alpha_1, t \sin\alpha_1, r_2 \cos\alpha_2, r_2 \sin\alpha_2, \dots, r_N \cos\alpha_N, r_N \sin\alpha_N)$$
where $r_1, \dots, r_N > 0$ are fixed, $0 \leq t \leq r_1$, and for each i we have $0 < \alpha_i \leq 2\pi$. Find the volume of T.

8. Find the volume of N-dimensional balls and spheres of radius r.

9. Show that the volume of a K-manifold is invariant under translations or orthogonal transformations.

6

K-VECTORS AND WEDGE PRODUCTS

6.1 K-Vectors in $\mathbb{R}^N$ and the Wedge Product

When it is first encountered, the wedge product may appear both artificial and exotic. This appearance is deceptive. The wedge product is something like a generalization of the cross product of 3-dimensional vectors, so it should not be surprising that things looking like wedge products should arise in natural phenomena. Certain wedge products, the simple K-vectors, have striking geometric interpretations with a strong connection to determinants and oriented volume, so that the wedge product is an excellent tool for analytic geometry in Euclidean spaces of arbitrary dimension. This connection with geometry leads in turn to an elegant and marvelously unified language for calculus not simply in Euclidean spaces but in manifolds. It is this last aspect of the theory of wedge products which draws us to its study.

Let us begin with heuristic considerations.

We are used to thinking of vectors as directed line segments (or, more precisely, as equivalence classes of directed line segments). These line segments have an orientation (a preferred direction), components associated with the axes of the coordinate system (handy for calculations), and a magnitude (the length of the vector). Let us call them 1-vectors because of their 1-dimensional nature.

Now suppose we construct a parallelogram P in $\mathbb{R}^N$ with one vertex at the origin. We want to think of this as representing a vector of a new sort which we will call a 2-*vector* (see Figure 6.1.1.). We attach components to the 2-vector by the following strategy:

Consider the x_i- and x_j-axes where $i \neq j$. These two axes determine a coordinate plane which is uniquely specified by writing the ordered pair (i, j) where, without

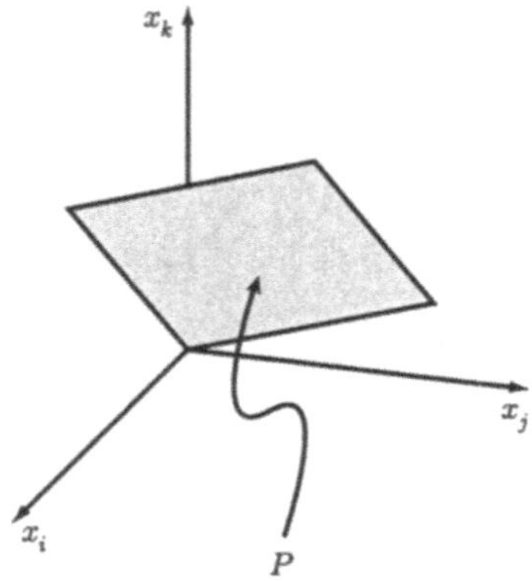

FIGURE 6.1.1.

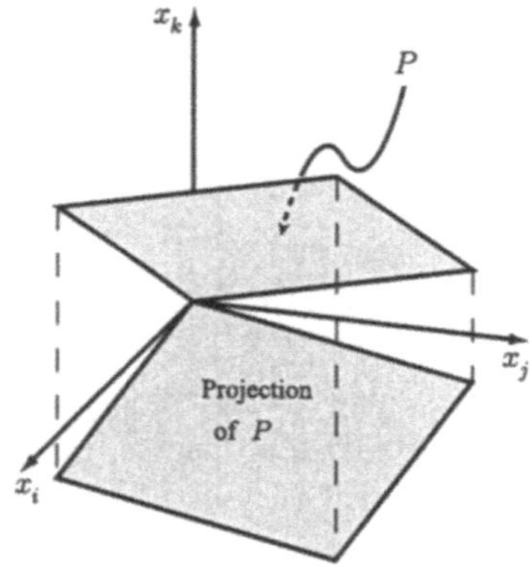

FIGURE 6.1.2.

loss of generality, we assume $i < j$. We want the area of the orthogonal projection of the parallelogram into this plane to be a component of the 2-vector (see Figure 6.1.2.). Since there are $\binom{N}{2}$ coordinate planes in $\mathbb{R}^N$, the 2-vector will have $\binom{N}{2}$ components.

However this is not quite the right way to do things. It will be convenient for 2-vectors (just as for 1-vectors) to sometimes have negative components, which means we will have to consider *oriented* areas. Looking again at the parallelogram P, we notice that it is determined by two 1-vectors, a and b, which emanate from the origin and constitute two edges of P. If $a = (\alpha_1, \alpha_2, \ldots, \alpha_N)$ and $b = (\beta_1, \beta_2, \ldots, \beta_N)$, we can take the oriented area of the projection onto the (i, j)-plane to be

$$(i, j)\text{-component} = \det \begin{pmatrix} \alpha_i & \beta_i \\ \alpha_j & \beta_j \end{pmatrix}.$$

(See Figure 6.1.3.) In this case we denote the 2-vector which P represents by the symbol $a \wedge b$.

This suggests the existence of another 2-vector associated with the same parallelogram P, namely $b \wedge a$. The components of the 2-vector $b \wedge a$ have the form

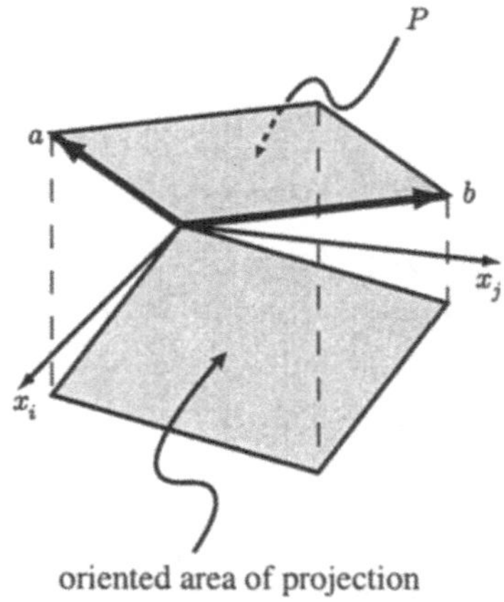

FIGURE 6.1.3.

$$(i, j)\text{-component} = \det\begin{pmatrix}\beta_i & \alpha_i\\ \beta_j & \alpha_j\end{pmatrix} = -\det\begin{pmatrix}\alpha_i & \beta_i\\ \alpha_j & \beta_j\end{pmatrix}.$$

Then we expect $b \wedge a = -a \wedge b$. The two forms for this 2-vector, $a \wedge b$ and $b \wedge a$, represent two different orientations. The distinction is much like that between right-handedness and left-handedness.

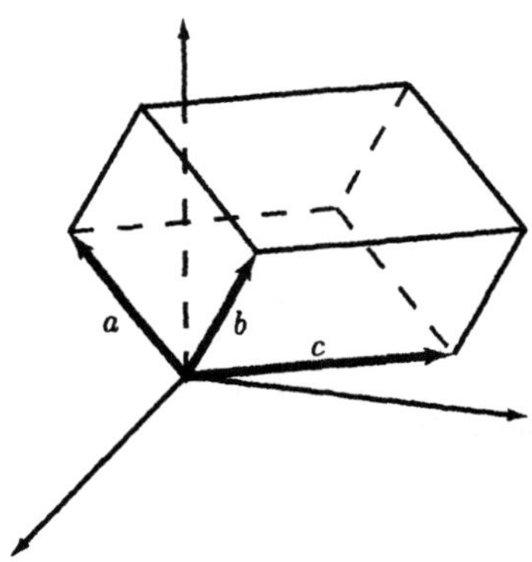

FIGURE 6.1.4.

Clearly this discussion extends to higher dimensions. Thus 1-vectors in $\mathbb{R}^N$, say a, b, and c, determine a parallelepiped P (Figure 6.1.4.). If $a = (\alpha_1, \alpha_2, \ldots, \alpha_N)$, $b = (\beta_1, \beta_2, \ldots, \beta_N)$, and $c = (\gamma_1, \gamma_2, \ldots, \gamma_N)$, we associate components with P by computing the oriented volume of the orthogonal projection of P onto every 3-dimensional subspace determined by distinct x_i-, x_j-, and x_k-axes, where $i < j < k$, thus

$$(i, j, k)\text{-component} = \det\begin{pmatrix}\alpha_i & \beta_i & \gamma_i\\ \alpha_j & \beta_j & \gamma_j\\ \alpha_k & \beta_k & \gamma_k\end{pmatrix}.$$

We think of P as representing a 3-*vector* which we designate $a \wedge b \wedge c$.

It is clear that whether we discuss 2-vectors, 3-vectors, or, more generally, K-vectors, the idea of *order* will play an important role. For example, for the 3-vector we just discussed, by looking at components, we expect

$$a \wedge b \wedge c = -b \wedge a \wedge c = b \wedge c \wedge a = -c \wedge b \wedge a = \text{etc.}$$

Also, if we return to our discussion of the 3-vector $a \wedge b \wedge c$ in $\mathbb{R}^N$ and assume $N = 1$ or 2, then for every component of the 3-vector we must have

$$(i, j, k)\text{-component} = \det \begin{pmatrix} \alpha_i & \beta_i & \gamma_i \\ \alpha_j & \beta_j & \gamma_j \\ \alpha_k & \beta_k & \gamma_k \end{pmatrix} = 0.$$

We cannot satisfy the condition that i, j, and k are distinct. So in dimensions 1 or 2, $a \wedge b \wedge c = 0$. Similarly, for a 2-vector $a \wedge b$ in $\mathbb{R}^N$ where $N = 1$, we always have

$$(i, j)\text{-component} = \det \begin{pmatrix} \alpha_i & \beta_i \\ \alpha_j & \beta_j \end{pmatrix} = 0$$

since i must equal j. Again we have $a \wedge b = 0$. These are particular instances of the following: If $K > N$, we expect a K-vector in $\mathbb{R}^N$ to be 0.

So far our discussion has been of an informal, motivational sort. Now we will become more careful. Our key to a rigorous development of K-vectors will be the idea that vectors can be specified by their components.

We can think of a typical $a = (\alpha_1, \alpha_2, \ldots, \alpha_N) \in \mathbb{R}^N$, a 1-vector in $\mathbb{R}^N$, as a function

$$f : \{1, 2, \ldots, N\} \to \mathbb{R}$$

defined by $f(i) = \alpha_i$. Each α_i (or $f(i)$) is just the signed length of the projection of a onto the ith axis. Basically then, vectors in an N-dimensional space are objects which are specified by associating a number with each of the N axes.

Definition 6.1.1 For $K = 1, 2, \ldots, N$, we define a K-*vector* in $\mathbb{R}^N$ to be a function f from the set of ordered K-tuples $(i_1, i_2, \ldots, i_K)$, such that each $i_j \in \{1, 2, \ldots, N\}$ and $i_1 < i_2 < \cdots < i_K$, into $\mathbb{R}$. We denote the space of K-vectors in $\mathbb{R}^N$ by the symbol $\wedge^K \mathbb{R}^N$. We also define $\wedge^0 \mathbb{R}^N = \mathbb{R}$ and $\wedge^K \mathbb{R}^N =$ a vector space containing only a zero vector if $K > N$.

It is convenient to identify $\mathbb{R}^N$ with the space of 1-vectors, $\wedge^1 \mathbb{R}^N$. Since $\wedge^K \mathbb{R}^N$ is a collection of real-valued functions on a common domain, it is closed under addition of functions and multiplication by real numbers. With these definitions we have the following result (which is left as an exercise):

Theorem 6.1.1 *$\wedge^K \mathbb{R}^N$ is a vector space over the reals.*

Recall that $e_1, e_2, \ldots, e_N$ is the standard basis for $\mathbb{R}^N$. We may consider each e_i as the function from $\{1, 2, \ldots, N\}$ into $\mathbb{R}$ defined by

$$e_i(j) = \begin{cases} 1 & \text{if } i = j \\ 0 & \text{otherwise.} \end{cases}$$

We want to construct K-vectors in $\mathbb{R}^N$, which we will designate $e_{i_1 i_2 \ldots i_K}$, and which play a role similar to the e_i's. Let us think of $e_{i_1 i_2 \ldots i_K}$ as being represented by the K-dimensional cube determined by the unit 1-vectors $e_{i_1}, e_{i_2}, \ldots, e_{i_K}$. (See Figure 6.1.5 for e_{13} in $\mathbb{R}^3$.) We would expect that the component of $e_{i_1 i_2 \ldots i_K}$ associated with the hyperplane determined by the x_{j_1}-, x_{j_2}-, $\ldots$, x_{j_K}-axes would be 1 if $(j_1, \ldots, j_K) = (i_1, \ldots, i_K)$ and 0 otherwise.

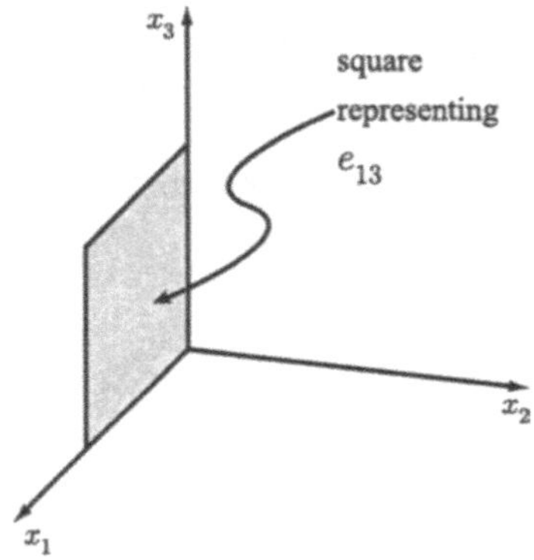

FIGURE 6.1.5.

This description is not quite accurate since it implicitly assumes the natural numbers in the sequence $i_1, \ldots, i_K$ are distinct; it turns out to be technically convenient to allow repetitions. It also neglects the importance of ordering. But it does motivate the following:

Definition 6.1.2 For $i_1,\ i_2,\ \ldots,\ i_K \in \{1, 2, \ldots, N\}$ we define a function $e_{i_1 i_2 \ldots i_K}$ from $\Big\{(j_1, j_2, \ldots, j_K) :\ j_1,\ j_2,\ \ldots,\ j_K \in \{1, 2, \ldots, N\}\Big\}$ into $\mathbb{R}$ by

$$e_{i_1 i_2 \ldots i_K}(j_1, j_2, \ldots, j_K) = \det \begin{pmatrix} e_{i_1}(j_1) & e_{i_2}(j_1) & \ldots & e_{i_K}(j_1) \\ e_{i_1}(j_2) & e_{i_2}(j_2) & \ldots & e_{i_K}(j_2) \\ \ldots & & & \\ e_{i_1}(j_K) & e_{i_2}(j_K) & \ldots & e_{i_K}(j_K) \end{pmatrix}.$$

It is easy to check that $e_{i_1 i_2 \ldots i_K}$ has the following properties:

(1) If the sequence $i_1, \ldots, i_K$ or the sequence $j_1, \ldots, j_K$ has any repetitions, then $e_{i_1 i_2 \ldots i_K}(j_1, j_2, \ldots, j_K) = 0$.

(2) If $K > N$, then $e_{i_1 i_2 \ldots i_K} = 0$. (This is consistent with our definition of $\wedge^K \mathbb{R}^N$ as a vector space with only a single element, 0.)

(3) If $(r_1, \ldots, r_K)$ is obtained from $(i_1, \ldots, i_K)$ by the permutation ρ, then

$$e_{r_1 r_2 \ldots r_K}(j_1, j_2, \ldots, j_K) = \operatorname{sgn}(\rho)\ e_{i_1 i_2 \ldots i_K}(j_1, j_2, \ldots, j_K).$$

If $(t_1, \ldots, t_K)$ is obtained from $(j_1, \ldots, j_K)$ by the permutation τ, then

$$e_{i_1 i_2 \ldots i_K}(t_1, t_2, \ldots, t_K) = \operatorname{sgn}(\tau)\ e_{i_1 i_2 \ldots i_K}(j_1, j_2, \ldots, j_K).$$

(This tells us that when $K \leq N$, we need only look at the case where $i_1 < \cdots < i_K$ and $j_1 < \cdots < j_K$.)

(4) If $K \leq N$ and $i_1 < \cdots < i_K$ and $j_1 < \cdots < j_K$, then

$$e_{i_1 i_2 \ldots i_K}(j_1, j_2, \ldots, j_K) = \begin{cases} 1 & \text{if } (i_1, \ldots, i_K) = (j_1, \ldots, j_K), \\ 0 & \text{otherwise.} \end{cases}$$

Theorem 6.1.2 *For $K = 1, 2, \ldots, N$ the set of K-vectors $e_{i_1 i_2 \ldots i_K}$ where $i_1 < i_2 < \cdots < i_K$ is a basis for $\wedge^K \mathbb{R}^N$. Therefore dim $(\wedge^K \mathbb{R}^N) = \binom{N}{K}$.*

Proof. Consider an arbitrary K-vector f. We can write

$$f = \sum_{i_1 < \cdots < i_K} f(i_1, \ldots, i_k)\, e_{i_1 \ldots i_K}$$

where

$$\sum_{i_1 < \cdots < i_K}$$

is understood to be the summation over all ordered K-tuples $(i_1, \ldots, i_K)$ such that $i_1 < \cdots < i_K$. Therefore every K-vector can be written as a linear combination of $\{e_{i_1 \ldots i_K}\}$ such that $i_1 < \cdots < i_K$.

Next suppose that

$$\sum_{i_1 < \cdots < i_K} \alpha_{i_1 \ldots i_K}\; e_{i_1 \ldots i_K} = 0$$

where each $\alpha_{i_1 \ldots i_K}$ is a real number. If we evaluate each side of the last equation at $(j_1, \ldots, j_K)$ where $j_1 < \cdots < j_K$, then we obtain

$$\alpha_{j_1 \ldots j_K} = 0.$$

This shows linear independence. Hence $\{e_{i_1 \ldots i_K}\}$ such that $i_1 < \cdots < i_K$ is a basis for $\wedge^K \mathbb{R}^N$. □

Because of this, if a is a K-vector in $\mathbb{R}^N$ with $K = 1, 2, \ldots, N$, we can write

$$a = \sum_{i_1 < \cdots < i_K} a(i_1, \ldots, i_K)\; e_{i_1 \ldots i_K}$$

where each $a(i_1, \ldots, i_K)$ is a uniquely determined real number. If, for example, we have the 2-vector a in $\mathbb{R}^3$ given by

$$a = 2\, e_{12} - 5\, e_{13} + \frac{1}{2}\, e_{23},$$

then by (4) of Definition 6.1.2, this amounts to saying that

$$a(1,2) = 2, \quad a(1,3) = -5, \quad \text{and} \quad a(2,3) = \frac{1}{2}.$$

We use this fact to define an interesting and very useful multiplication of K-vectors.

Definition 6.1.3 Let a and b be K- and L-vectors, respectively, in $\mathbb{R}^N$.

If $K + L \leq N$, then we have the unique expansions

$$a = \sum_{i_1 < \cdots < i_K} a(i_1, \ldots, i_K)\ e_{i_1 \ldots i_K}$$

and

$$b = \sum_{j_1 < \cdots < j_L} b(j_1, \ldots, j_L)\ e_{j_1 \ldots j_L}.$$

Then we define the *exterior product* or *wedge product* of a and b, the $(K + L)$-vector $a \wedge b$, by

$$a \wedge b = \sum_{i_1 < \cdots < i_K} \sum_{j_1 < \cdots < j_L} a(i_1, \ldots, i_K)\ b(j_1, \ldots, j_L)\ e_{i_1 \ldots i_K j_1 \ldots j_L}.$$

The order of the summation is clearly irrelevant. If we have $i_p = j_q$ for some p and q, then

$$e_{i_1 \ldots i_K j_1 \ldots j_L} = 0.$$

If, on the other hand, $i_1, \ldots, i_K, j_1, \ldots, j_L$ are all distinct, then by permuting them we can write

$$e_{i_1 \ldots i_K j_1 \ldots j_L} = \pm e_{r_1 r_2 \ldots r_{K+L}}$$

where $r_1 < r_2 < \cdots < r_{K+L}$. The $(K + L)$-vector $e_{r_1 r_2 \ldots r_{K+L}}$ is an element of the basis described in Theorem 6.1.2.

If $K + L > N$, then we define $a \wedge b$ to be the zero $(K + L)$-vector which is the single element of $\wedge^{K+L}\mathbb{R}^N$. In the case where $K = 0$, then $a \in \mathbb{R}$ and we take $a \wedge b$ to be simply multiplication of b by the scalar a, namely ab.

Exercises

1. Prove that $\wedge^K \mathbb{R}^N$ is a vector space over $\mathbb{R}$.

2. Prove that for $K = 1, 2, \ldots, N$, we have $\dim(\wedge^K \mathbb{R}^N) = \binom{N}{K}$.

3. List a set of basis elements for $\wedge^2 \mathbb{R}^4$.

4. Show that if $e_{j_1 j_2 \ldots j_K}$ is obtained from $e_{i_1 i_2 \ldots i_K}$ by interchanging any two indices, then $e_{j_1 j_2 \ldots j_K} = -e_{i_1 i_2 \ldots i_K}$.

5. (a) Show that if a and b are 1-vectors, then $a \wedge b = -(b \wedge a)$. (Hint: Consider basis vectors.) (b) Show that if a is a 1-vector and b is a 2-vector, then $a \wedge b = b \wedge a$. (c) What happens if a and b are both 3-vectors?

6. Show that if 0 is the zero K-vector and b is an L-vector, then $0 \wedge b = 0$, the zero $(K + L)$-vector.

6.2 Properties of $\wedge$

Theorem 6.2.1 *Suppose a is a K-vector in $\mathbb{R}^N$, b and c are L-vectors in $\mathbb{R}^N$, and $\lambda \in \mathbb{R}$. Then*

(1) $a \wedge b = (-1)^{KL}(b \wedge a)$.

(2) $a \wedge (b + c) = (a \wedge b) + (a \wedge c)$.

(3) $\lambda(a \wedge b) = (\lambda a) \wedge b = a \wedge (\lambda b)$.

Proof. Because of the way $\wedge$ is defined, it is necessary to prove these results only in the case where $K + L \leq N$ and only for basis elements of the form $e_{i_1 \dots i_K}$. We prove only (1).

Suppose $i_1 < \cdots < i_K$ and $j_1 < \cdots < j_L$. We know that

$$e_{i_1 \dots i_K} \wedge e_{j_1 \dots j_L} = e_{i_1 \dots i_K j_1 \dots j_L}.$$

By permuting j_1 successively with $i_K, i_{K-1}, \dots, i_1$, we obtain

$$e_{i_1 \dots i_K j_1 \dots j_L} = (-1)^K e_{j_1 i_1 \dots i_K j_2 \dots j_L}.$$

If we then carry out such permutations with $j_2, j_3, \dots, j_L$, we obtain

$$e_{i_1 \dots i_K j_1 \dots j_L} = (-1)^{KL} e_{j_1 \dots j_L i_1 \dots i_K}.$$

Since

$$e_{j_1 \dots j_L i_1 \dots i_K} = e_{j_1 \dots j_L} \wedge e_{i_1 \dots i_K},$$

we are done. □

Theorem 6.2.2 *$\wedge$ is associative.*

Proof. Let us consider K-, L-, and M-vectors in $\mathbb{R}^N$. It suffices to prove associativity for basis vectors. Choose

$$e_{i_1 \dots i_K}, \quad e_{j_1 \dots j_L}, \quad \text{and} \quad e_{l_1 \dots l_M}$$

where

$$i_1 < \cdots < i_K, \quad j_1 < \cdots < j_L, \quad \text{and} \quad l_1 < \cdots < l_M$$

and $K + L + M \leq N$. Let us further restrict ourselves to the case where $i_1, \dots, i_K$, $j_1, \dots, j_L, l_1, \dots, l_M$ are all distinct. Then

$$(e_{i_1 \dots i_K} \wedge e_{j_1 \dots j_L}) \wedge e_{l_1 \dots l_M} = e_{i_1 \dots i_K j_1 \dots j_L} \wedge e_{l_1 \dots l_M}.$$

By permuting $i_1, \dots, i_K, j_1, \dots, j_L$ we may write

$$e_{i_1 \dots i_K j_1 \dots j_L} = (-1)^R e_{r_1 r_2 \dots r_{K+L}}$$

where $r_1 < r_2 < \cdots < r_{K+L}$ and R is some integer. Then by the definition of $\wedge$,

$$e_{i_1 \ldots i_K j_1 \ldots j_L} \wedge e_{l_1 \ldots l_M} = (-1)^R e_{r_1 r_2 \ldots r_{K+L} l_1 \ldots l_M}.$$

Now if, in this last expression, we permute $r_1, \ldots, r_{K+L}$ back into the sequence $i_1, \ldots, i_K, j_1, \ldots, j_L$, we obtain

$$(-1)^R e_{r_1 r_2 \ldots r_{K+L} l_1 \ldots l_M} = e_{i_1 \ldots i_K j_1 \ldots j_L l_1 \ldots l_M}.$$

By a similar chain of reasoning we may also show that

$$e_{i_1 \ldots i_K} \wedge (e_{j_1 \ldots j_L} \wedge e_{l_1 \ldots l_M}) = e_{i_1 \ldots i_K j_1 \ldots j_L l_1 \ldots l_M}.$$

This gives us associativity for basis vectors, so it must hold in general. □

Corollary 6.2.1 $e_{i_1 i_2 \ldots i_K} = e_{i_1} \wedge e_{i_2} \wedge \cdots \wedge e_{i_K}$. *Therefore for* $K = 1, 2, \ldots, N$, *a basis for* $\wedge^K \mathbb{R}^N$ *is*

$$\{ e_{i_1} \wedge e_{i_2} \wedge \cdots \wedge e_{i_K} : i_1 < i_2 < \cdots < i_K \}.$$

This last result tells us that all K-vectors can be built up using 1-vectors, the wedge product, scalar multiplication, and addition.

Example 6.2.1 For $\wedge^0 \mathbb{R}^3$, a basis is the set consisting of the single number 1. Remember that we identify $\wedge^0 \mathbb{R}^3$ with $\mathbb{R}$. For $\wedge^1 \mathbb{R}^3$, which we identify with $\mathbb{R}^3$, a basis is the set consisting of e_1, e_2, and e_3. For $\wedge^2 \mathbb{R}^3$, we know that the set consisting of $e_1 \wedge e_2$, $e_1 \wedge e_3$, and $e_2 \wedge e_3$ is a basis. For $\wedge^3 \mathbb{R}^3$, we may take the single 3-vector $e_1 \wedge e_2 \wedge e_3$ as a basis.

Our next result shows an important link between K-vectors and determinants.

Theorem 6.2.3 *Suppose* $a_1, a_2, \ldots, a_K \in \mathbb{R}^N$ *where* $K = 1, 2, \ldots, N$. *If for each* i *we have*

$$a_i = \sum_{j=1}^{N} \alpha_{ij} e_j,$$

then

$$a_1 \wedge a_2 \wedge \cdots \wedge a_K = \sum_{i_1 < \cdots < i_K} \det \begin{pmatrix} \alpha_{1i_1} & \alpha_{2i_1} & \ldots & \alpha_{Ki_1} \\ \alpha_{1i_2} & \alpha_{2i_2} & \ldots & \alpha_{Ki_2} \\ \ldots & & & \\ \alpha_{1i_K} & \alpha_{2i_K} & \ldots & \alpha_{Ki_K} \end{pmatrix} e_{i_1} \wedge e_{i_2} \wedge \cdots \wedge e_{i_K}.$$

Proof. We have

$$\begin{aligned} a_1 \wedge a_2 \wedge \cdots \wedge a_K &= \left(\sum_{i_1=1}^{N} \alpha_{1i_1} e_{i_1} \right) \wedge \left(\sum_{i_2=1}^{N} \alpha_{2i_2} e_{i_2} \right) \wedge \cdots \wedge \left(\sum_{i_K=1}^{N} \alpha_{Ki_K} e_{i_K} \right) \\ &= \sum_{i_1, i_2, \ldots, i_K} \alpha_{1i_1} \alpha_{2i_2} \ldots \alpha_{Ki_K} \, (e_{i_1} \wedge e_{i_2} \wedge \cdots \wedge e_{i_K}). \end{aligned}$$

In this last summation every possible sequence $(i_1, \ldots, i_K)$ of elements chosen from $\{1, 2, \ldots, N\}$ occurs exactly once. We may ignore any sequence $(i_1, \ldots, i_K)$ in which some natural number occurs more than once, because then $e_{i_1} \wedge e_{i_2} \wedge \cdots \wedge e_{i_K} = 0$. Let us choose a sequence $(j_1, j_2, \ldots, j_K)$ such that $j_1 < j_2 < \cdots < j_K$ and by

$$\sum\nolimits^{*} \alpha_{1i_1}\alpha_{2i_2} \ldots \alpha_{Ki_K}(e_{i_1} \wedge e_{i_2} \wedge \cdots \wedge e_{i_K})$$

let us mean the summation over all sequences $(i_1, i_2, \ldots, i_K)$ which are permutations of $(j_1, j_2, \ldots, j_K)$. We may write

$$\sum\nolimits^{*} \alpha_{1i_1}\alpha_{2i_2} \ldots \alpha_{Ki_K}(e_{i_1} \wedge e_{i_2} \wedge \cdots \wedge e_{i_K})$$
$$= \sum_{\sigma \in \mathcal{P}_K} \alpha_{1j_{\sigma(1)}}\alpha_{2j_{\sigma(2)}} \ldots \alpha_{Kj_{\sigma(K)}}(e_{j_{\sigma(1)}} \wedge e_{j_{\sigma(2)}} \wedge \cdots \wedge e_{j_{\sigma(K)}})$$

where $\mathcal{P}_K$ is the set of permutations on $\{1, 2, \ldots, K\}$. This last expression can be rewritten

$$\left(\sum_{\sigma \in \mathcal{P}_K} \operatorname{sgn}(\sigma)\alpha_{1j_{\sigma(1)}}\alpha_{2j_{\sigma(2)}} \ldots \alpha_{Kj_{\sigma(K)}}\right) e_{j_1} \wedge e_{j_2} \wedge \cdots \wedge e_{j_K}$$

which amounts to

$$\det\begin{pmatrix} \alpha_{1j_1} & \ldots & \alpha_{Kj_1} \\ \ldots & & \\ \alpha_{1j_K} & \ldots & \alpha_{Kj_K} \end{pmatrix} e_{j_1} \wedge \cdots \wedge e_{j_K}.$$

We conclude that

$$\begin{aligned} a_1 \wedge a_2 \wedge \cdots \wedge a_K \\ &= \sum_{j_1 < \cdots < j_K} \sum_{\sigma \in \mathcal{P}_K} \alpha_{1j_{\sigma(1)}}\alpha_{2j_{\sigma(2)}} \ldots \alpha_{Kj_{\sigma(K)}}(e_{j_{\sigma(1)}} \wedge e_{j_{\sigma(2)}} \wedge \cdots \wedge e_{j_{\sigma(K)}}) \\ &= \sum_{j_1 < \cdots < j_K} \det\begin{pmatrix} \alpha_{1j_1} & \ldots & \alpha_{Kj_1} \\ \ldots & & \\ \alpha_{1j_K} & \ldots & \alpha_{Kj_K} \end{pmatrix} e_{j_1} \wedge \cdots \wedge e_{j_K}. \qquad \square \end{aligned}$$

Example 6.2.2 Suppose we want to compute $a_1 \wedge a_2 \wedge a_3$ where a_1, a_2, a_3 are vectors in $\mathbb{R}^4$,

$$\begin{aligned} a_1 &= 3\, e_1 - e_3 + 4\, e_4, \\ a_2 &= 2\, e_1 + 2\, e_2 + e_3 - e_4, \\ a_3 &= e_2 + e_3 + e_4. \end{aligned}$$

We can, of course, simply write the product of the three 1-vectors, use distributivity and the sign changes which occur when one commutes 1-vectors, and work out the answer a bit at a time. The last theorem tells us that another way to find the answer is to form the matrix

$$\begin{pmatrix} 3 & 2 & 0 \\ 0 & 2 & 1 \\ -1 & 1 & 1 \\ 4 & -1 & 1 \end{pmatrix}$$

containing the vectors as column vectors and then compute the determinants of all the 3×3 submatrices to obtain the coefficients for the basis vectors. For example, to obtain the coefficient in front of $e_1 \wedge e_3 \wedge e_4$, we take the submatrix consisting of the first, third, and fourth rows and compute

$$\det \begin{pmatrix} 3 & 2 & 0 \\ -1 & 1 & 1 \\ 4 & -1 & 1 \end{pmatrix} = 16.$$

The final result is

$$\begin{aligned} &a_1 \wedge a_2 \wedge a_3 \\ &\quad = e_1 \wedge e_2 \wedge e_3 \;+\; 17\, e_1 \wedge e_2 \wedge e_4 \;+\; 16\, e_1 \wedge e_3 \wedge e_4 \;+\; 7\, e_2 \wedge e_3 \wedge e_4. \end{aligned}$$

Note. In the proof of Theorem 6.2.3 it is not really important that the given vectors be written in terms of the basis vectors $e_1, e_2, \ldots, e_N$. We could replace $e_1, e_2, \ldots, e_N$ by $u_1, u_2, \ldots, u_M$ and the conclusion would still hold. We leave this as an exercise.

Corollary 6.2.2 *If* $a_1, a_2, \ldots, a_N \in \mathbb{R}^N$, *then*

$$a_1 \wedge a_2 \wedge \cdots \wedge a_N = \det(a_1^T, a_2^T, \ldots, a_N^T)\; e_1 \wedge e_2 \wedge \cdots \wedge e_N.$$

Definition 6.2.1 If a is a K-vector in $\mathbb{R}^N$ given by

$$a = \sum_{i_1 < \cdots < i_K} a(i_1, \ldots, i_K)\; e_{i_1} \wedge \cdots \wedge e_{i_K},$$

where $K = 1, 2, \ldots, N$, then the *norm* of a is

$$|a| = \sqrt{\sum_{i_1 < \cdots < i_K} a(i_1, \ldots, i_K)^2}.$$

If $K > N$, we set $|\, a \,| = 0$.

For 1-vectors this is simply the Euclidean norm on $\mathbb{R}^N$. We may also think of this as a generalization of the absolute value of real numbers since we identify real numbers with 0-vectors.

Example 6.2.3 Consider the 2-vector

$$a = 5\, e_1 \wedge e_2 \;-\; 6\, e_1 \wedge e_4 \;-\; 4\, e_2 \wedge e_3 \;+\; 2\, e_3 \wedge e_4$$

in $\mathbb{R}^4$. We have

$$|a| = \sqrt{5^2 + 6^2 + 4^2 + 2^2} = 9.$$

Theorem 6.2.4 *If $a_1, a_2, \ldots, a_K \in \mathbb{R}^N$ where $K \leq N$, then $\mid a_1 \wedge a_2 \wedge \cdots \wedge a_K \mid$ is the K-dimensional volume of the parallelepiped determined by $a_1, a_2, \ldots, a_K$.*

Proof. Let us imagine that we draw directed line segments to represent $a_1, a_2, \ldots, a_K$ and then draw the K-dimensional parallelepiped determined by these line segments. Now consider the K-dimensional subspace V determined by the $i_1-, i_2-, \ldots, i_K$-axes. If each $a_i = (\alpha_{i1}, \alpha_{i2}, \ldots, \alpha_{iN})$, then the "oriented" K-dimensional volume of the projection of the parallelepiped into V is given by

$$\det \begin{pmatrix} \alpha_{1i_1} & \alpha_{2i_1} & \ldots & \alpha_{Ki_1} \\ \alpha_{1i_2} & \alpha_{2i_2} & \ldots & \alpha_{Ki_2} \\ \ldots & & & \\ \alpha_{1i_K} & \alpha_{2i_K} & \ldots & \alpha_{Ki_K} \end{pmatrix}.$$

By Theorem 6.2.3 we know that $\mid a_1 \wedge a_2 \wedge \cdots \wedge a_K \mid^2$ is the sum of the squares of the K-dimensional volumes of all projections into such spaces V, and, by Section 1.9, this is the square of the desired K-dimensional volume of the parallelepiped.

The proofs of the next two results are left as exercises. □

Theorem 6.2.5 *If $f : \mathbb{R}^K \to \mathbb{R}^N$ is a linear transformation where $1 \leq K \leq N$, then $\mathcal{D}(f) = \mid f(e_1) \wedge f(e_2) \wedge \cdots \wedge f(e_K) \mid$.*

Theorem 6.2.6 *If $a_1, a_2, \ldots, a_K \in \mathbb{R}^N$, then $a_1, a_2, \ldots, a_K$ are linearly independent if and only if $a_1 \wedge a_2 \wedge \cdots \wedge a_K \neq 0$.*

We now come to an idea of extreme importance in the next chapter: There is a natural way in which linear transformations of vectors induce linear transformations of the associated K-vectors.

Definition 6.2.2 Let $f : \mathbb{R}^M \to \mathbb{R}^N$ be a linear transformation. For $K = 1, 2, \ldots, M$, we know that every K-vector a has a unique expansion of the form

$$a = \sum_{i_1 < \cdots < i_K} a(i_1, \ldots, i_K)\ e_{i_1} \wedge \cdots \wedge e_{i_K}.$$

We define a function

$$\wedge^K f : \wedge^K \mathbb{R}^M \to \wedge^K \mathbb{R}^N$$

by

$$(\wedge^K f)(a) = \sum_{i_1 < \cdots < i_K} a(i_1, \ldots, i_K) f(e_{i_1}) \wedge \cdots \wedge f(e_{i_K}).$$

If $K > M$, we define $(\wedge^K f)(a) = 0$ where 0 is the zero vector in $\wedge^K \mathbb{R}^N$.

Theorem 6.2.7 *If $f : \mathbb{R}^M \to \mathbb{R}^N$ is a linear transformation, then $\wedge^K f : \wedge^K \mathbb{R}^M \to \wedge^K \mathbb{R}^N$ is the unique linear transformation which satisfies*

$$(\wedge^K f)(a_1 \wedge a_2 \wedge \cdots \wedge a_K) = f(a_1) \wedge f(a_2) \wedge \cdots \wedge f(a_K)$$

for all $a_1, a_2, \ldots, a_K \in \mathbb{R}^M$.

Proof. The linearity of $\wedge^K f$ is trivial. Choose $a_1, \ldots, a_K \in \mathbb{R}^M$. We may suppose that $a_i = \sum_{j=1}^N \alpha_{ij} e_j$ for each i. Then

$$
\begin{aligned}
&(\wedge^K f)(a_1 \wedge \cdots \wedge a_K) \\
&\quad = (\wedge^K f)\left(\sum_{j_1<\cdots<j_K} \det \begin{pmatrix} \alpha_{1j_1} & \ldots & \alpha_{Kj_1} \\ \ldots & & \\ \alpha_{1j_K} & \ldots & \alpha_{Kj_K} \end{pmatrix} e_{j_1} \wedge \cdots \wedge e_{j_K} \right) \\
&\quad = \sum_{j_1<\cdots<j_K} \det \begin{pmatrix} \alpha_{1j_1} & \ldots & \alpha_{Kj_1} \\ \ldots & & \\ \alpha_{1j_K} & \ldots & \alpha_{Kj_K} \end{pmatrix} f(e_{j_1}) \wedge \cdots \wedge f(e_{j_K}) \\
&\quad = \left(\sum_{i_1=1}^N \alpha_{1i_1} f(e_{i_1}) \right) \wedge \left(\sum_{i_2=1}^N \alpha_{2i_2} f(e_{i_2}) \right) \wedge \cdots \wedge \left(\sum_{i_K=1}^N \alpha_{Ki_K} f(e_{i_K}) \right) \\
&\quad = f(a_1) \wedge f(a_2) \wedge \cdots \wedge f(a_K).
\end{aligned}
$$

The uniqueness of $\wedge^K f$ follows from the way it acts on basis vectors, namely,

$$(\wedge^K f)(e_{j_1} \wedge \cdots \wedge e_{j_K}) = f(e_{j_1}) \wedge \cdots \wedge f(e_{j_K}).$$

Hence $\wedge^K f$ is the unique linear transformation which "distributes" over the wedge product. □

Example 6.2.4 Let $f : \mathbb{R}^3 \to \mathbb{R}^3$ be the linear transformation defined by

$$f(e_1) = e_2, \quad f(e_2) = e_1, \quad \text{and} \quad f(e_3) = -e_3.$$

To fully understand the behavior of $\wedge^K f$, it is sufficient to evaluate it on basis vectors.

$$
\begin{aligned}
(\wedge^2 f)(e_1 \wedge e_2) &= e_2 \wedge e_1 = -e_1 \wedge e_2, \\
(\wedge^2 f)(e_1 \wedge e_3) &= e_2 \wedge (-e_3) = -e_2 \wedge e_3, \\
(\wedge^2 f)(e_2 \wedge e_3) &= e_1 \wedge (-e_3) = -e_1 \wedge e_3,
\end{aligned}
$$

and

$$(\wedge^3 f)(e_1 \wedge e_2 \wedge e_3) = e_2 \wedge e_1 \wedge (-e_3) = e_1 \wedge e_2 \wedge e_3.$$

We see from this last equation that f is orientation-preserving.

Example 6.2.5 Let $g : \mathbb{R}^3 \to \mathbb{R}^3$ be the linear transformation defined by

$$g(e_1) = ae_1, \quad g(e_2) = be_2, \quad \text{and} \quad g(e_3) = ce_3.$$

Then

$$
\begin{aligned}
(\wedge^2 g)(e_1 \wedge e_2) &= ab\, e_1 \wedge e_2, \\
(\wedge^2 g)(e_1 \wedge e_3) &= ac\, e_1 \wedge e_3, \\
(\wedge^2 g)(e_2 \wedge e_3) &= bc\, e_2 \wedge e_3,
\end{aligned}
$$

and

$$(\wedge^3 g)(e_1 \wedge e_2 \wedge e_3) = abc\, e_1 \wedge e_2 \wedge e_3.$$

Exercises

1. Prove (2) and (3) of Theorem 6.2.1.

2. Show that in the proof of Theorem 6.2.2, if

$$i_1 < \cdots < i_K, \quad j_1 < \cdots < j_L, \quad \text{and} \quad l_1 < \cdots < l_M$$

but not all of $i_1, \ldots, i_K, j_1, \ldots, j_L, l_1, \ldots, l_M$ are distinct, then

$$(e_{i_1 \ldots i_K} \wedge e_{j_1 \ldots j_L}) \wedge e_{l_1 \ldots l_M} = 0 = e_{i_1 \ldots i_K} \wedge (e_{j_1 \ldots j_L} \wedge e_{l_1 \ldots l_M}).$$

3. Give a proof of Theorem 6.2.2 in the case where $K + L + M > N$.

4. Show that every N-vector in $\mathbb{R}^N$ has the form $\lambda\, e_1 \wedge e_2 \wedge \cdots \wedge e_N$ where $\lambda \in \mathbb{R}$.

5. Suppose that $a_1, a_2, \ldots, a_L$ and $b_1, b_2, \ldots, b_M$ are vectors in $\mathbb{R}^N$ and for each i we have

$$b_i = \sum_{j=1}^{L} \alpha_{ij} a_j.$$

If $M \leq L$, show that

$$b_1 \wedge b_2 \wedge \cdots \wedge b_M$$
$$= \sum_{i_1 < \cdots < i_M} \det \begin{pmatrix} \alpha_{1i_1} & \alpha_{2i_1} & \cdots & \alpha_{Mi_1} \\ \alpha_{1i_2} & \alpha_{2i_2} & \cdots & \alpha_{Mi_2} \\ \cdots & & & \\ \alpha_{1i_M} & \alpha_{2i_M} & \cdots & \alpha_{Mi_M} \end{pmatrix} a_{i_1} \wedge a_{i_2} \wedge \cdots \wedge a_{i_M}.$$

If $M > L$, what can be said about $b_1 \wedge b_2 \wedge \cdots \wedge b_M$?

6. Show that if $a_1, a_2, \ldots, a_K$ are vectors in $\mathbb{R}^N$ and a vector occurs in this list more than once, then $a_1 \wedge a_2 \wedge \cdots \wedge a_K = 0$.

7. Show that if σ is a permutation of $\{1, 2, \ldots, K\}$, then we have that

$$e_{i_{\sigma(1)}} \wedge e_{i_{\sigma(2)}} \wedge \cdots \wedge e_{i_{\sigma(K)}} = \operatorname{sgn}(\sigma)\, e_{i_1} \wedge e_{i_2} \wedge \cdots \wedge e_{i_K}.$$

8. Show that if $f : \mathbb{R}^K \to \mathbb{R}^N$ is a linear transformation where $1 \leq K \leq N$, then $\mathcal{D}(f) = |\, f(e_1) \wedge f(e_2) \wedge \cdots \wedge f(e_K)\, |$.

9. Show that if $a_1, a_2, \ldots, a_K \in \mathbb{R}^N$, then $a_1, a_2, \ldots, a_K$ are linearly independent if and only if $a_1 \wedge a_2 \wedge \cdots \wedge a_K \neq 0$.

10. (a) If the linear transformation $f : \mathbb{R}^3 \to \mathbb{R}^3$ has the matrix

$$\begin{pmatrix} a_{11} & a_{12} & a_{13} \\ a_{21} & a_{22} & a_{23} \\ a_{31} & a_{32} & a_{33} \end{pmatrix},$$

then find matrices for $\wedge^2 f$ and $\wedge^3 f$.

(b) If the linear transformation $f : \mathbb{R}^3 \to \mathbb{R}^2$ has the matrix

$$\begin{pmatrix} a_{11} & a_{12} & a_{13} \\ a_{21} & a_{22} & a_{23} \end{pmatrix},$$

then find a matrix for $\wedge^2 f$.

(c) If the linear transformation $f : \mathbb{R}^2 \to \mathbb{R}^3$ has the matrix

$$\begin{pmatrix} a_{11} & a_{12} \\ a_{21} & a_{22} \\ a_{31} & a_{32} \end{pmatrix},$$

then find a matrix for $\wedge^2 f$.

11. If $f : \mathbb{R}^N \to \mathbb{R}^N$ is a linear transformation, show there must be a unique real number λ such that $(\wedge^N f)(e_1 \wedge \cdots \wedge e_N) = \lambda\, e_1 \wedge \cdots \wedge e_N$ and find the value of λ.

12. Show that if f and g are linear transformations such that $f \circ g$ is defined then, $\wedge^K (f \circ g) = (\wedge^K f) \circ (\wedge^K g)$.

6.3 Applications of the Wedge Product and a Characterization of Simple K-Vectors

The wedge product permits one to do linear algebra and analytic geometry in a more powerful and direct manner. We shall also show later that it is useful in unifying and extending the machinery of vector analysis from $\mathbb{R}^2$ and $\mathbb{R}^3$ to manifolds of arbitrary dimension. For now we shall exhibit some simple applications of the wedge product and show that there are some ways of thinking about "simple" K-vectors and addition of K-vectors which appeal to our geometric intuition.

Theorem 6.3.1 (Cramer's rule) *If*

$$\det \begin{pmatrix} \alpha_{11} & \dots & \alpha_{N1} \\ \dots & & \\ \alpha_{1N} & \dots & \alpha_{NN} \end{pmatrix} \neq 0,$$

then the solution of the system of linear equations

$$\alpha_{11}x_1 + \cdots + \alpha_{N1}x_N = \beta_1$$
$$\dots$$
$$\alpha_{1N}x_1 + \cdots + \alpha_{NN}x_N = \beta_N$$

is given by

$$x_i = \det(a_1^T, \dots, a_{i-1}^T, b^T, a_{i+1}^T, \dots, a_N^T) / \det(a_1^T, \dots, a_N^T)$$

where

$$a_i = (\alpha_{i1}, \alpha_{i2}, \ldots, \alpha_{iN})$$

and

$$b = (\beta_1, \beta_2, \ldots, \beta_N).$$

Proof. Note that the system of equations amounts to the single vector equation

$$a_1 x_1 + a_2 x_2 + \cdots + a_N x_N = b.$$

Suppose we want to solve for x_1. From our vector equation we can write

$$(a_1 x_1 + a_2 x_2 + \cdots + a_N x_N) \wedge a_2 \wedge \cdots \wedge a_N = b \wedge a_2 \wedge \cdots \wedge a_N.$$

Since

$$\begin{aligned} a_2 \wedge a_2 \wedge a_3 \wedge \cdots \wedge a_N &= 0, \\ a_3 \wedge a_2 \wedge a_3 \wedge \cdots \wedge a_N &= 0, \\ \ldots \\ a_N \wedge a_2 \wedge a_3 \wedge \cdots \wedge a_N &= 0, \end{aligned}$$

this reduces to

$$x_1 \,(a_1 \wedge a_2 \wedge \cdots \wedge a_N) = b \wedge a_2 \wedge \cdots \wedge a_N,$$

which in turn can be expressed as

$$x_1 \det(a_1^T, a_2^T, \ldots, a_N^T)\, e_1 \wedge \cdots \wedge e_N = \det(b^T, a_2^T, \ldots, a_N^T)\, e_1 \wedge \cdots \wedge e_N.$$

This yields the promised solution for x_1. The remaining $x_2, \ldots, x_N$ can be found in a similar way. □

Theorem 6.3.2 *Let V be a K-dimensional linear subspace of $\mathbb{R}^N$ and let $a_1, a_2, \ldots, a_K$ be a basis for V. Then*

$$x \wedge a_1 \wedge a_2 \wedge \cdots \wedge a_K = 0$$

is an equation for V where $x \in \mathbb{R}^N$.

Proof. We can have $x \wedge a_1 \wedge a_2 \wedge \cdots \wedge a_N = 0$ if and only if $x, a_1, \ldots, a_N$ are linearly dependent. But $a_1, a_2, \ldots, a_N$ are linearly independent, so this last condition amounts to saying that x must be a linear combination of $a_1, \ldots, a_N$. □

Definition 6.3.1 Let V be an M-dimensional linear subspace of $\mathbb{R}^N$. We say that a is *K-vector in V* if and only if a is a sum of K-vectors of the form $\alpha\, a_1 \wedge \cdots \wedge a_K$ where $\alpha \in \mathbb{R}$ and $a_1, \ldots, a_K \in V$. We denote the space of K-vectors in V by the symbol $\wedge^K V$.

Theorem 6.3.3 *Let V be a K-dimensional linear subspace of $\mathbb{R}^N$ and suppose $a_1, a_2, \ldots, a_K$ is a basis for V. Then every K-vector a in V has the form $\alpha\, a_1 \wedge \cdots \wedge a_K$ for some $\alpha \in \mathbb{R}$. Hence* $\dim(\wedge^K V) = 1$.

Proof. Let a be a K-vector in V. Then a must be a sum of K-vectors of the form $\beta\, b_1 \wedge \cdots \wedge b_K$ where $\beta \in \mathbb{R}$ and $b_1, \ldots, b_K \in V$. We need only show that $\beta\, b_1 \wedge \cdots \wedge b_K$ can be written in the form $\alpha\, a_1 \wedge \cdots \wedge a_K$ for some $\alpha \in \mathbb{R}$. Since $a_1, \ldots, a_K$ is a basis for V we can write

$$\beta\, b_1 \wedge \cdots \wedge b_K = \beta \left(\sum_{j=1}^{K} \alpha_{1j} a_j \right) \wedge \cdots \wedge \left(\sum_{j=1}^{K} \alpha_{Kj} a_j \right)$$

where each α_{ij} is a real number. Multiplying this out we obtain a linear combination of K-vectors of the form $a_{i_1} \wedge a_{i_2} \wedge \cdots \wedge a_{i_K}$. This last K-vector is 0 if any a_{i_j} occurs more than once. If no K-vector is repeated, then we may rewrite $a_{i_1} \wedge a_{i_2} \wedge \cdots \wedge a_{i_K}$ in the form $\pm a_1 \wedge a_2 \wedge \cdots \wedge a_K$. This establishes the theorem. □

It was remarked earlier that vectors in $\mathbb{R}^N$ may be thought of as equivalence classes of directed line segments. There is a geometric description of when two vectors a and b are equal. Starting from the origin, let us draw directed line segments which represent a and b. Then $a = b$ if and only if the following conditions hold:

(1) The two directed line segments must have the same length.

(2) Their directed line segments must determine the same linear subspace, that is, the same line passing through the origin.

(3) If the two directed line segments determine the same line through the origin, then within that line, they must both be oriented in the same direction.

It is perhaps a remarkable fact that a similar characterization may be given to certain very important K-vectors, the *simple* K-vectors.

Definition 6.3.2 If V is a K-dimensional linear subspace of $\mathbb{R}^N$ and a and b are nonzero K-vectors in V, we say that they have the *same orientation* if and only if there is a positive scalar γ such that $a = \gamma b$. Otherwise we must have $a = \gamma b$ where $\gamma < 0$, and we say that a and b have *opposite orientations.*

V has precisely two unit K-vectors. Let us denote them as w and $-w$. We call w and $-w$ the *orientations* of V. We shall call $e_1 \wedge e_2 \wedge \cdots \wedge e_N$ the *standard orientation* of $\mathbb{R}^N$.

Definition 6.3.3 If $1 \le K \le N$ and a is K-vector in $\mathbb{R}^N$, by the *subspace determined by a* we mean the set of 1-vectors $b \in \mathbb{R}^N$ that satisfy $b \wedge a = 0$.

Definition 6.3.4 By a *simple* K-vector we mean one of the form $a_1 \wedge a_2 \wedge \cdots \wedge a_K$ where each a_i is, of course, a 1-vector.

Note that all 1-vectors are trivially simple and that every K-vector is a sum of simple K-vectors.

The following theorem says, in effect, that we may think of simple K-vectors as equivalence classes of oriented parallelepipeds.

Theorem 6.3.4 *Let a and b be two nonzero, simple K-vectors in $\mathbb{R}^N$. Then $a = b$ if and only if*

(1) $|a| = |b|$,

(2) a and b both determine the same linear subspace V of $\mathbb{R}^N$, and

(3) a and b both have the same orientation.

Proof. Let us assume (1), (2), and (3) hold. Since a is simple, we can write it in the form $a_1 \wedge a_2 \wedge \cdots \wedge a_K$ for some $a_1, a_2, \ldots, a_K \in \mathbb{R}^N$. Notice that for each a_i we must have $a_i \wedge (a_1 \wedge a_2 \wedge \cdots \wedge a_K) = 0$, so that by the definition of V, each $a_i \in V$. Therefore a must be a K-vector in V, and by a similar argument, the same must be true for b. Since $a_1 \wedge \cdots \wedge a_K \neq 0$, it follows that $a_1, a_2, \ldots, a_K$ must be a basis for V. By Theorem 6.3.3, there must be some scalar γ such that $b = \gamma a$. From (3) we see that $\gamma > 0$, and from (1) it follows that $\gamma = 1$. Thus $a = b$. The implication in the other direction is trivial. □

The reader is almost certainly familiar with the practice of denoting the vector from A to B in $\mathbb{R}^N$ by AB. The two most important properties of such vectors are

$$AB = -BA$$

and

$$AB + BC = AC.$$

These two useful and highly geometric properties can be generalized to the setting of K-vectors.

Let $A_0, A_1, \ldots, A_K$ be points in $\mathbb{R}^N$ with $1 \leq K \leq N$. These points can be taken as the vertices of a (sometimes degenerate) K-dimensional simplex. We will think of $A_0 A_1 \ldots A_K$ as representing an equivalence class of oriented, K-dimensional simplices. Namely, two such simplices will be considered equivalent if they determine the same K-dimensional subspace, if they have the same orientation within that subspace, and if they have the same K-dimensional volume.

Formally, we define

$$A_0 A_1 \ldots A_K = \frac{1}{K!}(A_0 A_1) \wedge (A_0 A_2) \wedge \cdots \wedge (A_0 A_K).$$

The $K!$ is present because the K-dimensional volume of the simplex determined by $A_0,\ A_1,\ \ldots,\ A_K$ is

$$\frac{1}{K!}|(A_0 A_1) \wedge (A_0 A_2) \wedge \cdots \wedge (A_0 A_K)|.$$

Theorem 6.3.5 *Let $A_0, A_1, \ldots, A_K$ be points in $\mathbb{R}^N$ with $1 \le K \le N$.*
(a) If for distinct i and j we interchange A_i and A_j in $A_0 \ldots A_i \ldots A_j \ldots A_K$, then

$$A_0 \ldots A_j \ldots A_i \ldots A_K = -A_0 \ldots A_i \ldots A_j \ldots A_K.$$

(b)
$$\sum_{i=0}^{K} (-1)^i A_0 A_1 \ldots \overline{A_i} \ldots A_K = 0$$

where $\overline{A_i}$ means that A_i has been omitted.

Proof. For (a) we need only show that if we interchange A_i and A_{i+1} in $A_0 A_1 \ldots A_K$ there is a sign change. Since

$$A_0 A_1 \ldots A_K = \frac{1}{K!} (A_0 A_1) \wedge (A_0 A_2) \wedge \cdots \wedge (A_0 A_K),$$

this is obvious if $i \ge 1$. We need only consider what happens if we interchange A_0 and A_1. Observe that

$$\begin{aligned} A_0 A_1 \ldots A_K &= \frac{1}{K!} (A_0 A_1) \wedge (A_0 A_1 + A_1 A_2) \wedge \cdots \wedge (A_0 A_1 + A_1 A_K) \\ &= \frac{1}{K!} (A_0 A_1) \wedge (A_1 A_2) \wedge \cdots \wedge (A_1 A_K) \\ &= -\frac{1}{K!} (A_1 A_0) \wedge (A_1 A_2) \wedge \cdots \wedge (A_1 A_K) \\ &= -A_1 A_0 A_2 \ldots A_K. \end{aligned}$$

To prove (b), notice that

$$\begin{aligned} A_1 A_2 \ldots A_K &= \frac{1}{(K-1)!} (A_1 A_2) \wedge (A_1 A_3) \wedge \cdots \wedge (A_1 A_K) \\ &= \frac{1}{(K-1)!} (A_1 A_0 + A_0 A_2) \wedge (A_1 A_0 + A_0 A_3) \wedge \ldots \\ &\qquad \wedge (A_1 A_0 + A_0 A_K). \end{aligned}$$

When we multiply this out, because $(A_1A_0) \wedge (A_1A_0) = 0$, we must have

$$A_1A_2 \dots A_K = \frac{1}{(K-1)!}\Big\{(A_0A_2) \wedge (A_0A_3) \wedge \cdots \wedge (A_0A_K)$$

$$+ \sum_{i=2}^{K} (A_0A_2) \wedge \cdots \wedge (A_0A_{i-1}) \wedge (A_1A_0) \wedge (A_0A_{i+1}) \wedge \cdots \wedge (A_0A_K)\Big\}$$

$$= A_0A_2A_3 \dots A_K$$

$$+ \frac{1}{(K-1)!}\left\{\sum_{i=2}^{K} (-1)^{i-1} (A_0A_1) \wedge (A_0A_2) \wedge \cdots \wedge \overline{(A_0A_i)} \wedge \cdots \wedge (A_0A_K)\right\}$$

$$= \sum_{i=1}^{K} (-1)^{i-1} A_0A_1 \dots \overline{A_i} \dots A_K.$$

From this we immediately deduce the desired result. □

Notice that if $K = 2$, then (b) amounts to $A_0A_2 = A_0A_1 + A_1A_2$. This is the familiar fact that if we look at the sides of a triangle as vectors with the proper orientations, then two of the sides will sum to give us the third. To better see how (b) generalizes this relationship, consider the case $K = 3$ where $A_0, A_1, A_2, A_3 \in \mathbb{R}^3$. These four points determine a tetrahedron in $\mathbb{R}^3$. See Figure 6.3.1. The sides of this tetrahedron may be thought of as 2-vectors, $A_0A_1A_2$, $A_0A_2A_3$, etc. Then (b) says that, given the proper orientations, one of the sides may be thought of as the 2-vector sum of the other sides.

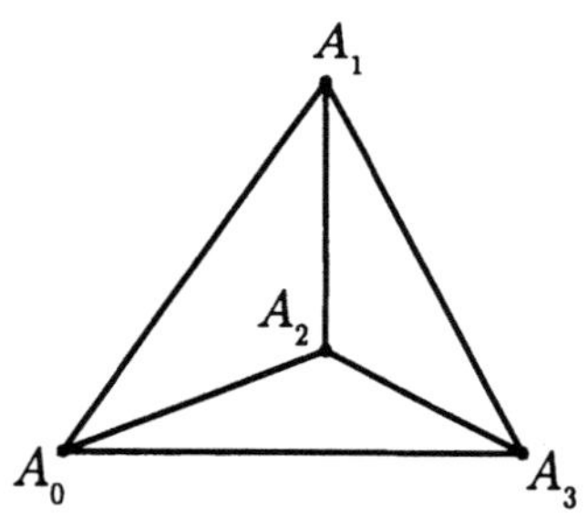

FIGURE 6.3.1.

Exercises

1. In the proof of Cramer's rule, show how one would solve for x_2.

2. Show that if V is a vector subspace of $\mathbb{R}^N$, then $\wedge^K V$ is a vector space.

3. Prove that a linear subspace of $\mathbb{R}^N$ has precisely two orientations and one must be the negative of the other.

4. $\mathbb{R}^3$ equipped with the standard orientation is sometimes called a right-handed coordinate system. If it is given the opposite orientation, it is called a left-

handed coordinate system. Draw some pictures and give an explanation of this terminology.

5. Show that in $\mathbb{R}^4$ the 2-vector $(e_1 \wedge e_2) + (e_3 \wedge e_4)$ is not simple.

6. Show that if V is the linear subspace of $\mathbb{R}^N$ determined by $a_1 \wedge \cdots \wedge a_K$ where, $a_1, \ldots, a_K$ are vectors in $\mathbb{R}^N$, and $a_1 \wedge \cdots \wedge a_K \neq 0$, then $\{a_1, \ldots, a_K\}$ must be a basis for V.

7. Let V be a 2-dimensional linear subspace of $\mathbb{R}^3$ and let $f : \mathbb{R}^3 \to \mathbb{R}^3$ be a linear transformation. Show that $f(V) = V$ if and only if there is a nonzero real number λ and a nonzero simple 2-vector v in V such that $\wedge^2 f(v) = \lambda\, v$.

6.4 The Dot Product and the Star Operator

We want first to extend to K-vectors the idea of a dot product.

Definition 6.4.1 Let $a, b \in \wedge^K \mathbb{R}^N$ where $1 \leq K \leq N$. If

$$a = \sum_{i_1 < \cdots < i_K} a(i_1, \ldots, i_K) \;\; e_{i_1} \wedge e_{i_2} \wedge \cdots \wedge e_{i_K}$$

and

$$b = \sum_{i_1 < \cdots < i_K} b(i_1, \ldots, i_K) \;\; e_{i_1} \wedge e_{i_2} \wedge \cdots \wedge e_{i_K},$$

then we define

$$a \cdot b = \sum_{i_1 < \cdots < i_K} a(i_1, \ldots, i_K) \; b(i_1, \ldots, i_K).$$

Since we identify $\wedge^0 \mathbb{R}^N$ with $\mathbb{R}$, we interpret $a \cdot b$ in this case as the multiplication of the real numbers a and b. For $\wedge^K \mathbb{R}^N$ with $K > N$, we of course take $a \cdot b = 0$.

Theorem 6.4.1 *For all $a, b, c \in \wedge^K \mathbb{R}^N$ and $\lambda \in \mathbb{R}$ we have*

(a) $a \cdot b = b \cdot a$,

(b) $a \cdot (b + c) = a \cdot b + a \cdot c$,

(c) $\lambda(a \cdot b) = (\lambda a) \cdot b = a \cdot (\lambda b)$,

(d) $a \cdot a = | a |^2$.

There is a nice formula for computing the dot product of simple K-vectors.

Theorem 6.4.2 *Suppose a and b are simple K-vectors in $\mathbb{R}^N$. If $a = a_1 \wedge \cdots \wedge a_K$ and $b = b_1 \wedge \cdots \wedge b_K$, then*

$$a \cdot b = \det \begin{pmatrix} a_1 \cdot b_1 & \ldots & a_K \cdot b_1 \\ \ldots & & \\ a_1 \cdot b_K & \ldots & a_K \cdot b_K \end{pmatrix}.$$

Proof. Assume that for each i we have

$$a_i = \sum_{j=1}^{N} \alpha_{ij} e_j \quad \text{and} \quad b_i = \sum_{j=1}^{N} \beta_{ij} e_j.$$

Then

$$\begin{aligned} a \cdot b &= (a_1 \wedge \cdots \wedge a_K) \cdot (b_1 \wedge \cdots \wedge b_K) \\ &= \sum_{i_1=1}^{N} \cdots \sum_{i_K=1}^{N} \beta_{1i_1} \ldots \beta_{Ki_K} (a_1 \wedge \cdots \wedge a_K) \cdot (e_{i_1} \wedge \cdots \wedge e_{i_K}). \end{aligned}$$

Similarly we can show that

$$\begin{aligned} &\det \begin{pmatrix} a_1 \cdot b_1 & \ldots & a_K \cdot b_1 \\ \ldots & & \\ a_1 \cdot b_K & \ldots & a_K \cdot b_K \end{pmatrix} \\ &= \sum_{i_1=1}^{N} \cdots \sum_{i_K=1}^{N} \beta_{1i_1} \ldots \beta_{Ki_K} \det \begin{pmatrix} a_1 \cdot e_{i_1} & \ldots & a_K \cdot e_{i_1} \\ \ldots & & \\ a_1 \cdot e_{i_K} & \ldots & a_K \cdot e_{i_K} \end{pmatrix}. \end{aligned}$$

By the definition of dot product and Theorem 6.2.3 we see that

$$(a_1 \wedge \cdots \wedge a_K) \cdot (e_{i_1} \wedge \cdots \wedge e_{i_K}) = \det \begin{pmatrix} \alpha_{1i_1} & \ldots & \alpha_{Ki_1} \\ \ldots & & \\ \alpha_{1i_K} & \ldots & \alpha_{Ki_K} \end{pmatrix}.$$

Trivially we have

$$\det \begin{pmatrix} a_1 \cdot e_{i_1} & \ldots & a_K \cdot e_{i_1} \\ \ldots & & \\ a_1 \cdot e_{i_K} & \ldots & a_K \cdot e_{i_K} \end{pmatrix} = \det \begin{pmatrix} \alpha_{1i_1} & \ldots & \alpha_{Ki_1} \\ \ldots & & \\ \alpha_{1i_K} & \ldots & \alpha_{Ki_K} \end{pmatrix},$$

and the theorem is established. □

One result of this theorem is that if a linear transformation f is orthogonal, then we may also think of $\wedge^K f$ as being "orthogonal".

Theorem 6.4.3 *If $f : \mathbb{R}^N \to \mathbb{R}^N$ is an orthogonal transformation and $1 \le K \le N$, then for all $a, b \in \wedge^K \mathbb{R}^N$ we have*

$$\Big((\wedge^K f)(a) \Big) \cdot \Big((\wedge^K f)(b) \Big) = a \cdot b.$$

Proof. We need only prove this for simple K-vectors. It is a straightforward calculation:

$$\begin{aligned}
&\Big((\wedge^K f)(a_1 \wedge \cdots \wedge a_K)\Big) \cdot \Big((\wedge^K f)(b_1 \wedge \cdots \wedge b_K)\Big) \\
&= \Big(f(a_1) \wedge \cdots \wedge f(a_K)\Big) \cdot \Big(f(b_1) \wedge \cdots \wedge f(b_K)\Big) \\
&= \det \begin{pmatrix} f(a_1)\cdot f(b_1) & \dots & f(a_K)\cdot f(b_1) \\ \dots & & \\ f(a_1)\cdot f(b_K) & \dots & f(a_K)\cdot f(b_K) \end{pmatrix} \\
&= \det \begin{pmatrix} a_1 \cdot b_1 & \dots & a_K \cdot b_1 \\ \dots & & \\ a_1 \cdot b_K & \dots & a_K \cdot b_K \end{pmatrix} \\
&= (a_1 \wedge \cdots \wedge a_K) \cdot (b_1 \wedge \cdots \wedge b_K).
\end{aligned}$$

□

The standard basis for $\wedge^K \mathbb{R}^N$ is the set of K-vectors of the form $e_{i_1} \wedge \cdots \wedge e_{i_K}$ where $i_1 < \cdots < i_K$. This is seen to be "orthonormal" in the sense that, for $i_1 < \cdots < i_K$ and $j_1 < \cdots < j_K$, we have

$$(e_{i_1} \wedge \cdots \wedge e_{i_K}) \cdot (e_{j_1} \wedge \cdots \wedge e_{j_K}) = \begin{cases} 1 & \text{if } (i_1, \dots, i_K) = (j_1, \dots, j_K), \\ 0 & \text{otherwise.} \end{cases}$$

It is now easy to see that any orthonormal basis of $\mathbb{R}^N$ generates an "orthonormal" basis of $\wedge^K \mathbb{R}^N$.

Theorem 6.4.4 *If $a_1, \dots, a_N$ is an orthonormal basis of $\mathbb{R}^N$ and $1 \le K \le N$, then the set of K-vectors of the form $a_{i_1} \wedge \cdots \wedge a_{i_K}$, where $i_1 < \cdots < i_K$, is a basis for $\wedge^K \mathbb{R}^N$ and satisfies*

$$(a_{i_1} \wedge \cdots \wedge a_{i_K}) \cdot (a_{j_1} \wedge \cdots \wedge a_{j_K}) = \begin{cases} 1 & \textit{if } (i_1, \dots, i_K) = (j_1, \dots, j_K), \\ 0 & \textit{otherwise.} \end{cases}$$

Proof. There is an orthogonal transformation $f : \mathbb{R}^N \to \mathbb{R}^N$ which has the property that $f(a_i) = e_i$ for all i.

Suppose we have scalars $\lambda_{i_1 \dots i_K}$ such that

$$\sum_{i_1 < \cdots < i_K} \lambda_{i_1 \dots i_K} (a_{i_1} \wedge \cdots \wedge a_{i_K}) = 0.$$

Applying f to both sides of this equation yields

$$\sum_{i_1 < \cdots < i_K} \lambda_{i_1 \dots i_K} (e_{i_1} \wedge \cdots \wedge e_{i_K}) = 0.$$

Since the K-vectors $e_{i_1} \wedge \cdots \wedge e_{i_K}$ are linearly independent, it follows that each $\lambda_{i_1 \dots i_K} = 0$, and hence the K-vectors $a_{i_1} \wedge \cdots \wedge a_{i_K}$ are also linearly independent.

Since there are $\binom{N}{K}$ of the K-vectors $a_{i_1} \wedge \cdots \wedge a_{i_K}$ where $i_1 < \cdots < i_K$,and since this is the dimension of $\wedge^K \mathbb{R}^N$, it follows that $\{a_{i_1} \wedge \cdots \wedge a_{i_K}\}$ is a basis for $\wedge^K \mathbb{R}^N$.

For $i_1 < \cdots < i_K$ and $j_1 < \cdots < j_K$ we have

$$\begin{aligned}
(a_{i_1} \wedge \cdots \wedge a_{i_K}) &\cdot (a_{j_1} \wedge \cdots \wedge a_{j_K}) \\
&= (f(a_{i_1}) \wedge \cdots \wedge f(a_{i_K})) \cdot (f(a_{j_1}) \wedge \cdots \wedge f(a_{j_K})) \\
&= (e_{i_1} \wedge \cdots \wedge e_{i_K}) \cdot (e_{j_1} \wedge \cdots \wedge e_{j_K}) \\
&= \begin{cases} 1 & \text{if } (i_1, \ldots, i_K) = (j_1, \ldots, j_K), \\ 0 & \text{otherwise.} \end{cases}
\end{aligned}$$

The theorem is proved. □

Recall that for a linear transformation $g\colon V \to W$ between vector spaces over the reals, assuming the presence of a dot product in V and W, the dual of g is the unique linear transformation $g^\diamond\colon W \to V$ satisfying $g(v) \cdot w = v \cdot g^\diamond(w)$ for $v \in V$ and $w \in W$. Theorem 6.4.2 makes possible a simple description of the dual to the linear transformation $\wedge^K f$.

Theorem 6.4.5 *If f is a linear transformation, then for simple K-vectors $a_1 \wedge \cdots \wedge a_K$ we have $\left(\wedge^K f\right)^\diamond (a_1 \wedge \cdots \wedge a_K) = f^\diamond(a_1) \wedge \cdots \wedge f^\diamond(a_K)$.*

Proof. For simple K-vectors $a_1 \wedge \cdots \wedge a_K$ and $b_1 \wedge \cdots \wedge b_K$ we have

$$\begin{aligned}
&\left[\left(\wedge^K f\right)^\diamond (a_1 \wedge \cdots \wedge a_K)\right] \cdot (b_1 \wedge \cdots \wedge b_K) \\
&= (a_1 \wedge \cdots \wedge a_K) \cdot \left[\left(\wedge^K f\right) (b_1 \wedge \cdots \wedge b_K)\right] \\
&= (a_1 \wedge \cdots \wedge a_K) \cdot \left(f(b_1) \wedge \cdots \wedge f(b_K)\right) \\
&= \det \begin{pmatrix} a_1 \cdot f(b_1) & \ldots & a_K \cdot f(b_1) \\ \ldots & & \\ a_1 \cdot f(b_K) & \ldots & a_K \cdot f(b_K) \end{pmatrix} \\
&= \det \begin{pmatrix} f^\diamond(a_1) \cdot b_1 & \ldots & f^\diamond(a_K) \cdot b_1 \\ \ldots & & \\ f^\diamond(a_1) \cdot b_K & \ldots & f^\diamond(a_K) \cdot b_K \end{pmatrix} \\
&= \left(f^\diamond(a_1) \wedge \cdots \wedge f^\diamond(a_K)\right) \cdot (b_1 \wedge \cdots \wedge b_K).
\end{aligned}$$

Since this holds for arbitrary $b_1, \ldots, b_K$, the desired conclusion follows. □

Example 6.4.1 Consider the linear transformation defined by

$$f(e_1) = e_1, \quad f(e_2) = e_2, \quad \text{and} \quad f(e_3) = e_1 + 2e_2.$$

The matrices of f and $f^\diamond$ are

$$[f] = \begin{pmatrix} 1 & 0 & 1 \\ 0 & 1 & 2 \end{pmatrix} \quad \text{and} \quad [f^\diamond] = \begin{pmatrix} 1 & 0 \\ 0 & 1 \\ 1 & 2 \end{pmatrix}.$$

Thus

$$f^{\diamond}(e_1) = e_1 + e_3,$$
$$f^{\diamond}(e_2) = e_2 + 2e_3,$$

and

$$\begin{aligned}(\wedge^2 f)^{\diamond}(e_1 \wedge e_2) &= (e_1 + e_3) \wedge (e_2 + 2e_3)\\ &= (e_1 \wedge e_2) + 2(e_1 \wedge e_3) - (e_2 \wedge e_3).\end{aligned}$$

Recall that a simple K-vector may be pictured as an oriented parallelepiped or simplex. With each such parallelepiped we may associate an $(N-K)$-dimensional parallelepiped which is its "orthogonal complement". For instance, given a 2-dimensional parallelogram in $\mathbb{R}^3$, we can find a vector (a 1-dimensional, oriented parallelepiped) which is perpendicular to the parallelogram. This vague description hints at a sort of duality between K-vectors and $(N-K)$-vectors in $\mathbb{R}^N$. This duality will be established by the (Hodge) star operator, a mapping $a \mapsto {}^*a$ which transforms K-vectors into $(N-K)$-vectors.

In order to define ${}^*a$ we first need a result about finite dimensional vector spaces. Suppose V is a finite dimensional vector space over the field of reals and $\{a_1, a_2, \ldots, a_N\}$ is a basis for V. It is simple to define a "dot product" on V in which $\{a_1, a_2, \ldots, a_N\}$ plays the role of an "orthonormal" basis. For $a = \sum_{i=1}^{N} \alpha_i a_i$ and $b = \sum_{i=1}^{N} \beta_i a_i$ we define

$$a \cdot b = \alpha_1\beta_1 + \alpha_2\beta_2 + \cdots + \alpha_N\beta_N.$$

This is exactly what was done in the case of $\wedge^K \mathbb{R}^N$. One can also easily show that

(a) $a \cdot b = b \cdot a$,

(b) $a \cdot (b + c) = a \cdot b + a \cdot c$,

(c) $\lambda(a \cdot b) = (\lambda a) \cdot b = a \cdot (\lambda b)$,

(d) $a_i \cdot a_j = \begin{cases} 1 & \text{if } i = j \\ 0 & \text{otherwise.} \end{cases}$

Theorem 6.4.6 *Let V be a finite dimensional vector space over the reals and suppose $(a, b) \mapsto a \cdot b$ is a "dot product" on V defined as indicated above. Then for every linear transformation $f : V \to \mathbb{R}$ there is a unique $a \in V$ such that $f(b) = a \cdot b$ for all $b \in V$.*

Proof. Set $a = f(a_1)\,a_1 + \cdots + f(a_N)\,a_N$. Since $a \cdot a_i = f(a_i)$ and the maps $b \mapsto f(b)$ and $b \mapsto a \cdot b$ are linear, it follows that $f(b) = a \cdot b$ for all $b \in V$. (Uniqueness of a is left as an exercise.) □

We are now ready to show the existence of the star operator.

Theorem 6.4.7 *Suppose $0 \le K \le N$. For every K-vector a in $\mathbb{R}^N$ there is a unique $(N-K)$-vector $^*a$ with the property that*

$$a \wedge b = (^*a \cdot b)\, e_1 \wedge \cdots \wedge e_N$$

for every $(N-K)$-vector b in $\mathbb{R}^N$.

Proof. Choose a K-vector a in $\mathbb{R}^N$. For every $(N-K)$-vector b in $\mathbb{R}^N$ there is a uniquely determined real number $f(b)$ such that $a \wedge b = f(b)\, e_1 \wedge \cdots \wedge e_N$. The mapping $b \mapsto f(b)$ is clearly linear. Therefore there must be a uniquely determined $(N-K)$-vector $^*a$ such that $^*a \cdot b = f(b)$ for all $(N-K)$-vectors b. □

Definition 6.4.2 For $0 \le K \le N$, the transformation of $\wedge^N\mathbb{R}^N$ into $\wedge^{N-K}\mathbb{R}^N$ defined by $a \mapsto {}^*a$ is the *Hodge star operator.*

Theorem 6.4.8 *For $0 \le K \le N$ the mapping $a \mapsto {}^*a$ is a linear transformation of $\wedge^K\mathbb{R}^N$ into $\wedge^{N-K}\mathbb{R}^N$.*

Example 6.4.2 Let us find $^*e_1$ in $\mathbb{R}^3$. This must be a 2-vector, and hence must have the form $\lambda_1(e_1 \wedge e_2) + \lambda_2(e_1 \wedge e_3) + \lambda_3(e_2 \wedge e_3)$. We see that

$$^*e_1 \cdot (e_1 \wedge e_2) = \Big(\lambda_1(e_1 \wedge e_2) + \lambda_2(e_1 \wedge e_3) + \lambda_3(e_2 \wedge e_3)\Big) \cdot (e_1 \wedge e_2) = \lambda_1.$$

Appealing to the equation $a \wedge b = (^*a \cdot b)\, e_1 \wedge \cdots \wedge e_N$, we then obtain

$$\lambda_1\, e_1 \wedge e_2 \wedge e_3 = \Big(^*e_1 \cdot (e_1 \wedge e_2)\Big)\, e_1 \wedge e_2 \wedge e_3 = e_1 \wedge (e_1 \wedge e_2) = 0.$$

Similarly, we find that

$$\lambda_2\, e_1 \wedge e_2 \wedge e_3 = (^*e_1 \cdot (e_1 \wedge e_3)) e_1 \wedge e_2 \wedge e_3 = e_1 \wedge (e_1 \wedge e_3) = 0,$$
$$\lambda_3\, e_1 \wedge e_2 \wedge e_3 = (^*e_1 \cdot (e_2 \wedge e_3)) e_1 \wedge e_2 \wedge e_3 = e_1 \wedge (e_2 \wedge e_3) = e_1 \wedge e_2 \wedge e_3.$$

Thus $\lambda_1 = \lambda_2 = 0$ and $\lambda_3 = 1$. This tells us that

$$^*e_1 = e_2 \wedge e_3.$$

We want to justify the claim that, for simple K-vectors, $^*a$ is an "orthogonal complement" for a.

Let $a_1, \ldots, a_N$ be an orthonormal basis for $\mathbb{R}^N$. We must have $a_1 \wedge \cdots \wedge a_N = \lambda\,(e_1 \wedge \cdots \wedge e_N)$ for some real number λ. Since $|\, e_1 \wedge \cdots \wedge e_N \,| = 1$, we must have $|\, a_1 \wedge \cdots \wedge a_N \,| = |\, \lambda \,|$. We also know that

$$\begin{aligned} |\, a_1 \wedge \cdots \wedge a_N \,|^2 &= (a_1 \wedge \cdots \wedge a_N) \cdot (a_1 \wedge \cdots \wedge a_N) \\ &= \det\,(a_i \cdot a_j) \\ &= 1, \end{aligned}$$

so that $\lambda = \pm 1$.

Theorem 6.4.9 *Let $a_1, \ldots, a_N$ be an orthonormal basis for $\mathbb{R}^N$. Let λ be the unique value, ± 1, which satisfies $a_1 \wedge \cdots \wedge a_N = \lambda\,(e_1 \wedge \cdots \wedge e_N)$. Then for $0 \leq K \leq N$ we have $^*(a_1 \wedge \cdots \wedge a_K) = \lambda\,(a_{K+1} \wedge a_{K+2} \wedge \cdots \wedge a_N)$.*

Proof. We consider only the case where $0 < K < N$. Recall that the set of $(N-K)$-vectors of the form $a_{i_1} \wedge \cdots \wedge a_{i_{N-K}}$ where $i_1 < \cdots < i_{N-K}$ is an "orthonormal" basis for $\wedge^{N-K}\mathbb{R}^N$. Therefore $^*(a_1 \wedge \cdots \wedge a_K)$ is completely determined by the coefficients $^*(a_1 \wedge \cdots \wedge a_K) \cdot (a_{i_1} \wedge \cdots \wedge a_{i_{N-K}})$. Assume $i_1 < \cdots < i_{N-K}$. Note that

$$
\begin{aligned}
&(a_1 \wedge \cdots \wedge a_K) \wedge (a_{i_1} \wedge \cdots \wedge a_{i_{N-K}}) \\
&\quad = \begin{cases} \lambda\,(e_1 \wedge \cdots \wedge e_N) & \text{if } i_1 = K+1, \;\; i_2 = K+2, \;\; \ldots, \;\; i_{N-K} = N \\ 0 & \text{otherwise.} \end{cases}
\end{aligned}
$$

The zero in the second case arises from the fact that some vector must occur twice in the wedge product. We also know that

$$
\begin{aligned}
&(a_1 \wedge \cdots \wedge a_K) \wedge (a_{i_1} \wedge \cdots \wedge a_{i_{N-K}}) \\
&\quad = {}^*(a_1 \wedge \cdots \wedge a_K) \cdot (a_{i_1} \wedge \cdots \wedge a_{i_{N-K}})\, e_1 \wedge \cdots \wedge e_N.
\end{aligned}
$$

Therefore

$$
\begin{aligned}
&{}^*(a_1 \wedge \cdots \wedge a_K) \cdot (a_{i_1} \wedge \cdots \wedge a_{i_{N-K}}) \\
&\quad = \begin{cases} \lambda & \text{if } (i_1, \ldots, i_{N-K}) = (K+1, K+2, \ldots, N) \\ 0 & \text{otherwise.} \end{cases}
\end{aligned}
$$

Thus $^*(a_1 \wedge \cdots \wedge a_K) = \lambda\,(a_{K+1} \wedge a_{K+2} \wedge \cdots \wedge a_N)$. □

Corollary 6.4.1 *For any K-vector a in $\mathbb{R}^N$ we have $^{**}a = (-1)^{K(N-K)}a$.*

Sketch of the proof. Divide the set of natural numbers $\{1, 2, \ldots, N\}$ into two disjoint sets, $\{i_1, \ldots, i_K\}$ and $\{j_1, \ldots, j_{N-K}\}$, and set $a = e_{i_1} \wedge \cdots \wedge e_{i_K}$ and $b = e_{j_1} \wedge \cdots \wedge e_{j_{N-K}}$. We must have $a \wedge b = (-1)^R e_1 \wedge \cdots \wedge e_N$ for some R. Now compute successively $^*a$ and $^{**}a$. The rest of this is left as an exercise.

Example 6.4.3 Suppose we want to find $^*(e_2 \wedge e_3)$ in $\mathbb{R}^4$. By Theorem 6.4.9, the answer must be $\lambda\,(e_1 \wedge e_4)$ where $\lambda = \pm 1$. We need only discover λ. Employing the properties of the dot product and the equation $a \wedge b = (^*a \cdot b)\, e_1 \wedge \cdots \wedge e_N$, we obtain

$$
\begin{aligned}
\lambda\, e_1 \wedge e_2 \wedge e_3 \wedge e_4 &= \left({}^*(e_2 \wedge e_3) \cdot (e_1 \wedge e_4)\right) e_1 \wedge e_2 \wedge e_3 \wedge e_4 \\
&= (e_2 \wedge e_3) \wedge (e_1 \wedge e_4) \\
&= e_1 \wedge e_2 \wedge e_3 \wedge e_4.
\end{aligned}
$$

Thus $^*(e_2 \wedge e_3) = e_1 \wedge e_4$.

The following result is straightforward and emphasizes the extent to which the star operator preserves important relations between K-vectors. We leave its proof as an exercise.

Theorem 6.4.10 *For $0 \leq K \leq N$, the following results hold for $\mathbb{R}^N$:*

(a) For all K-vectors a and b we have $(^*a) \cdot (^*b) = a \cdot b$.

(b) For each K-vector a we have $|^*a| = |a|$.

(c) For each K the transformation $a \mapsto {}^*a$ is continuous.

We will close by considering two ways in which the star operator touches material familiar to us from an earlier level of mathematics.

In discussing the cross product in $\mathbb{R}^3$, the usual notation replaces e_1, e_2, e_3 by $\imath$, $\jmath$, k, respectively. The cross product of two vectors, $a \times b$, can be defined by setting

$$\begin{aligned} \imath \times \jmath &= k, \\ \jmath \times k &= \imath, \\ k \times \imath &= \jmath, \end{aligned}$$

and assuming

$$a \times b = -b \times a$$

plus the usual laws of distributivity for multiplication and scalars. The cross product can also be described by the mnemonic

$$(\alpha_1 \imath + \alpha_2 \jmath + \alpha_3 k) \times (\beta_1 \imath + \beta_2 \jmath + \beta_3 k) = \det \begin{pmatrix} \imath & \jmath & k \\ \alpha_1 & \alpha_2 & \alpha_3 \\ \beta_1 & \beta_2 & \beta_3 \end{pmatrix}.$$

The property $a \times b = -b \times a$ suggests some kind of relationship to the wedge product. It is easily checked that

$$\begin{aligned} {}^*(\imath \wedge \jmath) &= k, \\ {}^*(\jmath \wedge k) &= \imath, \\ {}^*(k \wedge \imath) &= \jmath. \end{aligned}$$

It follows from this that the relation between the two kinds of multiplication is

$${}^*(a \wedge b) = a \times b.$$

A second application concerns $(N-1)$-dimensional linear subspaces of $\mathbb{R}^N$. (It may be helpful to picture a plane in $\mathbb{R}^3$ passing through the origin.) Such subspaces may be specified by writing an equation of the form

$$a \cdot x = 0,$$

where a is a given "normal" vector, or an equation of the form

$$b \wedge x = 0,$$

where b is a given simple $(N-1)$-vector. In each case x is simply an arbitrary point or vector in $\mathbb{R}^N$. If we start with a, we may construct a suitable b by setting $b =^* a$, or vice-versa as we wish.

Exercises

1. Prove Theorem 6.4.1.

2. Show that if $a = a_1 \wedge \cdots \wedge a_K$ and $b = b_1 \wedge \cdots \wedge b_K$ and a_1 is orthogonal to each b_i, then $a \cdot b = 0$.

3. Prove the uniqueness of the vector a in Theorem 6.4.6.

4. Prove that for $0 \le K \le N$ the mapping $a \mapsto {}^*a$ is a linear transformation of $\wedge^K \mathbb{R}^N$ into $\wedge^{N-K} \mathbb{R}^N$.

5. (a) Show that for $\mathbb{R}^2$ we have ${}^*e_1 = e_2$ and ${}^*e_2 = -e_1$.

   (b) Show that for every 1-vector a in $\mathbb{R}^2$ we have ${}^{**}a = -a$.

6. (a) For $\mathbb{R}^3$ compute ${}^*e_i$ and ${}^*(e_i \wedge e_j)$ where $i, j = 1, 2, 3$ and $i < j$.

   (b) Show that for every 1-vector $a \in \mathbb{R}^3$ we have ${}^{**}a = a$.

7. State and prove suitable versions of Theorem 6.4.9 for the cases $K = 0, N$. Note that for $K = 0$ we replace $a_1 \wedge \cdots \wedge a_K$ by 1.

8. Suppose that $1 < K < N$. (a) Show that if $a_1, \ \ldots, \ a_N$ is an orthonormal basis for $\mathbb{R}^N$ such that $a_1 \wedge \cdots \wedge a_N = \lambda\, e_1 \wedge \cdots \wedge e_N$, then ${}^*(a_{K+1} \wedge a_{K+2} \wedge \cdots \wedge a_N) = \lambda\, (-1)^{K(N-K)}\, a_1 \wedge a_2 \wedge \cdots \wedge a_K$. (b) Show that for any K-vector a in $\mathbb{R}^N$ we have ${}^{**}a = (-1)^{K(N-K)} a$.

9. Prove that for $0 \le K \le N$ the transformation $a \mapsto {}^*a$ maps $\wedge^K \mathbb{R}^N$ *onto* $\wedge^{N-K} \mathbb{R}^N$.

10. Show that if a is a simple K-vector in $\mathbb{R}^N$, then ${}^*a$ is also simple.

11. If $a, x \in \mathbb{R}^N$ and $a \ne 0$, show that $a \cdot x = 0$ if and only if ${}^*a \wedge x = 0$.

12. A plane in $\mathbb{R}^3$ is defined by the equation $a \cdot x = 0$ where $a = e_1 + e_2 + e_3$. Find a 2-vector b such that the same plane is defined by the equation $b \wedge x = 0$.

13. Show that for $0 \le K \le N$, the following results hold for $\mathbb{R}^N$: (a) For all K-vectors a and b we have $({}^*a) \cdot ({}^*b) = a \cdot b$. (b) For each K-vector a we have $|{}^*a| = |a|$. (c) For each K the transformation $a \mapsto {}^*a$ is continuous.

7

VECTOR ANALYSIS ON MANIFOLDS

7.1 Oriented Manifolds and Differential Forms

Two central ideas of this chapter are *orientation* and *vector field*. When we studied integrals of real-valued functions over manifolds, neither of these ideas were used. Yet orientations and vector fields often play important roles in integrals over curves, surfaces and higher dimensional manifolds. For example, when computing work done by a particle moving along a curve C through a potential field ϕ, we have

$$\int_C (\nabla\phi)\cdot T = \phi\,(\text{terminal point}) - \phi\,(\text{initial point})$$

where T is a unit tangent vector to C. Or perhaps the reader is familiar with the classical theorems of vector analysis, Green's theorem, Gauss' divergence theorem, and Stokes' theorem. He or she perhaps knows something of their importance in such fields as fluid mechanics and electromagnetism.

This chapter generalizes the concepts of orientation and vector field to a form appropriate for manifolds and proves a very central result known as the generalized Stokes' theorem (or Stokes' theorem for manifolds). We derive the classical theorems of vector analysis as corollaries of the generalized Stokes' theorem.

The machinery that we will develop, the theory of differential forms on manifolds, finds applications in many other areas such as differential geometry, Riemannian geometry, differential equations, fluid mechanics, electromagnetic field theory, and gravitation theory.

Warning: From this point on, whenever we mention K-manifolds, we shall mean C^r K-dimensional manifolds with $r \geq 1$. We shall similarly assume that all functions and all coordinate patches are C^r with $r \geq 1$.

Suppose p is a point on the K-manifold $\mathcal{M}$. Let $f : U \to \mathcal{M}$ be a coordinate patch with $f(q) = p$. In effect, f "pastes" the open set U onto $\mathcal{M}$ in a "smooth", continuously differentiable fashion. A vector of the form

$$\frac{f(q+\lambda y) - f(q)}{\lambda},$$

where λ is a real number and y is a vector in $\mathbb{R}^K$, can be pictured as a directed line segment beginning at $p = f(q)$ and can be thought of as becoming a tangent vector to $\mathcal{M}$ at the point p as $\lambda \to 0$. This leads to the idea that vectors of the form $f'(q)(y)$ must be tangent vectors to $\mathcal{M}$ at the point p. Let us make this official and generalize the idea to K-vectors.

Definition 7.1.1 Let p be a point on the K-manifold $\mathcal{M} \subseteq \mathbb{R}^N$. For $0 \le L \le K$ we say $w \in \wedge^L \mathbb{R}^N$ is a *tangent L-vector to $\mathcal{M}$ at p* if it has the form

$$w = \left(\wedge^L f'(q)\right)(v)$$

for some coordinate patch $f : U \to \mathcal{M}$ such that $f(q) = p$ and some $v \in \wedge^L \mathbb{R}^K$. Tangent 1-vectors will be called simply *tangent vectors.*

Example 7.1.1 Let $U = \mathcal{M} = \mathbb{R}^N$ and $f \colon U \to \mathbb{R}^N$ be the identity map. Then at every point $p \in \mathbb{R}^N$ we see that

$$e_1 \wedge \cdots \wedge e_N = \left(\wedge^N f'(p)\right)(e_1 \wedge \cdots \wedge e_N)$$

is a tangent N-vector.

Example 7.1.2 Recall that S^2 is the unit 2-sphere in $\mathbb{R}^3$. Define $g \colon \mathbb{R}^2 \to S^2$ by

$$g(\alpha, \beta) = (\cos\alpha\ \cos\beta\ , \cos\alpha\ \sin\beta\ ,\ \sin\alpha\ ).$$

It is straightforward to calculate $\mathcal{D}(g'(\alpha, \beta)) = |\cos\alpha| \neq 0$ provided α is not of the form $n\pi/2$ where n is an odd integer. If the domain of g is restricted to sets of the form $(-\frac{\pi}{2}, \frac{\pi}{2}) \times (\beta_0, \beta_1)$, then g can be used to define coordinate patches covering all of S^2 except the points $(0, 0, \pm 1)$.

For such coordinate patches, $[g'(\alpha, \beta)]\ (e_1)$ and $[g'(\alpha, \beta)]\ (e_2)$ will be tangent 1-vectors to S^2 at the point $p = g(\alpha, \beta)$. Since

$$[g'(\alpha, \beta)] = \begin{pmatrix} -\sin\alpha\ \cos\beta & -\cos\alpha\ \sin\beta \\ -\sin\alpha\ \sin\beta & \cos\alpha\ \cos\beta \\ \cos\alpha & 0 \end{pmatrix},$$

we can calculate that a tangent 2-vector at $p = (x, y, z) = g(\alpha, \beta)$ is given by

$$\begin{aligned} w &= [\wedge^2 g'(\alpha, \beta)]\ (e_1 \wedge e_2) \\ &= -\cos\alpha \left(z\ (e_1 \wedge e_2) - y\ (e_1 \wedge e_3) + x\ (e_2 \wedge e_3)\right). \end{aligned}$$

(The calculations are left for the reader.) See Figure 7.1.1.

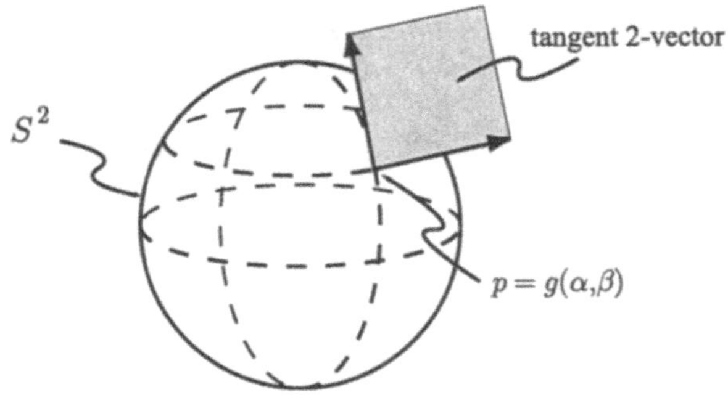

FIGURE 7.1.1.

For defining a tangent L-vector to $\mathcal{M}$ at p, one can use any coordinate patch which contains p; it does not have to be a particular one. To see this, let x be a tangent L-vector to $\mathcal{M}$ at p. There must exist some coordinate patch $f: U \to \mathcal{M}$ and some $y \in \wedge^L \mathbb{R}^K$ such that $f(q) = p$ and $x = \left(\wedge^L f'(q)\right)(y)$. Let $g : V \to \mathcal{M}$ be an arbitrary coordinate patch with the property that $g(r) = p$ for some r. If we set

$$z = \left(\wedge^L (g^{-1} \circ f)'(q)\right)(y),$$

then we easily obtain (this is a signal that this is left as an exercise)

$$x = \left(\wedge^L g'(r)\right)(z).$$

We now turn to the subject of orientable manifolds.

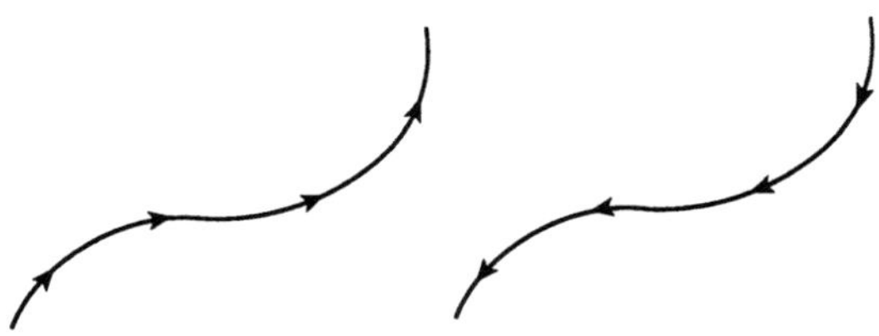

FIGURE 7.1.2.

By a *curve* in X we mean a function $f : J \to X$ where J is an interval in $\mathbb{R}$, where X is $\mathbb{R}^N$ or some manifold, and where f is a C^r function. (More precisely, we could say that f is C^r curve. We shall assume for all our curves that $r \geq 1$.) Intuitively one readily pictures the curve as a set of points in X which is traced by the point $f(t)$ as t increases. If $f'(t)$ is never zero, then the tracing point $f(t)$ can be thought of as moving steadily in a single direction along the curve. This action of $f(t)$, the action of moving in a single direction along the curve, assigns to the curve an orientation. Clearly there are only two directions in which a tracing point can move along a curve, so the curve can have one of two possible orientations. See Figure 7.1.2. An approach to orientation which is equivalent to this is to assign unit tangent vectors to every point of the curve in a *continuous* fashion. There are only two ways to do this. See Figure 7.1.3.

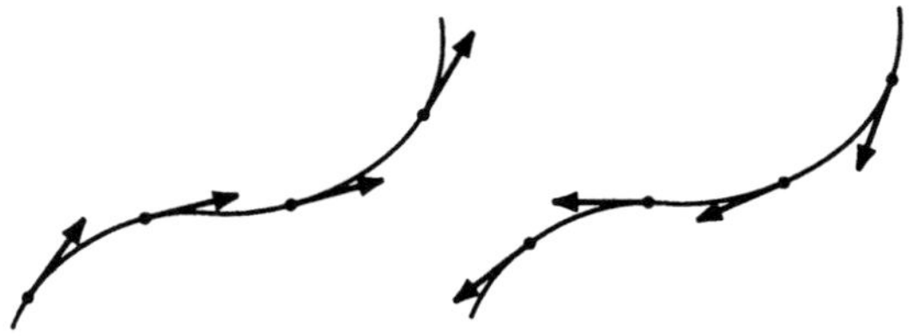

FIGURE 7.1.3.

This discussion will guide us when we consider the problem of assigning an orientation to a K-manifold.

Theorem 7.1.1 *A K-manifold $\mathcal{M}$ possesses exactly two unit tangent K-vectors at every point.*

Proof. Suppose u and v are unit tangent K-vectors to $\mathcal{M}$ at the point p. Let $f : U \to \mathcal{M}$ be a coordinate patch such that $f(q) = p$. There must exist $x, y \in \wedge^K \mathbb{R}^K$ such that

$$u = \left(\wedge^K f'(q)\right)(x) \quad \text{and} \quad v = \left(\wedge^K f'(q)\right)(y).$$

We know x and y must have the form

$$x = \alpha\,(e_1 \wedge \cdots \wedge e_K) \quad \text{and} \quad y = \beta\,(e_1 \wedge \cdots \wedge e_K)$$

where α and β are scalars and $\{e_1, \ldots, e_K\}$ is the standard basis for $\mathbb{R}^K$. It must be true that α and β are nonzero, since otherwise u or v or both would be zero. This means that

$$u = \frac{\alpha}{\beta} v.$$

But $|u| = |v| = 1$, so $\frac{\alpha}{\beta} = \pm 1$. Therefore, if u is a unit tangent K-vector to $\mathcal{M}$ at p, the only other unit tangent K-vector to $\mathcal{M}$ at p is $-u$. □

Definition 7.1.2 If $\mathcal{M}$ is a K-manifold in $\mathbb{R}^N$, by an *orientation* of $\mathcal{M}$ we mean a function $w : \mathcal{M} \to \wedge^K \mathbb{R}^N$ such that $w(p)$ is a unit tangent K-vector to $\mathcal{M}$ at p for all $p \in \mathcal{M}$ and w is continuous. If $\mathcal{M}$ possesses an orientation, it is called an *orientable* manifold. Otherwise it is *nonorientable*. By an *oriented manifold* we mean an orientable manifold $\mathcal{M}$ with a given orientation w.

If S is a subset of $\mathbb{R}^N$, then a function $f : S \to \wedge^K \mathbb{R}^N$ is a *K-vector field on S*. For $1 \le K \le N$, a K-vector field can be written in the form

$$f(x) = \sum_{i_1 < \cdots < i_K} f_{i_1 \ldots i_K}(x)\, e_{i_1} \wedge \cdots \wedge e_{i_K}$$

where the uniquely determined, real-valued functions $f_{i_1 \ldots i_K}(x)$ are the *component functions* of $f(x)$. We say f is continuous or a C^r function if and only if this is true for all its component functions. An orientation of $\mathcal{M}$ is a particular type of K-vector field on $\mathcal{M}$.

Example 7.1.3 The simplest example of an orientable manifold is $\mathbb{R}^N$. The standard orientation of $\mathbb{R}^N$ is taken to be $w(p) = e_1 \wedge e_2 \wedge \cdots \wedge e_N$. This is the same as the standard orientation of $\mathbb{R}^N$ considered as a vector space. Any subset of $\mathbb{R}^N$ which is an N-manifold, such as a closed ball or an open set, can also be given the orientation $w(p) = e_1 \wedge e_2 \wedge \cdots \wedge e_N$.

Example 7.1.4 Every K-dimensional sphere is an orientable manifold. Consider, for example, the unit sphere S^K in $\mathbb{R}^{K+1}$. Every point $p \in S^K$ may also be thought of as a vector, and if, starting at the point p, we draw a directed line segment to represent this vector, we see it will be a unit normal vector to the sphere. Choose $p \in S^K$ and let $\mathcal{T}$ be the set of all tangent vectors to S^K at the point p and let $\mathcal{P}$ be the set of all vectors that are orthogonal to p. We will show that $\mathcal{T} = \mathcal{P}$.

Suppose x is a tangent vector to S^K at the point p. Let $f : U \to S^K$ be a coordinate patch such that $f(q) = p$. For some u in $\mathbb{R}^K$ we must have $\left(f'(q)\right)(u) = x$. Because $f(r) \cdot f(r) = 1$ for all $r \in U$, it follows that $f(q) \cdot \left(f'(q)\right)(u) = 0$ (this is left as an exercise); that is, we have $p \cdot x = 0$. Thus $\mathcal{T} \subseteq \mathcal{P}$. Notice that $\mathcal{T}$ and $\mathcal{P}$ are linear subspaces of $\mathbb{R}^{K+1}$ and that $\left(f'(q)\right)(e_1), \ldots, \left(f'(q)\right)(e_K)$ are linearly independent vectors in $\mathcal{T}$. Since p is linearly independent of any vector in $\mathcal{P}$ and since $\mathbb{R}^{K+1}$ has dimension $K + 1$, it follows that $\mathcal{T} = \mathcal{P}$.

It is now easy to assign an orientation to S^K. The K-vector $^*p$ must have the form $a_1 \wedge \cdots \wedge a_K$ where each a_i is orthogonal to p, therefore $^*p$ will be a unit tangent K-vector to S^K at the point p. We may take $w(p) = ^* p$ as an orientation of S^K.

Example 7.1.5 Another example of an orientable manifold is the K-dimensional torus $T^K = S^1 \times \cdots \times S^1$ with K factors of the form S^1. We know S^1 has an orientation $w(p) = -y\, e_1 + x\, e_2$ where $p = (x, y)$ and $x^2 + y^2 = 1$. We use this to construct an orientation $t(p)$ for T^K as follows:

Each factor S^1 of T^K is a unit circle lying in a different copy of $\mathbb{R}^2$ so that T^K lies in $\mathbb{R}^{2K}$. Choose $p = (p_1, \ldots, p_K) \in T^K$ where each p_i lies in a different copy of S^1. For each i we can in turn write $p_i = (x_i, y_i)$ where $x_i^2 + y_i^2 = 1$. We define within $\mathbb{R}^{2K}$ an orientation of each factor S^1 of T^K by

$$\begin{aligned} w_1(p_1) &= -y_1\, e_1 + x_1\, e_2, \\ w_2(p_2) &= -y_2\, e_3 + x_2 e_4, \\ &\cdots \\ w_K(p_K) &= -y_K\, e_{2K-1} + x_K\, e_{2K}. \end{aligned}$$

Then an orientation for T^K is given by

$$t(p_1, p_2, \ldots, p_K) = w_1(p_1) \wedge w_2(p_2) \wedge \cdots \wedge w_K(p_K).$$

This illustrates a construction which leads to the following result (the proof being left as an exercise):

Theorem 7.1.2 *If $\mathcal{M}$ and $\mathcal{N}$ are orientable manifolds, at least one of which has an empty boundary, then $\mathcal{M} \times \mathcal{N}$ is also an orientable manifold.*

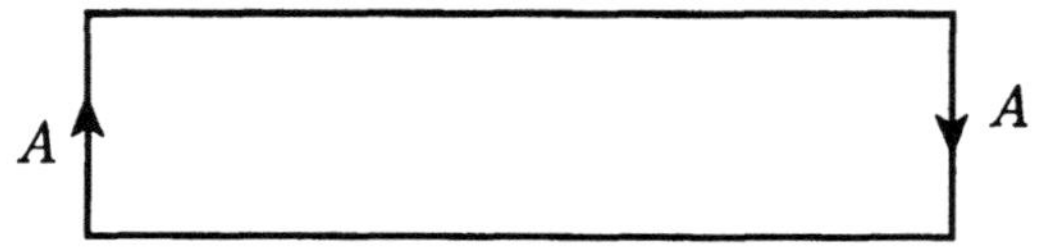

FIGURE 7.1.4.

Example 7.1.6 The standard example of a nonorientable manifold is a Möbius strip. This can be visualized as a rectangular strip of paper in which two opposite ends have been pasted together after a half twist. The way in which the two ends are pasted together is indicated by orientations of the ends labelled A in Figure 7.1.4. The result of this operation, the Möbius strip, is indicated in Figure 7.1.5.

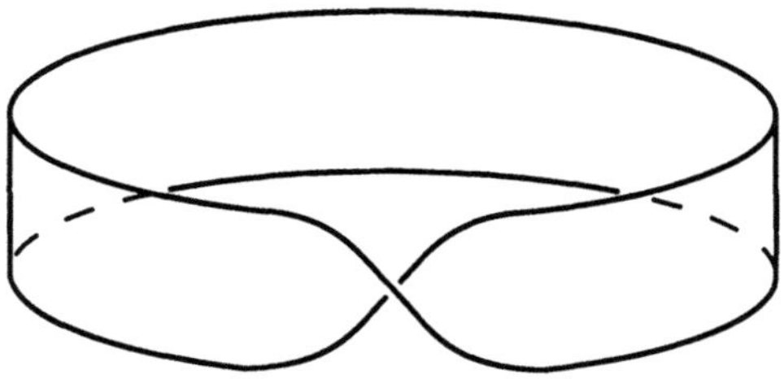

FIGURE 7.1.5.

To see why the Möbius strip is nonorientable, consider a "right-handed" ordered pair of vectors moving from left to right along the Möbius strip in Figure 7.1.6. Such an ordered pair represents a 2-vector. As a "right-handed" ordered pair moves off the right end of the strip in Figure 7.1.6, it comes back on the left end of the strip as a "left-handed" ordered pair of vectors. This implies that a continuous, unit 2-vector field cannot be defined on the Möbius strip since it would have to take on different values at the same point.

FIGURE 7.1.6.

We will say a manifold is *connected* provided that for every pair of points in the manifold there is a curve in the manifold that runs from one point to the other. That is, $\mathcal{M}$ is connected if for every $p, q \in \mathcal{M}$ there is a curve $f : [0, 1] \to \mathcal{M}$ such that $f(0) = p$ and $f(1) = q$. (Remember that f must be at least a C^1 function.)

Theorem 7.1.3 *Every orientable, connected manifold has exactly two orientations.*

Proof. Let w_1 and w_2 be two orientations of the orientable, connected manifold $\mathcal{M}$. Choose $p, q \in \mathcal{M}$. Let $f : [0, 1] \to \mathcal{M}$ be a curve such that $f(0) = p$ and $f(1) = q$. The function $h(t) = |w_1(f(t)) - w_2(f(t))|$ is continuous and real-valued.

Since for every r either $w_1(r) = w_2(r)$ or $w_1(r) = -w_2(r)$, it follows that $h(t)$ can take on only the values of 0 or 2. By continuity, we must have $h(0) = h(1)$. Since p and q were arbitrarily chosen points from the manifold, we see that $|w_1(r) - w_2(r)|$ must be a constant function on $\mathcal{M}$. Therefore either $w_1 = w_2$ or $w_1 = -w_2$. Thus $\mathcal{M}$ has exactly two orientations. □

Now let us consider two types of integrals which arise in vector analysis and the connection with orientable manifolds.

Line integrals in $\mathbb{R}^2$ or $\mathbb{R}^3$ have the form

$$\int_C f(x, y)\,dx + g(x, y)\,dy \quad \text{or} \quad \int_C f(x, y, z)\,dx + g(x, y, z)\,dy + h(x, y, z)\,dz.$$

These may both be rewritten in the form

$$\int_C F \cdot T\,ds$$

where F defines a vector field on the region of space through which the curve C passes and T is a function which attaches to every point of C a unit tangent vector pointing in the direction of the orientation of C. A reasonable interpretation is to think of F as a field of force and the integral as giving the amount of work done by a particle moving along the curve C in the direction indicated by the tangent vector field T.

Integrals over surfaces in $\mathbb{R}^3$ are often written in the form

$$\iint_S F \cdot n\,dA$$

where S is a surface and n is a unit normal vector to S defined in a continuous manner. One might think of $F(p)$ as representing the amount of fluid per unit of time flowing through a unit area in the direction given by $F(p)/|F(p)|$. Then the integral is the amount of flow per unit of time through S. This integral is equal to

$$\iint_S \left({}^*F \cdot {}^*n\right)\,dA$$

and ${}^*n$ is a unit tangent 2-vector at every point of S, in other words, an orientation of S. It is even, in a way, more logical to use ${}^*F$ than F because ${}^*F$, being a 2-vector, has the dimensions of an area, which is what one might expect of a vector representing fluid flow through a unit of area.

We generalize these ideas.

Definition 7.1.3 By a *differential K-form* (or *differential form* or *K-form*) we mean simply a K-vector field. That is, a K-form on U, a subset of $\mathbb{R}^N$, is a function $f : U \to \wedge^K \mathbb{R}^N$. If f is a differential K-form defined on $\mathcal{M}$ where $\mathcal{M}$ is an oriented K-manifold, then we define *the integral of f over the oriented manifold $\mathcal{M}$* by

$$\int_{\mathcal{M}} f = \int_{\mathcal{M}} f \cdot w$$

where w is the orientation of $\mathcal{M}$. (Of course, $f \cdot w$ must be integrable as a real-valued function.)

Note that in the integral on the left, it is important that $\mathcal{M}$ be oriented and that the orientation be known; one might almost prefer to replace $\mathcal{M}$ by some symbol such as $\mathcal{M}_w$ or $(\mathcal{M}, w)$. This is a new type of integral, different from integrals we have considered before. On the other hand, the integral on the right is the integral of a real-valued function over a manifold, a previously defined concept. The orientation of $\mathcal{M}$ occurs in the form of w as part of the integrand, not as information incorporated into the domain of integration. It should always be clear from context as to which type of integration is meant.

(One might also wonder why a K-form is described as a *differential* form. It turns out that some of the most basic 1-forms, those defined by $f(p) = e_i$ for all p, can be thought of as differentials of certain real-valued functions, namely those functions which assign p its ith coordinate. Further clarification of this comment must wait on the introduction of the idea of the differential operator.)

Let us look more closely at the evaluation of integrals of differential forms. Suppose f is a K-form on a K-manifold $\mathcal{M}$ in $\mathbb{R}^N$ where $\mathcal{M}$ has orientation w. Let $g: U \to \mathcal{M}$ be a coordinate patch in $\mathcal{M}$ and denote $g(U)$ as W. By the definitions of integrals of forms and integrals of real-valued functions on manifolds, we must have

$$\int_W f = \int_W f \cdot w = \int_U (f \circ g) \cdot (w \circ g)\, \mathcal{D}(g') .$$

U must be an open set in $\mathbb{R}^K$, and for arbitrary $x \in U$, the tangent space to $\mathcal{M}$ at $g(x)$ must be K-dimensional. It follows that

$$\left[\wedge^K g'(x)\right] (e_1 \wedge \cdots \wedge e_K) = \left(g'(x)\, e_1\right) \wedge \cdots \wedge \left(g'(x)\, e_K\right) = \pm\mathcal{D}(g'(x))\; w(g(x)).$$

(See Theorem 6.2.5.) Let us assume that we have the plus sign in the last expression, and in this case we say that g is *orientation-preserving*. We then have

$$\int_W f = \int_U ((f \circ g)(x)) \cdot \left\{\left(g'(x)\, e_1\right) \wedge \cdots \wedge \left(g'(x)\, e_K\right)\right\}\, dx .$$

Example 7.1.7 Let U be an open set in $\mathbb{R}^N$. We regard this as a manifold with the (standard) orientation $e_1 \wedge \cdots \wedge e_N$ at every point. If f is an N-form on U, we must be able to write $f(x) = h(x)\, e_1 \wedge \cdots \wedge e_N$ where h is a real-valued function. Then

$$\int_U f = \int_U h\, (e_1 \wedge \cdots \wedge e_N) \cdot (e_1 \wedge \cdots \wedge e_N) = \int_U h .$$

The last two integrals are simply integrals of real-valued functions over a set without regard to orientation.

Example 7.1.8 Suppose $f = f_1\, e_1 + f_2\, e_2 + f_3\, e_3$ where each of f_1, f_2, f_3 is a real-valued function, i.e., f is a 1-form on $\mathbb{R}^3$. Let

$$\mathcal{M} = \{\, (\cos t, \sin t, t) \in \mathbb{R}^3 \;:\; 0 \le t \le 2\pi \,\},$$

an arc in $\mathbb{R}^3$, thus a 1-manifold with boundary. ($\mathcal{M}$ is a portion of a helix which spirals about the x_3-axis.) Define $g\colon [0, 2\pi] \to \mathcal{M}$ by $g(t) = (\cos t, \sin t, t)$. g is (essentially) a coordinate patch. Let us assume that at each point $g(t)$, $\mathcal{M}$ has the orientation of the tangent vector $[g'(t)]e_1$. (That is, g is orientation-preserving.) We easily compute that

$$[g'(t)]e_1 = -\sin t\; e_1 + \cos t\; e_2 + e_3\,.$$

(Note that the e_1 on the left lives in $\mathbb{R}$ while the e_1 on the right lives in $\mathbb{R}^3$.) Then

$$\begin{aligned}\int_{\mathcal{M}} f &= \int_0^{2\pi} [(f_1\circ g)(t)\; e_1 + (f_2\circ g)(t)\; e_2 + (f_3\circ g)(t)\; e_3] \\ &\qquad \cdot(-\sin t\; e_1 + \cos t\; e_2 + e_3)\; dt \\ &= \int_0^{2\pi} [-(f_1\circ g)(t)\;\sin t + (f_2\circ g)(t)\;\cos t + (f_3\circ g)(t)]\; dt.\end{aligned}$$

Example 7.1.9 Let f be a 2-form on $\mathbb{R}^3$. We can write

$$f = f_{12}\; e_1\wedge e_2 + f_{13}\; e_1\wedge e_3 + f_{23}\; e_2\wedge e_3$$

where f_{12}, f_{13}, and f_{23} are real-valued functions on $\mathbb{R}^3$. Let $\mathcal{M}$ be the 2-manifold which is the graph in $\mathbb{R}^3$ of the equation $x_3 = x_1^2 + x_2^2$. This is a paraboloid opening upward from the x_1x_2-plane and centered on the x_3-axis. Define a coordinate patch $g\colon \mathbb{R}^2 \to \mathcal{M}$ by

$$g(x_1, x_2) = \left(x_1, x_2, x_1^2 + x_2^2\right).$$

We assume that $\mathcal{M}$ has the orientation at each point $g(x)$ of the tangent 2-vector $[\wedge^2 g'(x)]\,(e_1\wedge e_2)$ (where $x = (x_1, x_2)$, a point in $\mathbb{R}^2$). The matrix of $g'(x)$ is

$$[g'(x)] = \begin{pmatrix} 1 & 0 \\ 0 & 1 \\ 2x_1 & 2x_2 \end{pmatrix},$$

so we must have

$$\begin{aligned}\int_{\mathcal{M}} f &= \int_{\mathbb{R}^2} [(f\circ g)(x)]\cdot[\wedge^2 g'(x)]\,(e_1\wedge e_2)\; dx \\ &= \int_{\mathbb{R}^2} [(f\circ g)(x)]\cdot\{\,[g'(x)]\,(e_1)\wedge[g'(x)]e_2\,\}\; dx \\ &= \int_{\mathbb{R}^2} [(f\circ g)(x)]\cdot\{\,(e_1 + 2x_1\; e_3)\wedge(e_2 + 2x_2\; e_3)\,\}\; dx \\ &= \int_{\mathbb{R}^2} [(f_{12}\circ g)(x)\; e_1\wedge e_2 + (f_{13}\circ g)(x)\; e_1\wedge e_3 \\ &\qquad + (f_{23}\circ g)(x)\; e_2\wedge e_3]\cdot(e_1\wedge e_2 + 2x_2\; e_1\wedge e_3 - 2x_1\; e_2\wedge e_3)\; dx \\ &= \int_{\mathbb{R}^2} [(f_{12}\circ g)(x) + 2x_2\;(f_{13}\circ g)(x) - 2x_1\;(f_{23}\circ g)(x)]\; dx.\end{aligned}$$

This last integral is, of course, meant to be evaluated as an ordinary double integral.

Theorem 7.1.4 *If f and g are differential forms, λ is a real number, and $\mathcal{M}$ is an oriented manifold, then*

(a) $\int_{\mathcal{M}} f + g = \int_{\mathcal{M}} f + \int_{\mathcal{M}} g$.

(b) $\int_{\mathcal{M}} \lambda f = \lambda \int_{\mathcal{M}} f$.

Exercises

1. Let $T^K = S^1 \times \cdots \times S^1$ be our K-dimensional torus in $\mathbb{R}^{2K}$ as described in Example 7.1.5. Set $t = (t_1, \ldots, t_K)$, an arbitrary point in $\mathbb{R}^K$, and define $g \colon \mathbb{R}^K \to T^K$ by
$$g(t) = (\cos(t_1), \sin(t_1), \cos(t_2), \sin(t_2), \ldots, \cos(t_K), \sin(t_K)).$$
Show that at every point $p \in T^K$, g can be used to define a coordinate patch containing p and that $[\wedge^K g'(t)](e_1 \wedge \cdots \wedge e_K)$ agrees with the orientation assigned to that point in Example 7.1.5.

2. Let p be a point on the K-manifold $\mathcal{M} \subseteq \mathbb{R}^N$. Let $f : U \to \mathcal{M}$ and $g : V \to \mathcal{M}$ be coordinate patches such that $f(q) = p = g(r)$. Let $y \in \wedge^L \mathbb{R}^K$ (where $0 \le L \le K$) and set $x = \left(\wedge^L f'(q)\right)(y)$ and $z = \left(\wedge^L (g^{-1} \circ f)'(q)\right)(y)$. Show that we must have $x = \left(\wedge^L g'(r)\right)(z)$.

3. If $f : \mathbb{R}^K \rightarrowtail \mathbb{R}^N$ is a C^1 function such that $f(r) \cdot f(r) = 1$ for all r in the domain of f, show that $f(r) \cdot \left(f'(r)\right)(u) = 0$ for all $u \in \mathbb{R}^K$.

4. Show that if $f(p)$ and $g(p)$ are C^r K- and L-vector fields, respectively, then $f(p) \wedge g(p)$ is a C^r $(K + L)$-vector field.

5. Suppose $\mathbb{R}^N = \mathbb{R}^K \times \mathbb{R}^M$. Assume a is a K-vector in $\mathbb{R}^K$ in the sense that $a \in \wedge^K \mathbb{R}^N$ and a is constructed from vectors in $\mathbb{R}^K$ using the operations $+$ and $\wedge$ where $\mathbb{R}^K$ is regarded as a linear subspace of $\mathbb{R}^N$. Suppose similarly that b is an M-vector in $\mathbb{R}^M$. If $|a| = 1$ and $|b| = 1$, show that $|a \wedge b| = 1$.

6. Prove that if $\mathcal{M}$ and $\mathcal{N}$ are orientable manifolds, at least one of which has an empty boundary, then so is $\mathcal{M} \times \mathcal{N}$.

7. Here is a simple parametrization of a Möbius strip $\mathcal{M}$ in $\mathbb{R}^4$:
$$f(\rho, \theta) = (\cos(\theta),\ \sin(\theta),\ \rho\ \cos(\theta/2),\ \rho\ \sin(\theta/2))$$
where $-1 \le \rho \le 1$ and $0 \le \theta \le 2\pi$.

   (a) Show that $\mathcal{M}$ amounts to a rectangular strip with two of its opposite ends identified after a half twist by showing that, for distinct (ρ_1, θ_1) and (ρ_2, θ_2) in the rectangular domain of f, we have $f(\rho_1, \theta_1) = f(\rho_2, \theta_2)$ precisely when the two points have the form $(\rho, 0)$ and $(-\rho, 2\pi)$.

(b) Show that $\left(\wedge^2 f'(0,\theta)\right)(e_1 \wedge e_2)$ is a unit tangent 2-vector to $\mathcal{M}$ at the point $f(0,\theta)$.

(c) Prove that $\mathcal{M}$ is nonorientable.
(Hint: Suppose $w : \mathcal{M} \to \wedge^2\mathbb{R}^4$ is an orientation of $\mathcal{M}$, and consider $|w(f(0,\theta)) - \left(\wedge^2 f'(0,\theta)\right)(e_1 \wedge e_2)|$.)

8. If the differential form w is the orientation of the oriented manifold $\mathcal{M}$, then what is the geometric interpretation of $\int_{\mathcal{M}} w$?

9. Let $f = f_1\, e_1 + \cdots + f_N\, e_N$ be a 1-form in $\mathbb{R}^N$ and let $\mathcal{M}$ be a 1-manifold (curve) in $\mathbb{R}^N$ with orientation w. Suppose $g\colon (a,b) \to \mathcal{M}$ is a coordinate patch which covers all of $\mathcal{M}$ and is orientation-preserving, i.e., $[g'(t)]e_1 = \mathcal{D}(g'(t))\, w(g(t))$ for all $t \in (a,b)$. Show that

$$\int_{\mathcal{M}} f = \sum_{i=1}^{N} \int_a^b f_i(g(t))\, g_i'(t)\, dt$$

where $g = (g_1, \ldots, g_N)$.

10. Let $f = f_{12}\, e_1 \wedge e_2 + f_{13}\, e_1 \wedge e_3 + f_{23}\, e_2 \wedge e_3$ be a 2-form in $\mathbb{R}^3$ and let $\mathcal{M}$ be the 2-manifold defined by the equation $z = \sin(x+y)$. Let $g : \mathbb{R}^2 \to \mathcal{M}$ be the coordinate patch $g(x,y) = (x,\ y,\ \sin(x+y))$ and suppose $\mathcal{M}$ has an orientation such that g is orientation-preserving. Show that

$$\int_{\mathcal{M}} f = \int_{\mathbb{R}^2} \Big(f_{12}(x,\ y,\ \sin(x+y)) + \cos(x+y)\,(f_{13}(x,\ y,\ \sin(x+y)) - f_{23}(x,\ y,\ \sin(x+y)))\Big)\, dx\, dy.$$

11. Let $f = f_{12}\ e_1 \wedge e_2 + f_{13}\ e_1 \wedge e_3 + f_{23}\ e_2 \wedge e_3$ be a 2-form in $\mathbb{R}^3$ and let $\mathcal{M}$ be a 2-manifold in $\mathbb{R}^3$ with orientation w. Suppose $g\colon U \to \mathcal{M}$ is an orientation-preserving coordinate patch such that $U \subseteq \mathbb{R}^2$ and $g(U) = \mathcal{M}$. Show that

$$\int_{\mathcal{M}} f = \int_U \left\{ (f_{12} \circ g) \left(\frac{\partial g_1}{\partial x_1} \frac{\partial g_2}{\partial x_2} - \frac{\partial g_1}{\partial x_2} \frac{\partial g_2}{\partial x_1} \right) + (f_{13} \circ g) \left(\frac{\partial g_1}{\partial x_1} \frac{\partial g_3}{\partial x_2} - \frac{\partial g_1}{\partial x_2} \frac{\partial g_3}{\partial x_1} \right) + (f_{23} \circ g) \left(\frac{\partial g_2}{\partial x_1} \frac{\partial g_3}{\partial x_2} - \frac{\partial g_2}{\partial x_2} \frac{\partial g_3}{\partial x_1} \right) \right\}$$

where $g(x) = (g_1(x), g_2(x), g_3(x))$ and $x = (x_1, x_2)$.

12. Let $\mathcal{M}$ be the torus $S^1 \times S^1$ lying in $\mathbb{R}^4$ and let $g : [0, 2\pi] \times [0, 2\pi] \to \mathcal{M}$ be the map

$$g(\alpha, \beta) = (\cos\alpha, \sin\alpha, \cos\beta, \sin\beta).$$

We can, in effect, treat g as a coordinate patch covering all of $\mathcal{M}$; it satisfies all of the requirements of a coordinate patch except for a set of measure zero. Suppose further that $\mathcal{M}$ is an oriented manifold and that g is orientation-preserving. Show that if

$$f = \sum_{i<j} f_{ij} \quad e_i \wedge e_j$$

is a 2-form defined on $\mathcal{M}$, then

$$\int_{\mathcal{M}} f = \int_{[0,2\pi]^2} \Big(f_{13}(g(\alpha, \beta)) \sin\alpha \sin\beta - f_{14}(g(\alpha, \beta)) \sin\alpha \cos\beta$$
$$- f_{23}(g(\alpha, \beta)) \cos\alpha \sin\beta + f_{24}(g(\alpha, \beta)) \cos\alpha \cos\beta \Big) \, d\alpha \, d\beta.$$

13. Show that if f and g are differential forms, λ is a real number, and $\mathcal{M}$ is an oriented manifold, then

    (a) $\int_{\mathcal{M}} f + g = \int_{\mathcal{M}} f + \int_{\mathcal{M}} g$.
    (b) $\int_{\mathcal{M}} \lambda f = \lambda \int_{\mathcal{M}} f$.

14. If F is an integrable, real-valued function over the oriented manifold $\mathcal{M}$, then construct a differential form f on $\mathcal{M}$ such that

$$\int_{\mathcal{M}} F = \int_{\mathcal{M}} f.$$

15. Show that if p is a point on the K-manifold $\mathcal{M}$ in $\mathbb{R}^N$ and $\mathcal{T}_p$ = the set of tangent 1-vectors to $\mathcal{M}$ at p, then $\mathcal{T}_p$ is a K-dimensional linear subspace of $\mathbb{R}^N$.

16. Let $\mathcal{M}$ be an $(N-1)$-manifold in $\mathbb{R}^N$ defined by $f(x) = 0$ where f is a real-valued function.

    (a) If $p \in \mathcal{M}$ and $\mathcal{T}_p$ = the space of tangent 1-vectors to $\mathcal{M}$ at p, show that $v \cdot \nabla f(p) = 0$ for all $v \in \mathcal{T}_p$.
    (b) If ∇f is never zero on $\mathcal{M}$, show that $\mathcal{M}$ must be orientable.

7.2 Induced Orientation, the Differential Operator, and Stokes' Theorem; What We Can Learn From Simple Cubes

By I^N we mean the unit cube in $\mathbb{R}^N$, namely the set of points $(x_1, x_2, \ldots, x_N)$ such that $0 \leq x_i \leq 1$ for all i. These cubes are not manifolds, but they are enough alike that we shall feel no hesitation about attaching orientations to them. There are two choices:

$$w(p) = e_1 \wedge e_2 \wedge \cdots \wedge e_N \quad \text{or} \quad w(p) = -e_1 \wedge e_2 \wedge \cdots \wedge e_N.$$

(Later we shall assume the "standard" orientation of I^N is $w(p) = e_1 \wedge e_2 \wedge \cdots \wedge e_N$, but for now we permit ourselves the luxury of considering both possibilities.)

The unit cube has $2N$ faces of dimension $N-1$. Each of these is specified by setting some x_i equal to 0 or 1. Suppose p is a point in one of these $(N-1)$-dimensional faces, F. Let $n(p)$ be the unit vector directed outward from the unit cube, perpendicular to F. If I^N has orientation $e_1 \wedge \cdots \wedge e_N$, then we assign F to be the *induced orientation*

$$\partial w(p) = {}^*n(p).$$

If, on the other hand, I^N has orientation $-e_1 \wedge \cdots \wedge e_N$, we set

$$\partial w(p) = -{}^*n(p).$$

In this way we assign an induced orientation to each face of ∂I^N, the boundary of I^N.

Example 7.2.1 Consider I^2. Let its 1-dimensional faces be denoted as follows:

$$\begin{aligned} I_{0,1} &= \{(x_1, x_2) \colon\ x_1 = 0\} \\ I_{1,1} &= \{(x_1, x_2) \colon\ x_1 = 1\} \\ I_{0,2} &= \{(x_1, x_2) \colon\ x_2 = 0\} \\ I_{1,2} &= \{(x_1, x_2) \colon\ x_2 = 1\}. \end{aligned}$$

Suppose I^2 has orientation $w = e_1 \wedge e_2$. Then for each face we have the following:

$$\begin{aligned} I_{0,1} &:\ n = -e_1, \quad \partial w = {}^*n = -e_2 \\ I_{1,1} &:\ n = e_1, \quad \partial w = {}^*n = e_2 \\ I_{0,2} &:\ n = -e_2, \quad \partial w = {}^*n = e_1 \\ I_{1,2} &:\ n = e_2, \quad \partial w = {}^*n = -e_1. \end{aligned}$$

∂w amounts to a counterclockwise orientation of the boundary of I^2. If we had started with $w = -e_1 \wedge e_2$, then we would find that the induced orientation ∂w is clockwise. See Figure 7.2.1.

Example 7.2.2 Consider I^3. Let its faces be defined by the following equations:

$$\begin{array}{llll} I_{0,1}\colon & x_1 = 0 & I_{1,1}\colon & x_1 = 1 \\ I_{0,2}\colon & x_2 = 0 & I_{1,2}\colon & x_2 = 1 \\ I_{0,3}\colon & x_3 = 0 & I_{1,3}\colon & x_3 = 1. \end{array}$$

Suppose I^3 has the orientation $w = e_1 \wedge e_2 \wedge e_3$. Then each face must have the following induced orientation ∂w:

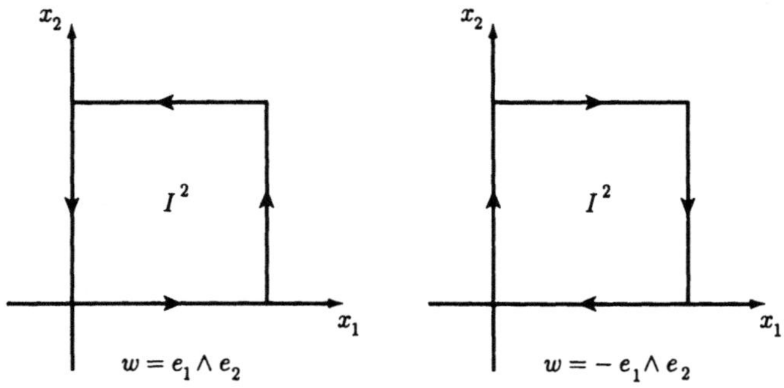

FIGURE 7.2.1.

$$
\begin{array}{lll}
I_{0,1}: & n = -e_1 & \partial w = {}^*n = -e_2 \wedge e_3 \\
I_{1,1}: & n = e_1 & \partial w = {}^*n = e_2 \wedge e_3 \\
I_{0,2}: & n = -e_2 & \partial w = {}^*n = e_1 \wedge e_3 \\
I_{1,2}: & n = e_2 & \partial w = {}^*n = -e_1 \wedge e_3 \\
I_{0,3}: & n = -e_3 & \partial w = {}^*n = -e_1 \wedge e_2 \\
I_{1,3}: & n = e_3 & \partial w = {}^*n = e_1 \wedge e_2.
\end{array}
$$

We may think of this induced orientation as imparting a certain "handedness" or "twist" (like a corkscrew) to each 2-dimensional face of I^3. See Figure 7.2.2.

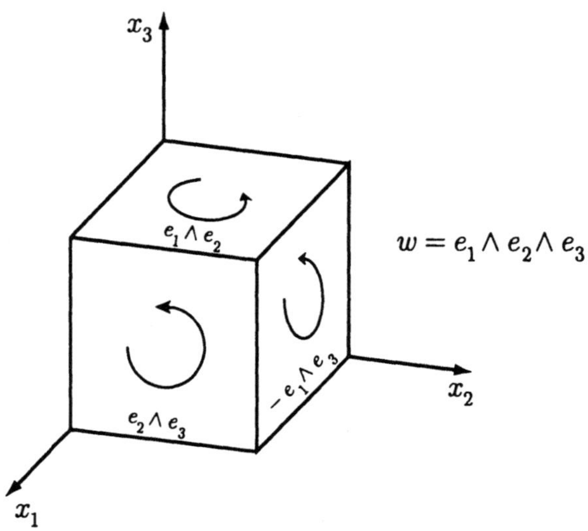

FIGURE 7.2.2.

We return for a moment to I^N, the unit cube in $\mathbb{R}^N$, and denote its $(N-1)$-dimensional faces as follows:

$$I_{0,i} = \{(x_1, \ldots, x_N) : x_i = 0\}$$
$$I_{1,i} = \{(x_1, \ldots, x_N) : x_i = 1\}$$

for $i = 1, \ldots, N$. It is straightforward to show that if I^N has orientation $w = e_1 \wedge \cdots \wedge e_N$, then each $I_{\alpha,i}$ has induced orientation

$$\partial w = (-1)^{i+\alpha}\, e_1 \wedge \cdots \wedge \overline{e_i} \wedge \cdots \wedge e_N,$$

where $\overline{e_i}$ indicates the omission of e_i. (For instance, if I^3 has orientation $w = e_1 \wedge e_2 \wedge e_3$, then $I_{0,2}$ has induced orientation

$$\partial w = (-1)^{2+0}\, e_1 \wedge \overline{e_2} \wedge e_3 = e_1 \wedge e_3.)$$

Without committing ourselves to a definition, we consider how one might assign an induced orientation to some simple squares embedded in $\mathbb{R}^N$.

Example 7.2.3 Let $C = I^2 \times \{c\}$ where c is a given constant. That is,

$$C = \{(x_1, x_2, x_3) : 0 \le x_1, x_2 \le 1 \quad \text{and} \quad x_3 = c\}.$$

Suppose we assign C the orientation $w = e_1 \wedge e_2$. What is a reasonable way to assign an induced orientation to the four arcs which make up ∂C?

Define $h\colon I^2 \to C$ by $h(x_1, x_2) = (x_1, x_2, c)$. Note that for every x the linear transformation $h'(x)$ has the matrix

$$[h'(x)] = \begin{pmatrix} 1 & 0 \\ 0 & 1 \\ 0 & 0 \end{pmatrix}.$$

If I^2 has orientation $e_1 \wedge e_2$, then h can be thought of as carrying the orientation of I^2 to that of C because

$$\left(\wedge^2 h'(x)\right)(e_1 \wedge e_2) = e_1 \wedge e_2.$$

(Remember that e_1 and e_2 on the left-hand side of the equation live in $\mathbb{R}^2$, but the e_1 and e_2 on the right-hand side live in $\mathbb{R}^3$.) To put it differently, h defines an orientation-preserving parametrization of C.

We take the induced orientation on ∂C to be that which h carries over from the boundary of I^2. The faces of C and their respective induced orientations are given as follows:

$$h(I_{0,1}) : \quad \partial w = \left(h'(x)\right)(-e_2) = -e_2$$
$$h(I_{1,1}) : \quad \partial w = \left(h'(x)\right)(e_2) = e_2$$
$$h(I_{0,2}) : \quad \partial w = \left(h'(x)\right)(e_1) = e_1$$
$$h(I_{1,2}) : \quad \partial w = \left(h'(x)\right)(-e_1) = -e_1.$$

See Figure 7.2.3.

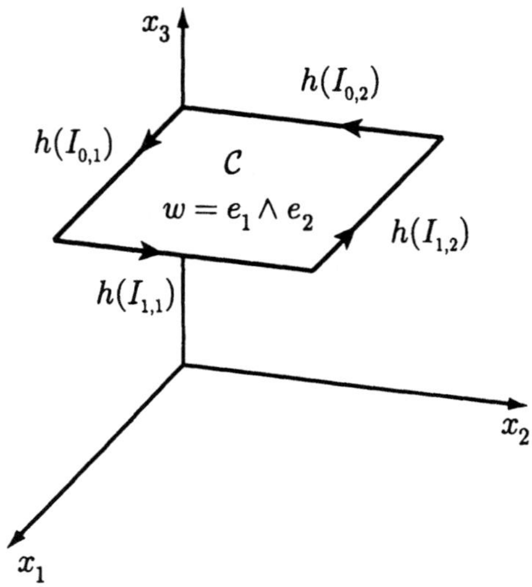

FIGURE 7.2.3.

Example 7.2.4 Let $C = \{(c_1, c_2)\} \times I^2$, that is

$$C = \{(x_1, x_2, x_3, x_4) \in \mathbb{R}^4 : x_1 = c_1, \quad x_2 = c_2, \quad \text{and } 0 \leq x_3, x_4 \leq 1\}$$

where c_1 and c_2 are given constants. Assume C has orientation $w = -e_3 \wedge e_4$. The boundary of this 2-dimensional square in $\mathbb{R}^4$ consists of four line segments:

$$\begin{aligned}
J_{0,3} &= \{(c_1, c_2, x_3, x_4) \in C : x_3 = 0\} \\
J_{1,3} &= \{(c_1, c_2, x_3, x_4) \in C : x_3 = 1\} \\
J_{0,4} &= \{(c_1, c_2, x_3, x_4) \in C : x_4 = 0\} \\
J_{1,4} &= \{(c_1, c_2, x_3, x_4) \in C : x_4 = 1\}.
\end{aligned}$$

We want to decide what would be a reasonable induced orientation for these line segments.

Define $h: I^2 \to C$ by $h(t_1, t_2) = (c_1, c_2, t_2, t_1)$ where I^2 has the usual orientation $e_1 \wedge e_2$. Since

$$[h'(x)] = \begin{pmatrix} 0 & 0 \\ 0 & 0 \\ 0 & 1 \\ 1 & 0 \end{pmatrix},$$

we see that

$$\left(\wedge^2 h'(x)\right)(e_1 \wedge e_2) = e_4 \wedge e_3 = -e_3 \wedge e_4,$$

so h is orientation-preserving.

To finish, we simply let h carry the induced orientation of ∂I^2 onto ∂C and thus find the induced orientation of each $J_{\alpha,i}$:

$$J_{0,3} = h(I_{0,2}) : \quad \partial w = \left(h'(x)\right)(e_1) = e_4$$

$$J_{1,3} = h(I_{1,2}) : \quad \partial w = \left(h'(x)\right)(-e_1) = -e_4$$

$$J_{0,4} = h(I_{0,1}) : \quad \partial w = \left(h'(x)\right)(-e_2) = -e_3$$

$$J_{1,4} = h(I_{1,1}) : \quad \partial w = \left(h'(x)\right)(e_2) = e_3.$$

(We leave it to the reader to sketch his or her own figure of this case.)

Now that we have the notion of induced orientation, we can begin to consider what amounts to a generalization of the fundamental theorem of calculus to higher dimensions. We will feel free to integrate differential forms over oriented K-dimensional cubes embedded in $\mathbb{R}^N$ in exactly the way we integrate them over oriented manifolds. But first some new notation.

Definition 7.2.1 Let $x_i : \mathbb{R}^N \to \mathbb{R}$ be the ith projection map, i.e., $x_i(t_1, \ldots, t_N) = t_i$. By dx_i we mean the constant 1-form $dx_i = e_i$.

This means that instead of writing a differential form as

$$f(x) = \sum_{i_1 < \cdots < i_K} f_{i_1 \ldots i_K}(x)\, e_{i_1} \wedge \cdots \wedge e_{i_K},$$

we shall write

$$f(x) = \sum_{i_1 < \cdots < i_K} f_{i_1 \ldots i_K}(x)\, dx_{i_1} \wedge \cdots \wedge dx_{i_K}.$$

This is the standard notation. As we shall see when we introduce the differential operator and differentials of real-valued functions, the differential of the projection map x_i will turn out to be the dx_i we introduced here. Some authors also leave out the wedge product symbols so that their differential forms look like this:

$$f(x) = \sum_{i_1 < \cdots < i_K} f_{i_1 \ldots i_K}(x)\, dx_{i_1} dx_{i_2} \ldots dx_{i_K}.$$

It may, however, be somewhat bewildering to students first encountering the theory of differential forms to see equations such as

$$dx_i dx_j = -dx_j dx_i.$$

We shall also feel free to use this notation when discussing orientations. For example, we can say the possible orientations of $\mathbb{R}^N$ are $dx_1 \wedge \cdots \wedge dx_N$ and $-dx_1 \wedge \cdots \wedge dx_N$.

It can be shown that for a $(K-1)$-form f defined on an oriented K-manifold $\mathcal{M}$ with boundary, there exists a K-form df such that

$$\int_{\partial\mathcal{M}} f = \int_{\mathcal{M}} df.$$

Accepting this for the moment, we would like to see how df is computed from f.

Example 7.2.5 Let $f = f_1 dx_1 + f_2 dx_2 + f_3 dx_3$, a 1-form on $\mathbb{R}^3$. Let $\mathcal{C} = I^2 \times \{c\}$ where c is a given constant. $\mathcal{C}$ is a square parallel to the x_1x_2-plane, and we assign it the orientation $dx_1 \wedge dx_2$. We write

$$\partial\mathcal{C} = I_{0,2} + I_{1,1} + I_{1,2} + I_{0,1}$$

and assign $\partial\mathcal{C}$ to be the induced orientation. See Figure 7.2.4. We then compute

$$\begin{aligned}\int_{\partial\mathcal{C}} f_1\,dx_1 &= \int_{I_{0,2}} f_1\,dx_1 + \int_{I_{1,1}} f_1\,dx_1 + \int_{I_{1,2}} f_1\,dx_1 + \int_{I_{0,1}} f_1\,dx_1 \\ &= \int_{I_{0,2}} f_1\,dx_1 + \int_{I_{1,2}} f_1\,dx_1\end{aligned}$$

(since $I_{1,1}$, $I_{0,1}$ have orientation $\pm dx_2$ and $dx_1 \cdot dx_2 = 0$)

$$= \int_0^1 f_1(x_1, 0, c)\,dx_1 - \int_0^1 f_1(x_1, 1, c)\,dx_1$$

(since $I_{1,2}$ has orientation $-dx_1$; note these are now integrals of functions over sets rather than integrals of differential forms)

$$\begin{aligned}&= -\int_0^1 \left(\int_0^1 \frac{\partial f_1}{\partial x_2}(x_1, x_2, c)\,dx_2\right) dx_1 \\ &= \int_{\mathcal{C}} \frac{\partial f_1}{\partial x_2}\,dx_2 \wedge dx_1\end{aligned}$$

(since $(dx_2 \wedge dx_1)\cdot(dx_1 \wedge dx_2) = -1$; note that we again deal with integrals of differential forms)

$$= \int_{\mathcal{C}} \left(\frac{\partial f_1}{\partial x_1}\,dx_1 + \frac{\partial f_1}{\partial x_2}\,dx_2 + \frac{\partial f_1}{\partial x_3}\,dx_3\right) \wedge dx_1$$

(since $dx_1 \wedge dx_1 = 0$ and $(dx_3 \wedge dx_1)\cdot(dx_1 \wedge dx_2) = 0$).

Similarly, one can show

$$\int_{\partial\mathcal{C}} f_2\,dx_2 = \int_{\mathcal{C}} \left(\frac{\partial f_2}{\partial x_1}\,dx_1 + \frac{\partial f_2}{\partial x_2}\,dx_2 + \frac{\partial f_2}{\partial x_3}\,dx_3\right) \wedge dx_2$$

and

$$\int_{\partial\mathcal{C}} f_3\,dx_3 = \int_{\mathcal{C}} \left(\frac{\partial f_3}{\partial x_1}\,dx_1 + \frac{\partial f_3}{\partial x_2}\,dx_2 + \frac{\partial f_3}{\partial x_3}\,dx_3\right) \wedge dx_3,$$

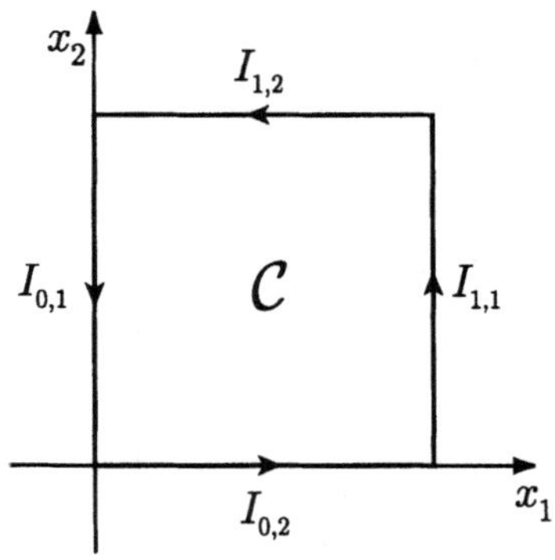

FIGURE 7.2.4.

though in these cases both integrals are zero. Putting these three equalities together, we obtain

$$\int_{\partial C} f = \int_{\partial C} \sum_{i=1}^{3} f_i \, dx_i = \int_C \sum_{i=1}^{3} \left(\sum_{j=1}^{3} \frac{\partial f_i}{\partial x_j} \, dx_j \right) \wedge dx_i .$$

This is very suggestive, but it is only natural to try some variations before forming a conclusion.

Example 7.2.6 If we take as before $C = I^2 \times \{c\}$ but this time endow C with the orientation $-dx_1 \wedge dx_2$, then it is straightforward to show

$$\int_{\partial C} f = \int_{\partial C} \sum_{i=1}^{3} f_i \, dx_i = \int_C \sum_{i=1}^{3} \left(\sum_{j=1}^{3} \frac{\partial f_i}{\partial x_j} \, dx_j \right) \wedge dx_i$$

still holds. (For example,

$$\begin{aligned} \int_{\partial C} f_1 \, dx_1 &= \int_0^1 \left(\int_0^1 \frac{\partial f_1}{\partial x_2}(x_1, x_2, c) \, dx_2 \right) dx_1 \\ &= \int_C \left(\frac{\partial f_1}{\partial x_1} dx_1 + \frac{\partial f_1}{\partial x_2} dx_2 + \frac{\partial f_1}{\partial x_3} dx_3 \right) \wedge dx_1 . \end{aligned}$$

Keep in mind, the orientation of ∂C is also reversed.)

Example 7.2.7 Suppose we take C to be a square parallel to the $x_1 x_3$-plane, say,

$$C = I \times \{c\} \times I = \{(x_1, x_2, x_3) : 0 \le x_1, x_3 \le 1 \text{ and } x_2 = c\}$$

where c is a given constant, and give it the orientation $dx_1 \wedge dx_3$. Then

$$\partial C = I_{0,3} + I_{1,1} + I_{1,3} + I_{0,1}$$

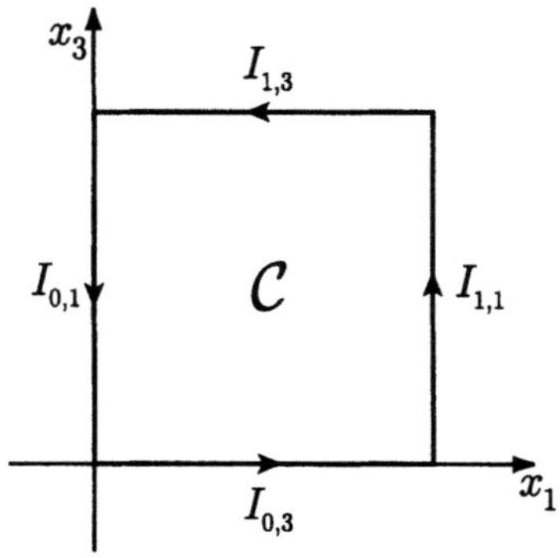

FIGURE 7.2.5.

where the induced orientation is as indicated in Figure 7.2.5. Then for a 1-form $f = f_1 dx_1 + f_2 dx_2 + f_3 dx_3$ we again have, by a straightforward calculation,

$$\int_{\partial C} f = \int_C \sum_{i=1}^{3} \left(\sum_{j=1}^{3} \frac{\partial f_i}{\partial x_j} \, dx_j \right) \wedge dx_i.$$

(For example,

$$\begin{aligned}\int_{\partial C} f_1 \, dx_1 &= -\int_0^1 \left(\int_0^1 \frac{\partial f_1}{\partial x_3}(x_1, c, x_3) \, dx_3 \right) dx_1 \\ &= \int_C \left(\frac{\partial f_1}{\partial x_1} dx_1 + \frac{\partial f_1}{\partial x_2} \, dx_2 + \frac{\partial f_1}{\partial x_3} \, dx_3 \right) \wedge dx_1 .)\end{aligned}$$

We see then that if $f = \sum_{i=1}^{3} f_i \, dx_i$, we want to set

$$df = \sum_{i=1}^{3} \left(\sum_{j=1}^{3} \frac{\partial f_i}{\partial x_j} \, dx_j \right) \wedge dx_i.$$

More generally, we make the following definition:

Definition 7.2.2 (The differential operator) If

$$f(x) = \sum_{i_1 < \cdots < i_K} f_{i_1 \ldots i_K}(x) \, dx_{i_1} \wedge \cdots \wedge dx_{i_K}$$

is a K-form defined in $\mathbb{R}^N$ and each $f_{i_1 \ldots i_K}$ is differentiable, then we define a $(K+1)$-form df by

$$df(x) = \sum_{i_1 < \cdots < i_K} \sum_{j=1}^{N} \left(\frac{\partial f_{i_1 \ldots i_K}}{\partial x_j} \right) (x) \;\; dx_j \wedge dx_{i_1} \wedge \cdots \wedge dx_{i_K}.$$

Example 7.2.8 If f is a 0-form, that is, if $f : \mathbb{R}^N \rightarrowtail \mathbb{R}$, then

$$df = \sum_{i=1}^{N} \frac{\partial f}{\partial x_i}\, dx_i = \nabla f.$$

Note in particular that if we interpret x_j as the function $x_j : \mathbb{R}^N \to \mathbb{R}$ which assigns each point its jth coordinate (that is, $x_j(p_1, p_2, \ldots, p_N) = p_j$), then we have

$$dx_j = \sum_{i=1}^{N} \frac{\partial x_j}{\partial x_i}\, dx_i$$

where

$$\frac{dx_j}{dx_i} = \begin{cases} 1 & \text{if } i = j \\ 0 & \text{otherwise.} \end{cases}$$

Thus we may interpret dx_j as either a special differential form or the result of d operating on x_j.

Example 7.2.9 If f is a 1-form in $\mathbb{R}^3$ having the form

$$f = f_1\, dx_1 + f_2\, dx_2 + f_3\, dx_3,$$

where each f_i is a real-valued function, then

$$df = \left(\frac{\partial f_3}{\partial x_2} - \frac{\partial f_2}{\partial x_3} \right) dx_2 \wedge dx_3 + \left(\frac{\partial f_3}{\partial x_1} - \frac{\partial f_1}{\partial x_3} \right) dx_1 \wedge dx_3 + \left(\frac{\partial f_2}{\partial x_1} - \frac{\partial f_1}{\partial x_2} \right) dx_1 \wedge dx_2.$$

Theorem 7.2.1 *Let f and g be differential forms with differentiable coefficients and let $\lambda \in \mathbb{R}$.*

(a) $d(f + g) = df + dg$ and $d(\lambda f) = \lambda df$.

(b) If f is a K-form, then $d(f \wedge g) = (df) \wedge g + (-1)^K f \wedge (dg)$.

(c) If f is C^2, then $d^2 f = 0$.

Proof of (c). By (a) we need consider only the case where $f(x) = h(x)\, dx_{i_1} \wedge \cdots \wedge dx_{i_K}$ defined on some open subset of $\mathbb{R}^N$ and h is a C^2 real-valued function. We know that

$$df(x) = \sum_{i=1}^{N} \frac{\partial h}{\partial x_i}\, dx_i \wedge dx_{i_1} \wedge dx_{i_2} \wedge \cdots \wedge dx_{i_K}.$$

Then

$$d^2 f(x) = \sum_{i=1}^{N} \sum_{j=1}^{N} \frac{\partial^2 h}{\partial x_j\, \partial x_i}\, dx_j \wedge dx_i \wedge dx_{i_1} \wedge dx_{i_2} \wedge \cdots \wedge dx_{i_K}.$$

Since $dx_i \wedge dx_i = 0$ and $dx_i \wedge dx_j = -dx_j \wedge dx_i$ for $i \neq j$ and $\frac{\partial^2 h}{\partial x_j \, \partial x_i} = \frac{\partial^2 h}{\partial x_i \, \partial x_j}$, it follows that $d^2 f(x) = 0$. □

We close this section by proving Stokes' theorem on I^N. In what follows, by $\int_{\partial I^N} g$, where g is a differential form, we mean

$$\sum_{i=1}^{N} \sum_{\alpha=0,1} \int_{I_{\alpha,i}} g$$

where each $I_{\alpha,i}$ has the orientation induced by the orientation of I^N.

Theorem 7.2.2 (Stokes' theorem for I^N) *Let f be a C^1 $(N-1)$-form on I^N. Assuming that I^N is oriented and ∂I^N has the induced orientation, we have*

$$\int_{I^N} df = \int_{\partial I^N} f.$$

Proof. Let us take the orientation of I^N to be $w = dx_1 \wedge \cdots \wedge dx_N$. (The proof is almost identical when the orientation is $-dx_1 \wedge \cdots \wedge dx_N$.) It is sufficient to prove the theorem for the case where $f = g \; dx_1 \wedge \cdots \wedge \overline{dx_i} \wedge \cdots \wedge dx_N$ and g is a real-valued C^1 function on I^N. In this case we have $df = \frac{\partial g}{\partial x_i} \, dx_i \wedge dx_1 \wedge \cdots \wedge \overline{dx_i} \wedge \cdots \wedge dx_N$. Then

$$\begin{aligned}
\int_{I^N} df &= \int_{I^N} \frac{\partial g}{\partial x_i} \; dx_i \wedge dx_1 \wedge dx_2 \wedge \cdots \wedge \overline{dx_i} \wedge \cdots \wedge dx_N \\
&= \int_{I^N} \frac{\partial g}{\partial x_i} \; (dx_i \wedge dx_1 \wedge dx_2 \wedge \cdots \wedge \overline{dx_i} \wedge \cdots \wedge dx_N) \\
&\quad \cdot (dx_1 \wedge \cdots \wedge dx_N) \\
&= (-1)^{i+1} \int_0^1 \cdots \int_0^1 \frac{\partial g}{\partial x_i}(x_1, \ldots, x_N) \, dx_1 \ldots dx_N \\
&= (-1)^{i+1} \int_0^1 \cdots \int_0^1 \Big\{ g(x_1, \ldots, x_{i-1}, 1, x_{i+1}, \ldots, x_N) \\
&\quad - g(x_1, \ldots, x_{i-1}, 0, x_{i+1}, \ldots, x_N) \Big\} \; dx_1 \ldots \overline{dx_i} \ldots dx_N
\end{aligned}$$

where the last step is a result of applying the fundamental theorem of calculus with respect to the ith variable and $\overline{dx_i}$ indicates that the ith integration is not performed. Now let us consider the integral over ∂I^N.

$$\begin{aligned}
\int_{\partial I^N} f &= \int_{\partial I^N} g \; dx_1 \wedge \cdots \wedge \overline{dx_i} \wedge \cdots \wedge dx_N \\
&= \sum_{i=1}^{N} \sum_{\alpha=0,1} \int_{I_{\alpha,i}} g \; dx_1 \wedge \cdots \wedge \overline{dx_i} \wedge \cdots \wedge dx_N \\
&= \sum_{i=1}^{N} \sum_{\alpha=0,1} \int_{I_{\alpha,i}} g \; (dx_1 \wedge \cdots \wedge \overline{dx_i} \wedge \cdots \wedge dx_N) \cdot \partial w \, .
\end{aligned}$$

We know that the induced orientation on $I_{\alpha,j}$ is given by

$$\partial w(p) = (-1)^{j+\alpha}\, dx_1 \wedge \cdots \wedge dx_{j-1} \wedge \overline{dx_j} \wedge dx_{j+1} \wedge \cdots \wedge dx_N.$$

It is clear there are only two terms of the series which do not vanish, namely the integrals over $I_{0,i}$ or $I_{1,i}$. We must have

$$\begin{aligned}\int_{\partial I^N} g\; dx_1 \wedge \cdots \wedge \overline{dx_i} \wedge \cdots \wedge dx_N &= (-1)^{i+1}\left\{\int_{I_{1,i}} g - \int_{I_{0,i}} g\right\}\\ &= (-1)^{i+1}\int_0^1 \cdots \int_0^1 \{g(x_1, \ldots, x_{i-1}, 1, x_{i+1}, \ldots, x_N)\\ &\quad -g(x_1, \ldots, x_{i-1}, 0, x_{i+1}, \ldots, x_N)\}\, dx_1 \ldots \overline{dx_i} \ldots dx_N.\end{aligned}$$

Thus the theorem is established. □

Exercises

1. Prove that in $\mathbb{R}^N$ we have

$$^*e_i = (-1)^{i+1}\; e_1 \wedge \cdots \wedge e_{i-1} \wedge \overline{e_i} \wedge e_{i+1} \wedge \cdots \wedge e_N.$$

2. Show that $(e_i \wedge e_1 \wedge e_2 \wedge \cdots \wedge \overline{e_i} \wedge \cdots \wedge e_N) \cdot (e_1 \wedge \cdots \wedge e_N) = (-1)^{i+1}$.

3. If $f = f_1 dx_1 + f_2 dx_2 + f_3 dx_3$, a 1-form on $\mathbb{R}^3$, go through the derivation of $\int_C df = \int_{\partial C} f$ where

$$C = \{c\} \times I^2 = \{(x_1, x_2, x_3) :\; x_1 = c \;\text{ and }\; 0 \le x_2, x_3 \le 1\},$$

c a given constant, assuming C has the orientation $w = -dx_2 \wedge dx_3$.

4. If $f = f_1 dx_1 + f_2 dx_2 + f_3 dx_3 + f_4\, dx_4$, a 1-form on $\mathbb{R}^4$, go through the derivation of $\int_C df = \int_{\partial C} f$ where

$$\begin{aligned}C = \{a\} \times I^2 \times \{b\} = \{\,&(x_1, x_2, x_3, x_4) :\; x_1 = a\\ &\text{and }\; 0 \le x_2, x_3 \le 1 \;\text{ and }\; x_4 = b\,\},\end{aligned}$$

a and b given constants, assuming C has the orientation $w = dx_2 \wedge dx_3$.

5. Let f and g be differential forms with differentiable coefficients and let $\lambda \in \mathbb{R}$. Show that

   (a) $d(f + g) = df + dg$ and $d(\lambda f) = \lambda df$.

   (b) If f is a K-form, then $d(f \wedge g) = (df) \wedge g + (-1)^K f \wedge (dg)$.

6. (a) Show that the set of K-forms

$$f(x) = \sum_{i_1<\cdots<i_K} f_{i_1\ldots i_K}(x)\, dx_{i_1} \wedge dx_{i_2} \wedge \cdots \wedge dx_{i_K}$$

defined on an open subset U of $\mathbb{R}^N$ is a vector space over the field of reals.

   (b) Show that the set of K-forms $dx_{i_1} \wedge dx_{i_2} \wedge \cdots \wedge dx_{i_K}$, where $i_1 < \cdots < i_K$, does not constitute a basis for this vector space.

7. Verify that if f is a 1-form in $\mathbb{R}^3$ having the form

$$f = f_1\, dx_1 + f_2\, dx_2 + f_3\, dx_3\,,$$

where each f_i is a real-valued function, then

$$df = \left(\frac{\partial f_3}{\partial x_2} - \frac{\partial f_2}{\partial x_3} \right) dx_2 \wedge dx_3 + \left(\frac{\partial f_3}{\partial x_1} - \frac{\partial f_1}{\partial x_3} \right) dx_1 \wedge dx_3 + \left(\frac{\partial f_2}{\partial x_1} - \frac{\partial f_1}{\partial x_2} \right) dx_1 \wedge dx_2\,.$$

8. If f is a 2-form in $\mathbb{R}^3$ having the form

$$f = f_1\, dx_2 \wedge dx_3 - f_2\, dx_1 \wedge dx_3 + f_3\, dx_1 \wedge dx_2\,,$$

then show that

$$df = \left(\frac{\partial f_1}{\partial x_1} + \frac{\partial f_2}{\partial x_2} + \frac{\partial f_3}{\partial x_3} \right) dx_1 \wedge dx_2 \wedge dx_3\,.$$

7.3 Integrals and Pullbacks

We want to generalize Stokes' theorem to manifolds. Before we can do this, we need something like the change-of-variables theorem for integrals, but adapted to differential forms on manifolds. We begin with some new notation.

Definition 7.3.1 If $g : \mathbb{R}^M \rightarrowtail \mathbb{R}^N$ is a C^1 function, then for every point x in the domain of g and every K such that $0 \le K \le M$, we let $g_*(x)$ stand for the induced linear transformation

$$\wedge^K g'(x) : \ \wedge^K \mathbb{R}^M \to \wedge^K \mathbb{R}^N.$$

There is, of course, a different $g_*(x)$ for every choice of K, but in every particular instance, it should be clear from the context which K is meant. For $K = 0$, the induced linear transformation $\wedge^0 g'(x)$ is simply the identity map of $\mathbb{R}$ to $\mathbb{R}$. (The K-vector $[g_*(x)]\, v$, where v is in $\wedge^K \mathbb{R}^M$, is sometimes referred to as the *push-forward* of v by g at x.)

With this definition, our formula for integrals of K-forms over coordinate patches now becomes

$$\int_{g(U)} f = \int_U (f \circ g(x)) \cdot \Big(g_*(x)\Big)\,(e_1 \wedge \cdots \wedge e_K)\,dx$$

where U is a subset of $\mathbb{R}^K$ and g is orientation-preserving.

Example 7.3.1 Let g be the map of $\mathbb{R}^3$ to $\mathbb{R}^3$ given by

$$g(x, y, z) = \left(\frac{x}{a}, \frac{y}{b}, \frac{z}{c}\right)$$

where a, b, c are positive constants. Note that g is a linear transformation so that we must have $g'(x, y, z) = g$. We can describe $g_*(x, y, z)$ by showing how it transforms basis vectors. For 1-vectors we have

$$\begin{aligned}
\Big(g_*(x, y, z)\Big)\,(e_1) &= e_1/a,\\
\Big(g_*(x, y, z)\Big)\,(e_2) &= e_2/b,\\
\Big(g_*(x, y, z)\Big)\,(e_3) &= e_3/c.
\end{aligned}$$

Passing to 2-vectors, note that

$$\Big(g_*(x, y, z)\Big)\,(e_1 \wedge e_2) = \Big(g'(x, y, z)\Big)\,(e_1) \;\wedge\; (g'(x, y, z))\,(e_2) = (e_1 \wedge e_2)/ab.$$

Similarly,

$$\Big(g_*(x, y, z)\Big)\,(e_1 \wedge e_3) = (e_1 \wedge e_3)/ac$$

and

$$\Big(g_*(x, y, z)\Big)\,(e_2 \wedge e_3) = (e_2 \wedge e_3)/bc.$$

Of course for 3-vectors we have

$$\Big(g_*(x, y, z)\Big)\,(e_1 \wedge e_2 \wedge e_3) = (e_1 \wedge e_2 \wedge e_3)/abc.$$

Example 7.3.2 Suppose that g is the rotation of $\mathbb{R}^3$ about the x-axis by an angle of θ. Then

$$g(x, y, z) = (\,x, \quad y\,\cos(\theta) - z\,\sin(\theta), \quad y\,\sin(\theta) + z\,\cos(\theta)\,).$$

Since g is a linear transformation, we have $g'(x, y, z) = g$, so it follows trivially that

$$\begin{aligned}
\Big(g_*(x, y, z)\Big)\,(e_1) &= e_1,\\
\Big(g_*(x, y, z)\Big)\,(e_2) &= \cos(\theta)\,e_2 \;+\; \sin(\theta)\,e_3,\\
\Big(g_*(x, y, z)\Big)\,(e_3) &= -\sin(\theta)\,e_2 \;+\; \cos(\theta)\,e_3.
\end{aligned}$$

For 2-vectors we have

$$
\begin{aligned}
\Big(g_*(x,y,z)\Big)(e_1 \wedge e_2) &= e_1 \wedge (\cos(\theta)e_2 + \sin(\theta)e_3)\\
&= \cos(\theta)e_1 \wedge e_2 + \sin(\theta)e_1 \wedge e_3\,,\\
\Big(g_*(x,y,z)\Big)(e_1 \wedge e_3) &= e_1 \wedge (-\sin(\theta)\ e_2\ +\ \cos(\theta)\ e_3)\\
&= -\sin(\theta)e_1 \wedge e_2 + \cos(\theta)e_1 \wedge e_3\,,\\
\Big(g_*(x,y,z)\Big)(e_2 \wedge e_3) &= (\cos(\theta)e_2 + \sin(\theta)e_3) \wedge (-\sin(\theta)e_2 + \cos(\theta)e_3)\\
&= e_2 \wedge e_3\,.
\end{aligned}
$$

Finally, for 3-vectors, we have

$$
\begin{aligned}
&\Big(g_*(x,y,z)\Big)(e_1 \wedge e_2 \wedge e_3)\\
&\quad = e_1 \wedge (\cos(\theta)e_2 + \sin(\theta)e_3) \wedge (-\sin(\theta)e_2 + \cos(\theta)e_3)\\
&\quad = e_1 \wedge e_2 \wedge e_3\,.
\end{aligned}
$$

(This last equation also tells us that g, considered as a linear transformation, is orientation preserving.)

We return to our quest for a change-of-variables formula.

Suppose $\mathcal{M}$ and $\mathcal{N}$ are oriented K-manifolds and $g : \mathcal{M} \to \mathcal{N}$ is an "orientation-preserving diffeomorphism" of $\mathcal{M}$ onto $\mathcal{N}$. (We will define this in a moment.) If f is a differential K-form defined on $\mathcal{N}$, then we claim there is a differential K-form $g^*(f)$ on $\mathcal{M}$, the pullback of f by g, which satisfies

$$
\int_{\mathcal{M}} g^*(f) = \int_{\mathcal{N}} f\,.
$$

Definition 7.3.2 Let $\mathcal{M}$ and $\mathcal{N}$ be K-manifolds. We say a map g of $\mathcal{M}$ onto $\mathcal{N}$ is a C^r *diffeomorphism* of $\mathcal{M}$ onto $\mathcal{N}$ provided that for all coordinate patches $u : U \to \mathcal{M}$ and $v : V \to \mathcal{N}$ such that $(g \circ u)(U)$ and $v(V)$ overlap, the map $v^{-1} \circ g \circ u$ is a C^r diffeomorphism between open subsets of $\mathbb{R}^K$. (See Figure 7.3.1.) If we simply describe g as a diffeomorphism, we shall assume it is at least C^1. If $\mathcal{M}$ and $\mathcal{N}$ are oriented manifolds with orientations w_0 and w_1 respectively, we say g is *orientation-preserving* provided that for every $x \in \mathcal{M}$ there is a positive real number λ such that $g_*(x)\Big(w_0(x)\Big) = \lambda\ w_1(g(x))$. If this last equation holds for a negative number λ, then g is *orientation-reversing*.

The simplest pullbacks are connected with coordinate patches. Let $g : U \to \mathcal{M}$ be an orientation-preserving coordinate patch in a K-manifold $\mathcal{M}$. Let $W = g(U)$ and suppose f is a K-form defined on some neighborhood of W. We treat U, an open subset of $\mathbb{R}^K$, as a K-manifold with orientation $e_1 \wedge \cdots \wedge e_K$. Then

$$
\int_W f = \int_U (f \circ g(x)) \cdot (g_*(x))(e_1 \wedge \cdots \wedge e_K)\, dx.
$$

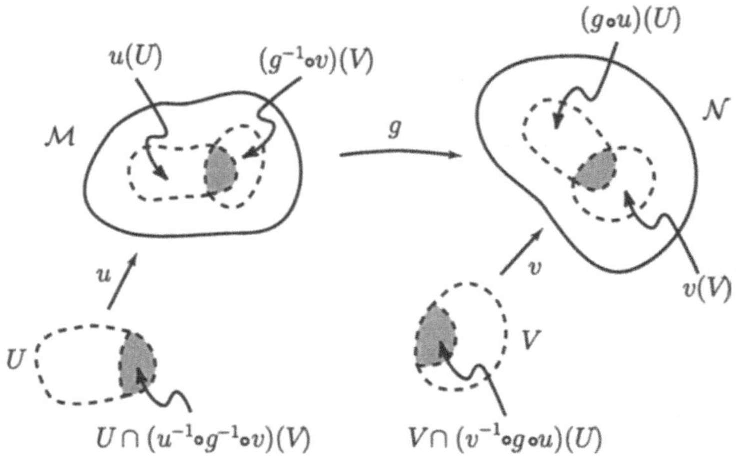

FIGURE 7.3.1.

(The last integral is that of a real-valued function, and dx is included not as a differential form but to indicate the variable of integration.) Taking the dual of the linear transformation $g_*(x)$, we write

$$\int_W f = \int_U \{g_*(x)\}^\diamond (f \circ g(x)) \cdot (e_1 \wedge \cdots \wedge e_K)\, dx\,.$$

Since f is a K-vector field on W, we see that $\{g_*(x)\}^\diamond (f \circ g(x))$ defines a K-vector field on U. We denote this $g^*(f)$ and have

$$\int_W f = \int_U g^*(f),$$

where both integrals are of differential forms and take into account the orientations of W and U respectively.

We generalize this idea to a definition.

Definition 7.3.3 Suppose $g : \mathbb{R}^M \rightarrowtail \mathbb{R}^N$ is a C^1 map and f is a differential K-form with domain in $\mathbb{R}^N$. Assume the range of g is contained in the domain of f. We define the *pullback* of f by g to be the unique differential K-form $g^*(f)$ with domain in $\mathbb{R}^M$ which satisfies

$$f(q) \cdot (\{g_*(p)\}\alpha) = (\{g^*(f)\}(p)) \cdot \alpha$$

for all p in the domain of g and all K-vectors α in $\mathbb{R}^N$ where $q = g(p)$. The value of the pullback at p is given by

$$(g^*(f))(p) = \{g_*(p)\}^\diamond((f \circ g)(p))$$

where $\{g_*(p)\}^\diamond$ is the dual of the linear transformation $g_*(p)$. In the case where $K = 0$, we define $g^*(f) = f \circ g$.

We next generalize the chain rule and use this generalization to see how to compute pullbacks. The proof of this theorem and parts of the next one are left as exercises.

Theorem 7.3.1 *If we are given maps $g : \mathbb{R}^M \rightarrowtail \mathbb{R}^N$ and $h : \mathbb{R}^K \rightarrowtail \mathbb{R}^M$, then at every point x at which the maps are defined, we have*

$$(g \circ h)_*(x) = g_*(h(x)) \circ h_*(x).$$

Theorem 7.3.2 *Suppose we have C^1 maps $g : \mathbb{R}^M \rightarrowtail \mathbb{R}^N$ and $h : \mathbb{R}^K \rightarrowtail \mathbb{R}^M$. Then*

(a) $g^*(\lambda f) = \lambda g^*(f)$ *where f is a differential form and α is a scalar.*

(b) $g^*(f_1 + f_2) = g^*(f_1) + g^*(f_2)$ *where f_1 and f_2 are differential forms.*

(c) $(g \circ h)^* = h^* \circ g^*$.

(d) $g^*(f_1 \wedge f_2) = g^*(f_1) \wedge g^*(f_2)$ *where f_1 and f_2 are differential forms.*

(e) $g^*(dx_i) = \sum_{j=1}^{M} \left(\frac{\partial g_i}{\partial x_j}\right) dx_j$ *where $g = (g_1, g_2, \ldots, g_N)$. (In this last equation, dx_i and dx_j are defined on different vector spaces, $\mathbb{R}^N$ and $\mathbb{R}^M$, respectively.)*

Proof of (c): Let f be a differential form and let p be a point in the domain of $f \circ g \circ h$. Now we use both (a) and (b) and the fact that for linear transformations F and G, when we find duals, we have $(F \circ G)^\diamond = G^\diamond \circ F^\diamond$. We obtain

$$\begin{aligned}
[(g \circ h)^*(f)](p) &= \left\{ \Big((g \circ h)_*(p) \Big)^\diamond \right\} (f \circ g \circ h)\,(p) \\
&= \left\{ \Big(g_*(h(p)) \circ h_*(p) \Big)^\diamond \right\} (f \circ g \circ h)\,(p) \\
&= \left\{ (h_*(p))^\diamond \circ \Big(g_*(h(p)) \Big)^\diamond \right\} (f \circ g)(h(p)) \\
&= \Big(h_*(p) \Big)^\diamond \left\{ g^*(f)\Big(h(p)\Big) \right\} \\
&= \left\{ h^*\Big(g^*(f)\Big) \right\} (p)\,.
\end{aligned}$$

Hint for (d): This is essentially Theorem 6.4.5. (If f_1 is real-valued, a 0-form, then we interpret $f_1 \wedge f_2$ as $f_1 f_2$, the product of a scalar and a K-vector. Then $g^*(f_1 \wedge f_2) = (f_1 \circ g)\, g^*(f_2) = g^*(f_1) \wedge g^*(f_2)$; the result still holds.)

Proof of (e): Suppose p is in the domain of g. We see that

$$[\,g^*(dx_i)\,](p) = \Big(g'(p) \Big)^\diamond \Big(dx_i(g(p)) \Big) = \Big(g'(p) \Big)^\diamond (e_i).$$

Applying the transpose of the matrix of $g'(p)$ to e_i, we obtain,

$$[g^*(dx_i)](p) = \sum_{j=1}^{M} \left(\left(\frac{\partial g_i}{\partial x_j} \right)(p) \right)(e_j) = \left\{ \sum_{j=1}^{M} \left(\frac{\partial g_i}{\partial x_j} \right) dx_j \right\}(p),$$

and the result is established. □

This theorem shows us how to compute $g^*(f)$. If $f = h\, dx_1 \wedge \cdots \wedge dx_K$ where h is a real-valued function, then

$$g^*(f) = (h \circ g)\, g^*(dx_1) \wedge \ldots \wedge g^*(dx_K).$$

Therefore the crucial step is to compute $g^*(dx_i)$.

Example 7.3.3 Suppose again that g is the rotation of $\mathbb{R}^3$ about the x-axis by an angle of θ, so that

$$g(x, y, z) = (x, \quad y \cos(\theta) - z \sin(\theta), \quad y \sin(\theta) + z \cos(\theta)).$$

Then using (e) of our theorem we see that

$$g^*(dx) = \frac{\partial g_1}{\partial x}\, dx + \frac{\partial g_1}{\partial y}\, dy + \frac{\partial g_1}{\partial z}\, dz = dx,$$
$$g^*(dy) = \frac{\partial g_2}{\partial x}\, dx + \frac{\partial g_2}{\partial y}\, dy + \frac{\partial g_2}{\partial z}\, dz = \cos(\theta)\, dy - \sin(\theta)\, dz\,,$$

and

$$g^*(dz) = \frac{\partial g_3}{\partial x}\, dx + \frac{\partial g_3}{\partial y}\, dy + \frac{\partial g_3}{\partial z}\, dz = \sin(\theta)\, dy + \cos(\theta)\, dz.$$

If we wish to compute the pullbacks of 2-forms, we appeal to (d):

$$\begin{aligned} g^*(dx \wedge dy) &= dx \wedge (\cos(\theta)\, dy - \sin(\theta)\, dz) \\ &= \cos(\theta)\, dx \wedge dy - \sin(\theta)\, dx \wedge dz, \\ g^*(dx \wedge dz) &= dx \wedge (\sin(\theta)\, dy + \cos(\theta)\, dz) \\ &= \sin(\theta)\, dx \wedge dy + \cos(\theta)\, dx \wedge dz, \end{aligned}$$

and

$$\begin{aligned} g^*(dy \wedge dz) &= (\cos(\theta)\, dy - \sin(\theta)\, dz) \wedge (\sin(\theta)\, dy + \cos(\theta)\, dz) \\ &= dy \wedge dz. \end{aligned}$$

It is now trivial that

$$g^*(dx \wedge dy \wedge dz) = dx \wedge dy \wedge dz.$$

Intuitively, this last equation says that g is both orientation- and volume-preserving.

The next example illustrates a useful abuse of notation.

Example 7.3.4 Consider the familiar transformation from polar to Cartesian coordinates:

$$x = r\cos(\theta),\ y = r\sin(\theta).$$

We define $g : \mathbb{R}^2 \to \mathbb{R}^2$ by

$$(x, y) = g(r, \theta) = (r\cos(\theta),\ r\ \sin(\theta)).$$

We must be aware at this point that there is more than one way in which we can interpret x and y.

One way is to look at them as the projection maps $x : \mathbb{R}^2 \to \mathbb{R}$ and $y : \mathbb{R}^2 \to \mathbb{R}$ which give us the first and second coordinates of points in $\mathbb{R}^2$. (This interpretation is in accord with the spirit in which the student is introduced to the Cartesian coordinate system: A plane $\mathcal{P}$ is supposed given, two number lines are drawn on this plane at right angles to one another, and a method is described for using these number lines to assign coordinates to points in the plane.) In this case dx and dy are the constant 1-vector fields $dx = e_1$ and $dy = e_2$. In this case, using part (e) of Theorem 7.3.2, we compute

$$g^*(dx) = \frac{\partial g_1}{\partial r}\, dr + \frac{\partial g_1}{\partial \theta}\, d\theta = \cos(\theta)\, dr - r\ \sin(\theta)\, d\theta$$

and

$$g^*(dy) = \frac{\partial g_2}{\partial r}\, dr + \frac{\partial g_2}{\partial \theta}\, d\theta = \sin(\theta)\, dr + r\ \cos(\theta)\, d\theta.$$

Another possible interpretation of x and y is as the first and second components of the function g, that is, $x = g_1$ and $y = g_2$. In that case, by the definition of the differential operator, we have

$$dx = dg_1 = \frac{\partial g_1}{\partial r}\, dr + \frac{\partial g_1}{\partial \theta}\, d\theta = \cos(\theta)\, dr - r\ \sin(\theta)\, d\theta$$

and

$$dy = dg_2 = \frac{\partial g_2}{\partial r}\, dr + \frac{\partial g_2}{\partial \theta}\, d\theta = \sin(\theta)\, dr + r\ \cos(\theta)\, d\theta.$$

We shall abuse notation by not bothering to distinguish between these two cases. Whether we are talking about dx and dy or about $g^*(dx)$ and $g^*(dy)$, we shall write dx and dy. It should be clear from context which interpretation we have in mind.

From this point on, we shall feel free to abuse notation in this manner without comment.

Theorem 7.3.3 (The Change-of-Variables Theorem) *Suppose $\mathcal{M}$ and $\mathcal{N}$ are oriented K-manifolds and $g : \mathcal{M} \to \mathcal{N}$ is an orientation-preserving diffeomorphism of $\mathcal{M}$ onto $\mathcal{N}$. If f is a differential K-form on $\mathcal{N}$, then*

$$\int_{\mathcal{M}} g^*(f) = \int_{\mathcal{N}} f.$$

If g is orientation-reversing, we have

$$\int_{\mathcal{M}} g^*(f) = -\int_{\mathcal{N}} f.$$

Proof. Assume that g is orientation-preserving.

Case 1. Suppose that $\mathcal{M} = U$, an open subset of $\mathbb{R}^K$ with orientation $dx_1 \wedge \cdots \wedge dx_K$ and $g\colon U \to \mathcal{N}$ is a C^1 coordinate patch. From Theorem 5.3.3, g must be a diffeomorphism. We have already shown in this case that

$$\int_{\mathcal{M}} g^*(f) = \int_{\mathcal{N}} f\,.$$

Case 2. We want to show the validity of the change-of-variables formula when g is restricted to a coordinate patch. Suppose $u : U \to \mathcal{M}$ is a C^1 coordinate patch. Notice that $g \circ u$ is a coordinate patch in $\mathcal{N}$. The main difficulty in seeing this is to check that $\mathcal{D}\left((g \circ u)'\right)$ is never zero. To verify this, first let $v : V \to \mathcal{N}$ be a C^1 coordinate patch in $\mathcal{N}$ such that $(g \circ u)(U)$ and $v(V)$ overlap. Since g is a diffeomorphism between manifolds, the map $v^{-1} \circ g \circ u$ must be a diffeomorphism between the subsets U and V of $\mathbb{R}^K$. This implies that

$$\det(v^{-1} \circ g \circ u) \neq 0.$$

Then by Theorem 1.9.3, if $z = (v^{-1} \circ g \circ u)(x)$, we have

$$\begin{aligned}\mathcal{D}\Big((g \circ u)'(x)\Big) &= \mathcal{D}\Big((v \circ v^{-1} \circ g \circ u)'(x)\Big)\\ &= \mathcal{D}\Big(v'(z)\Big)\,|\det(v^{-1} \circ g \circ u)'(x)|.\end{aligned}$$

Since v is a coordinate patch, $\mathcal{D}\left(v'(z)\right)$ cannot be zero and hence neither can $\mathcal{D}\left((g \circ u)'(x)\right)$.

We now treat $u(U)$ and $(g \circ u)(U)$ as manifolds (submanifolds of $\mathcal{M}$ and $\mathcal{N}$ respectively). Appealing to case 1 and properties of pullbacks, we obtain

$$\int_{(g\circ u)(U)} f = \int_U (g \circ u)^*(f) = \int_U (u^* \circ g^*)(f) = \int_{u(U)} g^*(f).$$

Case 3. The general case. Let $\{(A_i, u_i)\}_{i=1}^{\infty}$ be a measurable decomposition of $\mathcal{M}$. Recall that each u_i is a coordinate patch and that if U_i is the domain of u_i, then $A_i \subseteq u_i(U_i)$. We see that $\{(g(A_i), g \circ u_i)\}_{i=1}^{\infty}$ is a measurable decomposition of $\mathcal{N}$. Then

$$
\begin{aligned}
\int_{\mathcal{N}} f &= \sum_{i=1}^{\infty} \int_{(g \circ u_i)(U_i)} f \; \chi_{g(A_i)} \text{ (where } \chi_{g(A_i)} \text{ is the characteristic} \\
&\qquad\qquad \text{function of } g(A_i)) \\
&= \sum_{i=1}^{\infty} \int_{u_i(U_i)} g^*(f \; \chi_{g(A_i)}) \\
&= \sum_{i=1}^{\infty} \int_{u_i(U_i)} g^*(f \wedge \chi_{g(A_i)}) \text{ (by definition of the wedge} \\
&\qquad\qquad \text{product of a 0-form)} \\
&= \sum_{i=1}^{\infty} \int_{u_i(U_i)} g^*(f) \wedge g^*(\chi_{g(A_i)}) \\
&= \sum_{i=1}^{\infty} \int_{u_i(U_i)} g^*(f) \; (\chi_{g(A_i)} \circ g) \\
&= \sum_{i=1}^{\infty} \int_{u_i(U_i)} g^*(f) \; \chi_{A_i} \\
&= \int_{\mathcal{M}} g^*(f).
\end{aligned}
$$

We leave to the reader the case where g is orientation-reversing. □

Example 7.3.5 Again we consider the familiar transformation from polar to Cartesian coordinates: $(x, y) = g(r, \theta) = (r \cos(\theta), r \sin(\theta))$. Let us compute the area of

$$S = \{(x, y) : 1 \le x^2 + y^2 \le 4 \text{ and } x, y \ge 0\}.$$

Among other methods, this can be found by evaluating $\int_S dx \wedge dy$. We know that g is a diffeomorphism of the set $T = [1, 2] \times [0, \pi/2]$ onto S. Since

$$dx \wedge dy = (\cos(\theta)\, dr - r \sin(\theta)\, d\theta) \wedge (\sin(\theta)\, dr + r \cos(\theta)\, d\theta) = r\, dr \wedge d\theta,$$

we have

$$\int_S dx \wedge dy = \int_T r\, dr \wedge d\theta = \int_0^{\pi/2} \int_1^2 r\, dr\, d\theta = \frac{3\pi}{4}.$$

Exercises

1. Prove that if we are given maps $g : \mathbb{R}^M \rightarrowtail \mathbb{R}^N$ and $h : \mathbb{R}^K \rightarrowtail \mathbb{R}^M$, then at every point x at which the maps are defined, we have

$$(g \circ h)_*(x) = g_*(h(x)) \circ h_*(x).$$

2. Compute the values of $g_*(x, y)$ on basis vectors of $\mathbb{R}^2$ and $\wedge^2\mathbb{R}^2$ where $g(x, y) = (x^2 - y^2, 2xy)$.

3. For Theorem 7.3.2, supply the details of the proofs of (a), (b), and (d).

4. Compute $\int_{\mathcal{M}} x^2 dx + y^2 dy$ where $\mathcal{M}$ is the portion of the right-hand branch of the hyperbola $x^2 - y^2 = 1$ running from $(2, -\sqrt{3})$ to $(2, \sqrt{3})$. (Hint: Use the parametrization $(x, y) = (\cosh(\alpha), \sinh(\alpha))$.)

5. Recall that S^1 is the unit circle in $\mathbb{R}^2$. If S^1 has the counterclockwise orientation, compute

   (a) $\int_{S^1} dx$.

   (b) $\int_{S^1} x\, dy - y\, dx$.

6. (a) Show that the function $(u, v) = g(x, y)$ defined by

$$u = x^2 - y^2 \quad \text{and} \quad v = 2xy$$

   maps S^1 onto itself.

   (b) Now show that if f is any 1-form defined on S^1, we must have

$$\int_{S^1} g^*(f) = 2 \int_{S^1} f.$$

   Why does this not conflict with Theorem 7.3.3?
   (Hint: We can parametrize S^1 by setting

$$(x, y) = g(\alpha) = (\cos(\alpha), \sin(\alpha)).)$$

7. (a) If $x, y : \mathbb{R}^N \rightarrowtail \mathbb{R}$, show that $d(xy) = x\, dy + y\, dx$.

   (b) If $x : \mathbb{R}^N \rightarrowtail \mathbb{R}$, show that $d(e^x) = e^x dx$.

   (c) More generally, if $x : \mathbb{R}^N \rightarrowtail \mathbb{R}$ and $f : \mathbb{R} \rightarrowtail \mathbb{R}$, show that $d(f(x)) = f'(x)\, dx$.

8. (a) Let

$$(x, y, z) = g(\alpha, \beta) = (\cos(\alpha), \sin(\alpha)\ \cos(\beta), \sin(\alpha)\ \sin(\beta)).$$

   Compute the pullbacks of dx, dy, dz, $dx \wedge dy$, etc.

   (b) Find an orientation w for the sphere $x^2 + y^2 + z^2 = 1$ in terms of x, y, z, dx, *etc.* and compute the pullback $g^*(w)$ of this orientation in $\mathbb{R}^2$.

9. Consider the map $(x, y, z) \mapsto (u, v, w)$ of $\mathbb{R}^3$ to $\mathbb{R}^3$ defined by

$$u = yz, \quad v = xz, \quad \text{and} \quad w = xy.$$

Show that

$$\begin{aligned}
du &= y\,dz + z\,dy, \\
dv &= x\,dz + z\,dx, \\
dw &= x\,dy + y\,dx, \\
du \wedge dv &= -z(-x\,dy \wedge dz + y\,dx \wedge dz + z\,dx \wedge dy), \\
du \wedge dw &= -y(x\,dy \wedge dz + y\,dx \wedge dz + z\,dx \wedge dy), \\
dv \wedge dw &= -x(x\,dy \wedge dz + y\,dx \wedge dz - z\,dx \wedge dy), \\
du \wedge dv \wedge dw &= 2xyz\,dx \wedge dy \wedge dz.
\end{aligned}$$

10. Let $T =$ the torus $S^1 \times S^1 \subseteq \mathbb{R}^4$.
   (a) Give an orientation w of T in terms of $x_1,\ x_2,\ x_3,\ x_4,\ dx_1,$ etc. Give a two variable parametrization of T and compute the pullback of w under this parametrization.
   (b) Use the parametrization of T and pullbacks to compute $\int_T dx_1 \wedge dx_3 + dx_2 \wedge dx_4$.

11. Suppose $\mathcal{M}$ and $\mathcal{N}$ are C^1 manifolds and $g : \mathcal{M} \to \mathcal{N}$. We assume g satisfies the following two requirements:
   1) g is C^1. (It may be necessary to think of the domain of g as being extended to some neighborhood of $\mathcal{M}$ in order for the partial derivatives to exist.)
   2) For all coordinate patches $u : U \to \mathcal{M}$ and $v : V \to \mathcal{N}$ for which the compositions are defined, $v^{-1} \circ g \circ u$ is C^1.

   (a) Show that if x is a tangent vector to $\mathcal{M}$ at p, then $(g\prime(p))x$ must be a tangent vector to $\mathcal{N}$ at $g(p)$.
   (b) Show that if x is a tangent K-vector to $\mathcal{M}$ at p, then $(g_*(p))x$ must be a tangent K-vector to $\mathcal{N}$ at $g(p)$

12. Show that if $\mathcal{M}$ and $\mathcal{N}$ are diffeomorphic K-manifolds, then $\mathcal{M}$ is orientable if and only if $\mathcal{N}$ is.

7.4 Stokes' Theorem for Chains

We want to extend Stokes' theorem from cubes to *cells* and from cells to *chains*. A K-dimensional cell in $\mathbb{R}^N$ (where $K \le N$) may be visualized as a curved version of a rectangle. (See Figure 7.4.1.) It is often useful to think of manifolds as collections

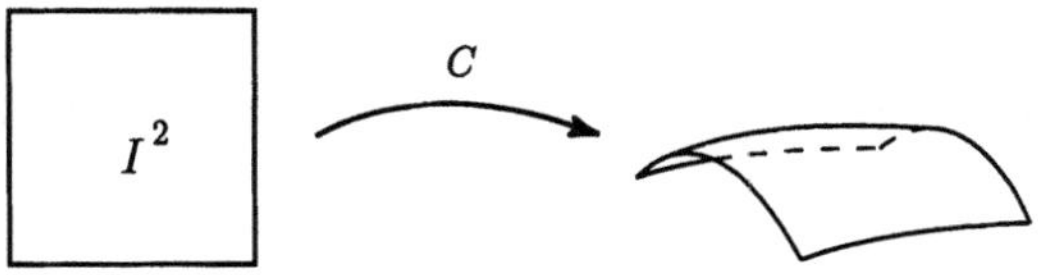

FIGURE 7.4.1.

of cells which are "sewn together" in a nice way. This is an example of a chain of cells. In Figure 7.4.2 we see a rectangle divided into four subrectangles. The large rectangle is to be thought of as having opposite ends "glued" together in such a way as to form the torus shown in the figure. Thus a torus can be formed by four 2-cells.

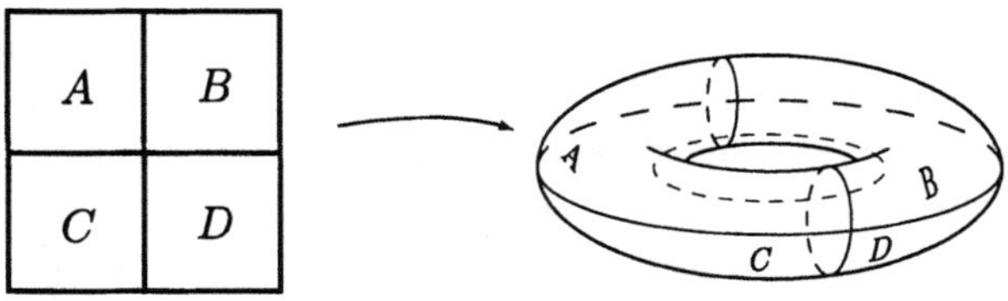

FIGURE 7.4.2.

Definition 7.4.1 Recall that I^K is the unit cube in $\mathbb{R}^K$. By a C^r *K-cell* in $\mathbb{R}^N$ (or, more briefly, a *cell* or a *K-cell*) we mean a C^r map $c : I^K \to \mathbb{R}^N$. (We will deal only with cells which are at least C^1 and will usually not explicitly state this condition.) We shall further restrict ourselves to cells such that

c is one-to-one,
c^{-1} is continuous,

and

$\mathcal{D}(c')$ is never zero.

The *orientation* of c is then taken to be

$$w(p) = \frac{c_*(q)\left(e_1 \wedge e_2 \wedge \cdots \wedge e_K\right)}{|c_*(q)\left(e_1 \wedge e_2 \wedge \cdots \wedge e_K\right)|}$$

or, equivalently,

$$w(p) = \frac{1}{\mathcal{D}(c'(q))}\ c_*(q)\left(e_1 \wedge e_2 \wedge \cdots \wedge e_K\right)$$

where $p = c(q)$. If f is any differential K-form defined over an open set containing the range of c, then we define the integral of f over c by

$$\int_c f = \int_{I^K} c^*(f)$$

where I^K is taken to have the standard orientation $dx_1 \wedge dx_2 \wedge \cdots \wedge dx_K$. By a *K-chain* (or, more briefly, a *chain*) we mean a formal sum

$$\alpha_1 c_1 + \alpha_2 c_2 + \cdots + \alpha_r c_r$$

where $\alpha_1, \ldots, \alpha_r$ are integers and $c_1, \ldots, c_r$ are K-cells. We may add and subtract chains and multiply them by integers in an obvious fashion. For example,

$$(\alpha_1 c_1 + \cdots + \alpha_r c_r) + (\beta_1 c_1 + \cdots + \beta_r c_r) = (\alpha_1 + \beta_1) c_1 + \cdots + (\alpha_r + \beta_r) c_r .$$

If the differential K-form f is defined on some open set containing $c_1(I^K) \cup c_2(I^K) \cup \cdots \cup c_r(I^K)$, then the integral of f over the chain $\alpha_1 c_1 + \alpha_2 c_2 + \cdots + \alpha_r c_r$ is defined by

$$\int_{\alpha_1 c_1 + \cdots + \alpha_r c_r} f = \sum_{i=1}^{r} \alpha_i \int_{c_i} f .$$

The reader may find the expression "formal sum" a bit vague. The important thing, of course, is that one can integrate over such things, over chains, and that the integration is carried out in an obvious and natural and—above all—unambiguous way. Here is one way to formalize this idea: Suppose $\mathfrak{C}_K$ is the set of all K-cells in $\mathbb{R}^N$. Now let $\mathfrak{F}_K$ be the set of all functions from $\mathfrak{C}_K$ into $\mathbb{Z}$, the set of integers, which are zero for all but a finite number of K-cells. Note that we are dealing with well-defined objects. Suppose F is one of the functions in $\mathfrak{F}_K$. We can find distinct K-cells $c_1, \ldots, c_r$ such that $F(c_i) = a_i \neq 0$ for each i and $F(c) = 0$ for every other K-cell c. Clearly we can denote F by the symbol $\alpha_1 c_1 + \cdots + \alpha_r c_r$, and the cells which make up this chain and their coefficients are unambiguously specified. It makes sense therefore to add chains and multiply them by integers. For example, for arbitrary $F, G \in \mathfrak{F}_K$, we define $F + G$ by $(F + G)(c) = F(c) + G(c)$ where $c \in \mathfrak{C}_K$. The set of K-chains has a natural algebraic structure.

Example 7.4.1 Suppose we wish to compute the work done by a particle of unit mass moving through a gravitational field f as the particle traverses the unit circle in the plane twice in the counterclockwise direction. We define the cells $c_1, c_2 : [0, 1] \to \mathbb{R}^2$ by

$$c_1(t) = (\cos(\pi t), \sin(\pi t))$$

and

$$c_2(t) = (\cos[\pi(t + 1)], \sin[\pi(t + 1)]).$$

Both c_1 and c_2 have counterclockwise orientation with c_1 covering the top half and c_2 the bottom half of the unit circle. Clearly $c_1 + c_2$ represents one complete traversal of the circle and $2c_1 + 2c_2$ represents two such traversals. The work for two traversals is given by

$$\int_{2c_1 + 2c_2} f$$

where we treat f as a 1-form.

We can use our understanding of the boundary of I^K to define the boundary of a cell.

Definition 7.4.2 Suppose c is a K-cell in $\mathbb{R}^N$ where $K \geq 1$. Then c is a function, $c : I^K \to \mathbb{R}^N$. Recall that the $(K-1)$-dimensional faces of I^K are the sets

$$I_{0,i} = \{ (x_1, \ldots, x_K) \in I^K \ : \ x_i = 0 \},$$
$$I_{1,i} = \{ (x_1, \ldots, x_K) \in I^K \ : \ x_i = 1 \}.$$

We define the corresponding faces of c, namely $c_{\alpha,i} : I^{K-1} \to \mathbb{R}^N$, by

$$c_{\alpha,i}(x_1, \ldots, x_{i-1}, x_{i+1}, \ldots, x_K) = c(x_1, \ldots, x_{i-1}, \alpha, x_{i+1}, \ldots, x_K).$$

Then by the *boundary* of the cell c we mean the chain

$$\partial c = \sum_{i=1}^{K} \sum_{\alpha=0,1} (-1)^{i+\alpha} \, c_{\alpha,i}.$$

This formula is inspired by our earlier discussion of induced orientation for the boundaries of cubes. Figure 7.4.3 shows an example of a 2-cell in $\mathbb{R}^2$ with the orientation of the boundary cells indicated, assuming $\det(c') > 0$.

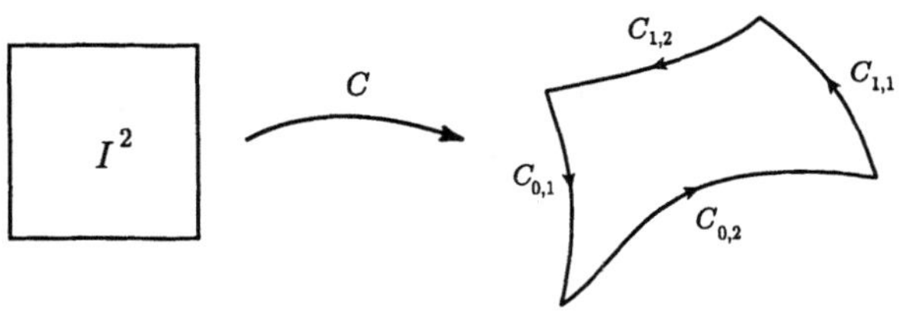

FIGURE 7.4.3.

Lemma 7.4.1 *If c is a K-cell and f is a $(K-1)$-form defined on some open set containing $c(\partial I^K)$, then*

$$\int_{\partial c} f = \int_{\partial I^K} c^*(f) .$$

Proof. For the case where $K = 1$ we may interpret both integrals as $f(c(1)) - f(c(0))$. Let us consider the case where $K \geq 2$.

Recall the definitions of $I_{\alpha,i}$ and $c_{\alpha,i}$ from the definition of ∂c. We define $p_{\alpha,i}$: $I^{K-1} \to I_{\alpha,i}$ by

$$p_{\alpha,i}(x_1, \ldots, x_{K-1}) = (x_1, \ldots, x_{i-1}, \alpha, x_i, \ldots, x_{K-1}) .$$

Then $c_{\alpha,i} = c \circ p_{\alpha,i}$. We know that

$$\int_{\partial I^K} c^*(f) = \sum_{i=1}^{K} \sum_{\alpha=0,1} \int_{I_{\alpha,i}} c^*(f)$$

and

$$\int_{\partial c} f = \sum_{i=1}^{K} \sum_{\alpha=0,1} (-1)^{i+\alpha} \int_{c_{\alpha,i}} f .$$

Recall also that if $g : \mathcal{M} \to \mathcal{N}$ is an orientation-preserving diffeomorphism between two N-manifolds and h is a N-form on $\mathcal{N}$, then

$$\int_{\mathcal{M}} g^*(h) = \int_{\mathcal{N}} h .$$

If g is orientation-reversing, then

$$\int_{\mathcal{M}} g^*(h) = -\int_{\mathcal{N}} h .$$

Now consider the maps

$$I^{K-1} \xrightarrow{p_{\alpha,i}} I_{\alpha,i} \xrightarrow{c} c(I_{\alpha,i}) .$$

We may think of the maps $p_{\alpha,i}$ and c as being extended to open sets that contain I^{K-1} and $I_{\alpha,i}$ in such a way that $p_{\alpha,i}$ and c are diffeomorphisms between manifolds and we may use the theorem about the transformation of integrals. The orientation of I^{K-1} is $dx_1 \wedge dx_2 \wedge \cdots \wedge dx_{K-1}$ and that of $I_{\alpha,i}$ is

$$w_{\alpha,i} = (-1)^{i+\alpha}\, dx_1 \wedge \cdots \wedge \overline{dx_i} \wedge \cdots \wedge dx_K .$$

We know that $c_{\alpha,i}$ is orientation-preserving because we define the orientation of $c(I_{\alpha,i})$ to be that induced by $c_{\alpha,i}$. It is easily calculated that

$$\Big((p_{\alpha,i})_*(x)\Big)\,(e_1 \wedge e_2 \wedge \cdots \wedge e_{K-1}) = e_1 \wedge \cdots \wedge \overline{e_i} \wedge \cdots \wedge e_K ,$$

so that $p_{\alpha,i}$ is seen to be orientation-preserving or reversing depending on whether $(-1)^{i+\alpha}$ is positive or negative.

Suppose that $(-1)^{i+\alpha} = 1$ so that $p_{\alpha,i}$ is orientation-preserving. We then have

$$\begin{aligned}
\int_{I_{\alpha,i}} c^*(f) &= \int_{I^{K-1}} (p_{\alpha,i})^* \Big(c^*(f)\Big) \\
&= (-1)^{i+\alpha} \int_{I^{K-1}} (c_{\alpha,i})^* (f) \\
&= (-1)^{i+\alpha} \int_{c_{\alpha,i}} f .
\end{aligned}$$

Now suppose that $(-1)^{i+\alpha} = -1$ so that $p_{\alpha,i}$ is orientation-reversing. We then have

$$\begin{aligned}
\int_{I_{\alpha,i}} c^*(f) &= -\int_{I^{K-1}} (p_{\alpha,i})^* \Big(c^*(f)\Big) \\
&= (-1)^{i+\alpha} \int_{I^{K-1}} (c_{\alpha,i})^* (f) \\
&= (-1)^{i+\alpha} \int_{c_{\alpha,i}} f .
\end{aligned}$$

Thus we are done. □

Definition 7.4.3 If $c = \alpha_1 c_1 + \cdots + \alpha_r c_r$, a chain, where each c_i is a K-cell, then we define the *boundary* of c by $\partial c = \alpha_1 \partial c_1 + \cdots + \alpha_r \partial c_r$.

Corollary 7.4.1 *If $c = \alpha_1 c_1 + \cdots + \alpha_r c_r$ is a chain of K-cells and f is a $(K-1)$-form with domain such that the integrals below exist, then*

$$\int_{\partial c} f = \sum_{i=1}^{r} \alpha_i \int_{\partial I^K} c_i^*(f).$$

To be in a position to prove Stokes' theorem for chains, we need only show that pullbacks commute with the differential operator.

Theorem 7.4.1 *If $g : \mathbb{R}^M \rightarrowtail \mathbb{R}^N$ is C^2 and f is a differential form on $\mathbb{R}^N$, then*

$$dg^*(f) = g^*(df).$$

Proof. We show this by an induction. Suppose first that f is a 0-form and $g = (g_1, \ldots, g_N)$. Then

$$\begin{aligned} dg^*(f) &= d(f \circ g) \\ &= \sum_{i=1}^{M} \frac{\partial}{\partial x_i}(f \circ g)\, dx_i \\ &= \sum_{i=1}^{M} \sum_{j=1}^{N} \left(\left(\frac{\partial f}{\partial x_j} \right) \circ g \right) \frac{\partial g_j}{\partial x_i}\, dx_i \\ &= \sum_{j=1}^{N} \left(\left(\frac{\partial f}{\partial x_j} \right) \circ g \right) \left(\sum_{i=1}^{M} \frac{\partial g_j}{\partial x_i}\, dx_i \right) \\ &= \sum_{j=1}^{N} \left(\left(\frac{\partial f}{\partial x_j} \right) \circ g \right) \left\{ g^*(dx_j) \right\} \quad \text{(by 7.3.2(e))} \\ &= g^* \left(\sum_{j=1}^{N} \frac{\partial f}{\partial x_j}\, dx_j \right) \quad \text{(by the remark after the proof of Theorem 7.3.2)} \\ &= g^*(df). \end{aligned}$$

Now let us suppose that it has been established for all K-forms h on $\mathbb{R}^N$ that we have $dg^*(h) = g^*(dh)$. To extend this result to $(K+1)$-forms, it is sufficient to show that $dg^*(f \wedge dx_i) = g^*(d(f \wedge dx_i))$ where f is a K-form on $\mathbb{R}^N$. As before, $g = (g_1, \ldots, g_N)$. Then

$$\begin{aligned} dg^*(f \wedge dx_i) &= d(g^*(f) \wedge g^*(dx_i)) \\ &= d(g^*(f)) \wedge g^*(dx_i) + (-1)^K g^*(f) \wedge dg^*(dx_i). \end{aligned}$$

Notice that

$$\begin{aligned} dg^*(dx_i) &= d\left(\sum_{j=1}^{M} \frac{\partial g_i}{\partial x_j}\, dx_j\right) \\ &= \sum_{j=1}^{M} \sum_{r=1}^{M} \frac{\partial^2 g_i}{\partial x_r \partial x_j}\, dx_r \wedge dx_j \\ &= 0 \end{aligned}$$

since $dx_r \wedge dx_j = -dx_j \wedge dx_r$. Therefore

$$\begin{aligned} dg^*(f \wedge dx_i) &= d(g^*(f)) \wedge g^*(dx_i) \\ &= g^*(df) \wedge g^*(dx_i) \\ &= g^*(df \wedge dx_i). \end{aligned}$$

Because

$$\begin{aligned} d(f \wedge dx_i) &= df \wedge dx_i + (-1)^K f \wedge d^2 x_i \\ &= df \wedge dx_i\,, \end{aligned}$$

we see that

$$dg^*(f \wedge dx_i) = g^*\Big(d(f \wedge dx_i)\Big),$$

and we are done. □

Theorem 7.4.2 (Stokes' theorem for chains) *If f is a $(K-1)$-form defined over the K-chain c, where c is C^2, then*

$$\int_c df = \int_{\partial c} f\,.$$

Proof. We need only consider the case where c is a cell. In the case where $K = 1$, we can show both integrals amount to $f(c(1)) - f(c(0))$, so let us consider the case where $K \geq 2$.

Since we have proved the theorem for I^K, we have

$$\begin{aligned} \int_c df &= \int_{I^K} c^*(df) \\ &= \int_{I^K} dc^*(f) \\ &= \int_{\partial I^K} c^*(f) \\ &= \int_{\partial c} f\,. \end{aligned}$$

The result is established. □

Exercises

1. Suppose $\mathfrak{C}_K$ is the set of all K-cells in $\mathbb{R}^N$. Let $\mathfrak{F}_K$ be the set of all functions from $\mathfrak{C}_K$ into $\mathbb{R}$ which are zero for all but a finite number of K-cells. Show that $\mathfrak{F}_K$ is a vector space over the field of real numbers.

2. Sketch the image of the cell $c(x, y) = (x/2,\ x^2(1 - y) + y)$ where, of course, $(x, y) \in I^2$.

3. Show that if c_1 and c_2 are K-chains with $K \geq 1$, then $\partial(c_1 + c_2) = \partial c_1 + \partial c_2$.

4. Let $\mathcal{A}$ be the annulus in $\mathbb{R}^2$ described by $1 \leq x^2 + y^2 \leq 4$. Let $\mathcal{A}^+$ and $\mathcal{A}^-$ be the right and left halves respectively of $\mathcal{A}$, that is, let
$$\mathcal{A}^+ = \{(x, y) \in \mathcal{A} :\ x \geq 0\} \text{ and } \mathcal{A}^- = \{(x, y) \in \mathcal{A} :\ x \leq 0\}.$$
Write analytic descriptions of cells c^+ and c^- such that $\mathcal{A}^+$ and $\mathcal{A}^-$ are their respective images. If f and g are, respectively, a 2-form and a 1-form defined on $\mathcal{A}$, what can be said about the relation between $\int_{\mathcal{A}} f$ and $\int_{c^+ + c^-} f$? Between $\int_{\partial\mathcal{A}} g$ and $\int_{\partial(c^+ + c^-)} g$?

5. (a) Find f such that $df = x\, dx + y\, dy$.
   (b) Evaluate $\int_{\mathcal{C}} x\, dx + y\, dy$ where $\mathcal{C}$ is the curve
$$(x, y) = (\cos(t) + t,\ \sin(t) - t), \quad 0 \leq t \leq 2\pi.$$

6. Give details of the proof of Stokes' theorem for chains in the case where c is a 1-cell.

7. In $\mathbb{R}^N$ let f be the $(N - 1)$-form
$$\frac{1}{N} \sum_{i=1}^{N} (-1)^{i+1}\, x_i\, dx_1 \wedge \cdots \wedge \overline{dx_i} \wedge \cdots \wedge dx_N.$$
If c is an N-cell in $\mathbb{R}^N$, show that $\int_c f =$ the N-dimensional volume of $c(I^N)$.

8. In $\mathbb{R}^3$ let $\mathcal{M}$ be the 2-manifold $x^2 + y^2 - z^2 = 1$, where $-1 \leq z \leq 1$.
   (a) Sketch the graph of $\mathcal{M}$.
   (b) Let us parametrize $\mathcal{M}$ by the map
$$(x, y, z) = g(\alpha, \beta) = (\cosh(\alpha)\ \cos(\beta),\ \cosh(\alpha)\ \sin(\beta),\ \sinh(\alpha)),$$
where $-\alpha_0 \leq \alpha \leq \alpha_0$ and $0 \leq \beta \leq 2\pi$, the number α_0 being the unique value for which $\sinh(\alpha_0) = 1$. Use this parametrization to define a chain of four cells, $c = c_1 + c_2 + c_3 + c_4$, with the (intuitively clear) property that for 2-forms f and 1-forms F we have
$$\int_c f = \int_{\mathcal{M}} f \quad \text{and} \quad \int_{\partial c} F = \int_{\partial\mathcal{M}} F.$$
   (c) Evaluate $\int_{\partial c} x\, dx + y\, dy$.
   (d) Find the area of $\mathcal{M}$ by evaluating $\int_{\partial c} \frac{1}{2}(y\, dx - x\, dy)$.

7.5 Stokes' Theorem for Oriented Manifolds

Much of what works for chains can be carried over to manifolds. This depends on the fact that we can think of every point in a K-manifold as lying inside a K-cell which in turn lies inside the manifold in a "nice way".

Let $\mathcal{M}$ be a K-manifold in some $\mathbb{R}^N$. We say that a C^r K-cell c is a *C^r K-cell in $\mathcal{M}$* provided that $c(I^K) \subseteq \mathcal{M}$, and for every coordinate patch $g\colon U \to \mathcal{M}$ such that $g(U)$ intersects $c(I^K)$, we have that $g^{-1} \circ c$ and $c^{-1} \circ g$ are C^r maps between subsets of $\mathbb{R}^K$. (Note that if we restrict c to the interior points of I^K, the restricted c is a coordinate patch in $\mathcal{M}$.) If $\mathcal{M}$ has an orientation w, we say the orientation of c agrees with that of $\mathcal{M}$ at $x = c(p)$ provided

$$[c_*(p)]\,(e_1 \wedge \cdots \wedge e_K) = \alpha w(x)$$

for some scalar $\alpha > 0$. (Of course, since I^K is connected, if the orientation of c agrees with that of $\mathcal{M}$ at one point, it agrees at all points; if it disagrees at one point, it disagrees at all points.) If $x = c(p)$ and p is an interior point of I^K, we shall call x an *interior point of* c; if p lies on the boundary of I^K, we shall say x belongs to ∂c, that is, x is a *boundary point of* c.

It is important that for a K-cell in a K-manifold, an integral over the cell can be transformed to an integral over the manifold. If f is a differential K-form on an oriented K-manifold $\mathcal{M}$, then for every K-cell c in $\mathcal{M}$ whose orientation agrees with that of $\mathcal{M}$, we have

$$\int_c f = \int_{\mathcal{M}} f\,\chi_A$$

where χ_A is the characteristic function of the set $A = c(I^K)$. This is because both integrals reduce to

$$\int_{I^K} (f \circ c) \cdot (w \circ c)\,\mathcal{D}(c')$$

where w is orientation of $\mathcal{M}$. Of course, if we know f is zero outside A (an important case for us later), we simply have

$$\int_c f = \int_{\mathcal{M}} f.$$

We show how to use cells to define induced orientation in a manifold.

Let $\mathcal{M}$ be a C^1 K-manifold with nonempty boundary in $\mathbb{R}^N$ such that $\mathcal{M}$ has orientation w. Suppose $x \in \partial\mathcal{M}$. There exists a C^1 K-cell $c\colon I^K \to \mathcal{M}$ such that:

c agrees with the orientation of $\mathcal{M}$,
$c(I^K)$ intersects $\partial\mathcal{M}$ precisely in the set $c(I_{1,1})$ while the rest of $c(I^K)$ lies in the interior of $\mathcal{M}$, and
$x = c(p)$ for some point p in the interior of $I_{1,1}$.

(Recall that $I_{1,1}$ is the set of $(u_1, u_2, \ldots, u_K) \in I^K$ such that $u_1 = 1$.) The existence of such a cell c is easily seen by starting with a coordinate patch $g\colon U \to \mathcal{M}$ which covers x (where U is open in the half-space, $\mathbb{H}^K$) and constructing an appropriate

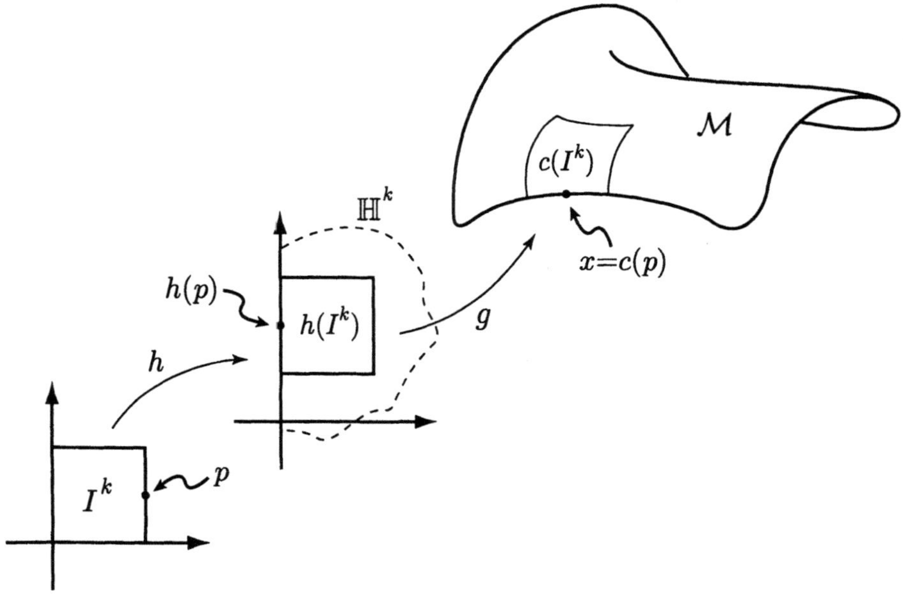

FIGURE 7.5.1.

map $h\colon I^K \to \mathbb{H}^K$ such that c can be taken to be $g \circ h$. See Figure 7.5.1. We know the induced orientation on ∂I^K at the point p is $e_2 \wedge e_3 \wedge \cdots \wedge e_K$, so we define the induced orientation on $\mathcal{M}$ at x to be

$$\partial w(x) = \alpha\ [c_*(p)]\ (e_2 \wedge \cdots \wedge e_K)$$

where α is a positive scalar chosen in such a way as to ensure $\partial w(p)$ will be a unit vector. (It can be shown that $\alpha = \frac{1}{\mathcal{D}((c_{1,1})'(p))}$.)

We need to show this definition of induced orientation at $x \in \partial\mathcal{M}$ is independent of our choice of c. Let c_1 and c_2 be two K-cells in $\mathcal{M}$ with $c_1(p_1) = c_2(p_2) = x$ where p_1 and p_2 are interior points of $I_{1,1}$ and otherwise having all the properties imposed on c in the preceding paragraph. The transformation $c_2^{-1} \circ c_1$ is orientation-preserving since c_1 and c_2 separately have this property. Also, $c_2^{-1} \circ c_1$ carries points in a neighborhood of p_1 in the face $I_{1,1}$ of I^K into the face $I_{1,1}$. Because of this, $[(c_2^{-1} \circ c_1)_*(p_1)]\ (e_2 \wedge \cdots \wedge e_K) = \beta\ (e_2 \wedge \cdots \wedge e_K)$ for some real scalar β. The fact that $c_2^{-1} \circ c_1$ is orientation-preserving forces β to be positive. Applying $(c_2)_*(p_2)$ to both sides of the last equation yields $[(c_1)_*(p_1)]\ (e_2 \wedge \cdots \wedge e_K) = \beta\ [(c_2)_*(p_2)](e_2 \wedge \cdots \wedge e_K)$ where $\beta > 0$. That is, c_1 and c_2 both induce the same orientation in $\partial\mathcal{M}$ at the point x; thus ∂w is well defined.

Example 7.5.1 Let $\mathcal{M} = \{(x_1, x_2) \in \mathbb{R}^2\colon x_1^2 + x_2^2 \le 1\}$, a unit disk, and we endow this 2-manifold with the orientation $dx_1 \wedge dx_2$. Then $\partial\mathcal{M} = \{(x_1, x_2) \in \mathbb{R}^2\colon x_1^2 + x_2^2 = 1\}$. We want to find the induced orientation at a point $x_0 = (x_{01}, x_{02}) \in \partial\mathcal{M}$.

There exists θ_0 such that $(x_{01}, x_{02}) = (\cos(\theta_0), \sin(\theta_0))$. There exist $\delta > 0$ and $r_0 \in (0, 1)$ such that the map $g(r, \theta) = (r\ \cos\theta, r\ \sin\theta)$ is a diffeomorphism on the rectangle $[r_0, 1] \times [\theta_0 - \delta, \theta_0 + \delta]$ which carries the rectangle into $\mathcal{M}$, carries

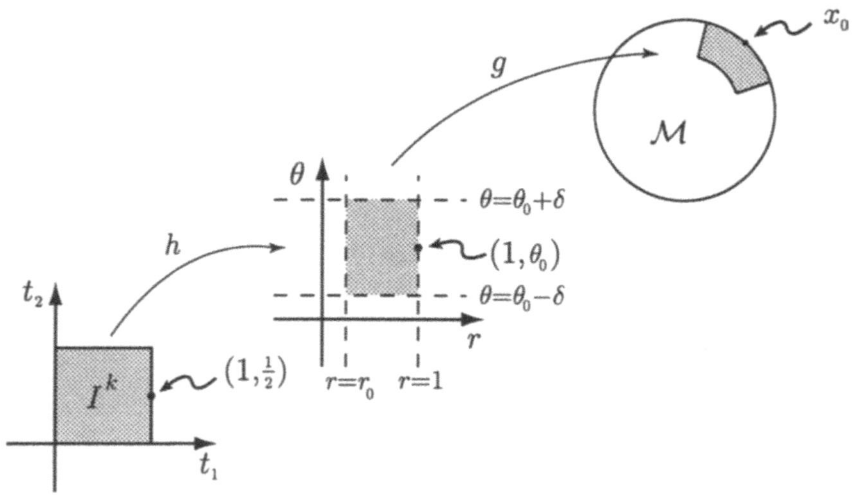

FIGURE 7.5.2.

one edge of the rectangle into $\partial\mathcal{M}$, and carries $(1, \theta_0)$ to x_0. The map $h(t_1, t_2) = (r_0 + t_1(1 - r_0), \theta_0 + \delta(2t_2 - 1))$ is a diffeomorphism which carries the unit square $[0, 1]^2$ onto the rectangle $[r_0, 1] \times [\theta_0 - \delta, \theta_0 + \delta]$ and takes $(1, 1/2)$ to the point $(1, \theta_0)$.

The composition of these two maps defines a cell $c\colon [0, 1]^2 \to \mathcal{M}$ so that one edge of the unit square goes into $\partial\mathcal{M}$ and $(1, 1/2)$ maps to x_0. See Figure 7.5.2.

It is straightforward to calculate that

$$[c'(t_1, t_2)]\ (e_1 \wedge e_2) = 2\delta(r_0 + t_1(1 - r_0))(1 - r_0)\ (e_1 \wedge e_2),$$

and since $2\delta(r_0+t_1(1-r_0))(1-r_0) > 0$, we see that c is orientation-preserving. Now the induced orientation of ∂I^2 at $(1, 1/2)$ is e_2, so we can find the induced orientation of $\partial\mathcal{M}$ at x_0 by calculating $[c'(1, 1/2)]\ (e_2) = -2\delta\ \sin(\theta_0)\ e_1 + 2\delta\ \cos(\theta_0)\ e_2$. Hence $\partial w(x_0) = -x_{02}e_1 + x_{01}e_2$.

Note: In this particular example, since $\mathcal{M}$ is a 2-dimensional manifold in $\mathbb{R}^2$ and $\mathcal{M}$ has the same orientation as $\mathbb{R}^2$, we could have found this induced orientation this way: Choose $x = (x_1, x_2) \in \partial\mathcal{M}$. The outward unit normal vector at this point is $n = x = x_1e_1 + x_2e_2$. Then $\partial w(x) = {}^*n = -x_2e_1 + x_1e_2$. This gives us the right answer because it is essentially how we define induced orientation on ∂I^K. It only works if the manifold and the Euclidean space which contain it have the same dimension and the same orientation. In the case of a 2-manifold in $\mathbb{R}^2$, this also means that if the manifold has the "standard" orientation of $\mathbb{R}^2$, then the boundary of the manifold will have "counterclockwise" orientation.

Before proving Stokes' theorem for manifolds, we need to show the existence of C^∞ real-valued functions on $\mathbb{R}^N$ with arbitrarily small support. Recall that if $f\colon \mathbb{R}^N \to \mathbb{R}$, the support of f, denoted $\operatorname{supp} f$, is the closure of $\{x \in \mathbb{R}^N\colon f(x) \neq 0\}$.

Define $E\colon \mathbb{R} \to \mathbb{R}$ by

$$E(x) = \begin{cases} e^{-1/(1-x^2)} & \text{if } -1 < x < 1 \\ 0 & \text{otherwise.} \end{cases}$$

Notice that $\lim_{x\to -1} E(x) = \lim_{x\to 1} E(x) = 0$ so that E is continuous. It can be shown by induction that for $k = 1, 2, 3, \ldots$ the kth derivative of E satisfies

$$E^{(k)}(x) = \begin{cases} \frac{p_k(x)}{u^m} E(x) & \text{for } |x| < 1 \\ 0 & \text{for } |x| \geq 1 \end{cases}$$

where $p_k(x)$ is a polynomial, $u = 1 - x^2$, and m is a natural number. (It is helpful to show, by appealing to L'Hopital's rule, that $\lim_{x\to -1^+} u^m e^{1/u} = \lim_{x\to 1^-} u^m e^{1/u} = \infty$.) The details of these claims are left as an exercise. Therefore $E(x)$ is a C^∞ function on $\mathbb{R}$ which is positive on $(-1, 1)$ and has support $[-1, 1]$.

Theorem 7.5.1 *If $R = [a_1, b_1] \times [a_2, b_2] \times \cdots \times [a_N, b_N]$, a nondegenerate N-dimensional rectangle in $\mathbb{R}^N$, there is a C^∞ real-valued, nonnegative function with domain $\mathbb{R}^N$ which is positive on the open set $(a_1, b_1) \times (a_2, b_2) \times \cdots \times (a_N, b_N)$ and has the closed set R as its support.*

Proof. Consider first a single interval $[a, b]$ in $\mathbb{R}$. The function $g(x) = (2x - (a + b))/(b - a)$ is a C^∞ function on $\mathbb{R}$ which takes $[a, b]$ onto $[-1, 1]$. Therefore $E \circ g$ is a C^∞ function which is positive on (a, b) and has $[a, b]$ as its support.

Now for $i = 1, 2, \ldots, N$, let E_i be a C^∞ real-valued function on $\mathbb{R}$ which is positive on (a_i, b_i) and has $[a_i, b_i]$ as its support. The desired function is $F(x_1, x_2, \ldots, x_N) = E_1(x_1)\ E_2(x_2)\ \ldots\ E_N(x_N)$. □

Theorem 7.5.2 (Stokes' theorem for manifolds) *If $\mathcal{M}$ is a C^2, compact, oriented K-manifold and f is a C^2 differential $(K - 1)$-form on $\mathcal{M}$, then*

$$\int_{\mathcal{M}} df = \int_{\partial \mathcal{M}} f.$$

Proof. Suppose $K = 1$. Any point x of $\mathcal{M}$ must have a neighborhood in $\mathcal{M}$ which is diffeomorphic to an interval J in $\mathbb{R}$. Because of this and because $\mathcal{M}$ is compact, $\mathcal{M}$ can be thought of as a finite collection of 1-cells, $c_1, \ldots, c_m$, none of which has points in common with any other. Since Stokes' theorem holds for chains, we are done.

Now suppose $K > 1$.

We will construct a finite number of C^∞ functions on $\mathcal{M}$, let us call them ϕ_1, $\phi_2, \ldots, \phi_n$, such that

1) each $\phi_i \geq 0$,

2) $\sum_{i=1}^n \phi_i(x) = 1$ for all $x \in \mathcal{M}$,

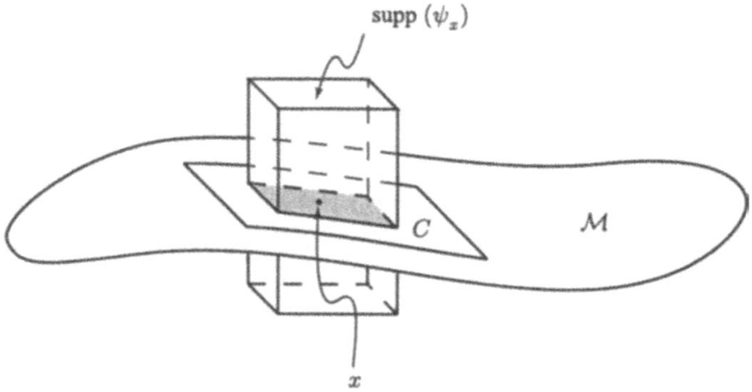

FIGURE 7.5.3.

and

3) $\int_{\mathcal{M}} d(\phi_i f) = \int_{\partial\mathcal{M}} \phi_i f$ for all i.

(Such a construction, if it satistfies 1) and 2), is an example of a *partition of unity* for $\mathcal{M}$.)

Choose $x \in \mathcal{M}$.

Suppose x is not a boundary point of $\mathcal{M}$. Then there is a K-cell c in $\mathcal{M}$ such that x is an interior point of c and $c(I^K)$ lies entirely in the interior of $\mathcal{M}$. By Theorem 7.5.1, there is a real-valued, nonnegative C^∞ function ψ_x on $\mathbb{R}^N$ which is positive on an open set U_x and has the property that $\mathcal{M} \cap (\text{supp}(\psi_x))$ lies in the interior of c. Let us denote $\mathcal{M} \cap U_x$ as O_x. (See Figure 7.5.3; the shaded portion is O_x.) This is an open subset of $\mathcal{M}$. The product $\psi_x f$ is a C^2 differential $(K-1)$-form on $\mathcal{M}$. Appealing to Stokes' theorem for chains, we have

$$\int_{\mathcal{M}} d(\psi_x f) = \int_c d(\psi_x f) = \int_{\partial c} \psi_x f.$$

This last integral is zero since O_x lies in the interior of c. We must also have $\int_{\partial\mathcal{M}} \psi_x f = 0$ by essentially the same reasoning. Therefore

$$\int_{\mathcal{M}} d(\psi_x f) = \int_{\partial\mathcal{M}} \psi_x f.$$

Now suppose $x \in \partial\mathcal{M}$. Then there is a K-cell c in $\mathcal{M}$ such that the intersection of $c(I^K)$ with $\partial\mathcal{M}$ is $c(I_{1,1})$ and x is an interior point of $c_{1,1}$. By Theorem 7.5.1, there is a real-valued, nonnegative C^∞ function ψ_x on $\mathbb{R}^N$ which is positive on an open set U_x and has the property that $\mathcal{M} \cap (\text{supp}(\psi_x))$ lies in the union of the interior of c and the interior of $c_{1,1}$. As before, we denote $\mathcal{M} \cap U_x$ as O_x and note that this is an open subset of $\mathcal{M}$. (See Figure 7.5.4; the shaded portion is O_x.) Again the product $\psi_x f$ is a C^2 differential $(K-1)$-form on $\mathcal{M}$, and, by Stokes' theorem for chains,

$$\int_{\mathcal{M}} d(\psi_x f) = \int_c d(\psi_x f) = \int_{\partial c} \psi_x f.$$

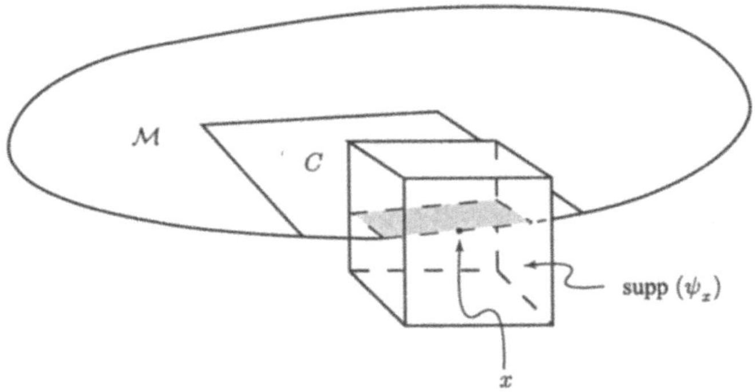

FIGURE 7.5.4.

When integrating over ∂c, we know ψ_x is nonzero only on $c_{1,1}$, so $\int_{\partial c} \psi_x f = \int_{c_{1,1}} \psi_x f$. Since $c_{1,1}$ lies in $\partial\mathcal{M}$ and ψ_x, when restricted to $\partial\mathcal{M}$, is zero outside $c_{1,1}$, we find that $\int_{c_{1,1}} \psi_x f = \int_{\partial\mathcal{M}} \psi_x f$. Therefore, as in our first case, we have

$$\int_{\mathcal{M}} d(\psi_x f) = \int_{\partial\mathcal{M}} \psi_x f.$$

Since $\mathcal{M}$ is compact, there exists a finite number of points, $x_1, x_2, \ldots, x_n$, such that $O_{x_1}, O_{x_2}, \ldots, O_{x_n}$ is a covering for $\mathcal{M}$, i.e., $\mathcal{M} \subseteq O_{x_1} \cup O_{x_2} \cup \cdots \cup O_{x_n}$. We define $\psi = \psi_{x_1} + \cdots + \psi_{x_n}$ and $\phi_i = \psi_{x_i}/\psi$ for $i = 1, 2, \ldots, n$. Since ψ_{x_i} is positive on O_{x_i} and every point of $\mathcal{M}$ lies in some O_{x_i}, the function ψ is positive at every point of $\mathcal{M}$. Therefore each ϕ_i is defined and C^∞ on $\mathcal{M}$. For every $x \in \mathcal{M}$ we have $\sum_{i=1}^n \phi_i(x) = 1$, and for $i = 1, 2, \ldots, n$, by the same argument we gave for ψ_x, we have $\int_{\mathcal{M}} d(\phi_i f) = \int_{\partial\mathcal{M}} \phi_i f$.

Therefore,

$$\begin{aligned}
\int_{\mathcal{M}} df &= \int_{\mathcal{M}} d\left(\left(\sum_{i=1}^n \phi_i\right) f\right) \\
&= \sum_{i=1}^n \int_{\mathcal{M}} d(\phi_i f) \\
&= \sum_{i=1}^n \int_{\partial\mathcal{M}} \phi_i f \\
&= \int_{\partial\mathcal{M}} \left(\sum_{i=1}^n \phi_i\right) f \\
&= \int_{\partial\mathcal{M}} f,
\end{aligned}$$

and we are done. □

Example 7.5.2 Let $\mathcal{M} = \{(x, y) \in \mathbb{R}^2 \colon x^2 + y^2 \leq 1\}$, the closed unit disk centered at the origin, with orientation $w = dx \wedge dy$. Then $\partial w = -y\ dx + x\ dy$. Let $f =$

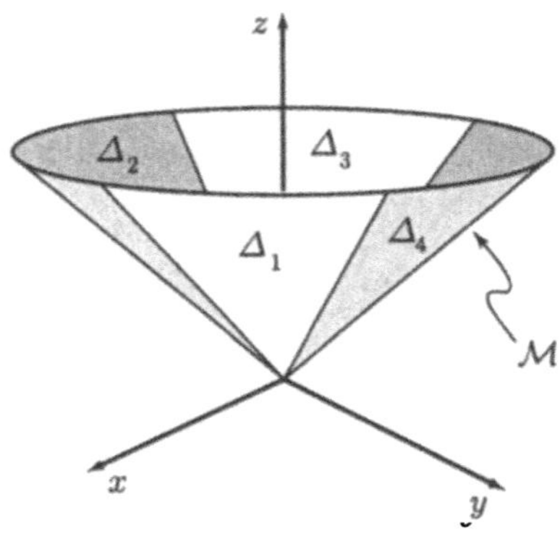

FIGURE 7.5.5.

$\frac{1}{2}(-y\,dx + x\,dy)$. One easily calculates $df = dx \wedge dy$, so that $\int_{\mathcal{M}} df = \pi$, the area of $\mathcal{M}$. Using coordinate patches of the form $g(\theta) = (\cos\theta, \sin\theta)$, one shows that we also have $\int_{\partial\mathcal{M}} f = \pi$.

Example 7.5.3 The restriction in Stokes' theorem that $\mathcal{M}$ be C^2 can be relaxed. Consider, for example,

$$\mathcal{M} = \{(x, y, z) \in \mathbb{R}^3 \colon z^2 = x^2 + y^2 \text{ and } 0 \le z \le 1\}.$$

This is a cone which sits above the disk

$$D = \{(x, y, 0) \colon x^2 + y^2 \le 1\}$$

in the xy-plane and fails to be smooth at the origin. We can orient $\mathcal{M}$ and its boundary by taking the standard orientation of D and its boundary and projecting them onto $\mathcal{M}$. This gives us

$$w = \frac{1}{\sqrt{2}}\left((dx \wedge dy + \frac{y}{r}\,(dx \wedge dz) + \frac{x}{r}\,(dy \wedge dz)\right),$$

where $r = \sqrt{x^2 + y^2}$, and

$$\partial w = -y\,dx + x\,dy.$$

We can think of $\mathcal{M}$ as being composed of a finite number of triangles, $\Delta_1, \ldots, \Delta_n$, and each triangle can be thought of as a sum of three 2-cells, $\Delta_i = c_1 + c_2 + c_3$. See Figures 7.5.5 and 7.5.6. Therefore $\mathcal{M}$ can be identified with a chain c such that ∂c and $\partial\mathcal{M}$ amount to the same thing. Since Stokes' theorem holds for chains, it holds for $\mathcal{M}$. (It may be objected that the orientation w is undefined at the origin, but since this is a set of measure zero, this does not matter.)

Example 7.5.4 The restriction in Stokes' theorem that $\mathcal{M}$ be compact is crucial. For instance, if we take $\mathcal{M}$ to be the open disk $\{(x, y) \in \mathbb{R}^2 \colon x^2 + y^2 < 1\}$ with the standard orientation, then $\mathcal{M}$ is certainly an oriented manifold, but $\int_{\partial\mathcal{M}} f$ cannot

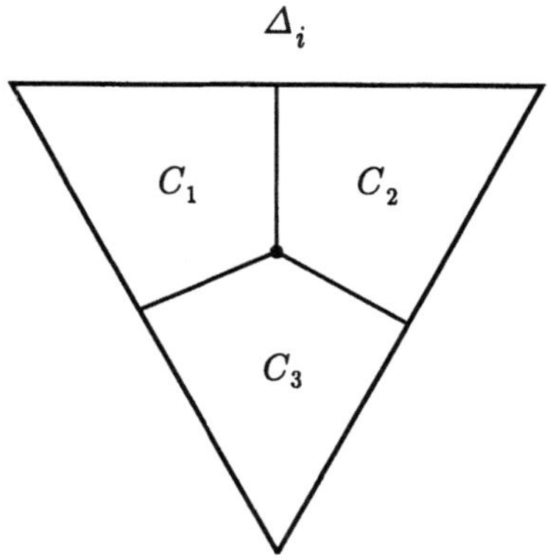

FIGURE 7.5.6.

be evaluated since $\partial\mathcal{M} = \emptyset$. Compact sets in $\mathbb{R}^N$ are those which are closed and bounded. Stokes' theorem fails here because this $\mathcal{M}$ is not closed. We leave as an exercise the case of a closed but unbounded manifold on which Stokes' theorem fails.

Exercises

1. Review the argument that the induced orientation ∂w of the boundary of a manifold is well defined and fill in any gaps you might find in the reasoning. Pay particular attention to the assertion $\beta > 0$.

2. Show that in Example 7.5.3, the orientation induced on $\mathcal{M}$ and its boundary by projecting upward from D is indeed given by

$$w = \frac{1}{\sqrt{2}}\left((dx \wedge dy) + \frac{y}{r}\,(dx \wedge dz) - \frac{x}{r}\,(dy \wedge dz)\right),$$

where $r = \sqrt{x^2 + y^2}$, and

$$\partial w = -y\,dx + x\,dy.$$

3. Let

$$\mathcal{M} = \{(x, y, z) \in \mathbb{R}^3 : x^2 + y^2 = 1 \quad \text{and} \quad z \geq 0\}.$$

   (a) Define an orientation w for $\mathcal{M}$ and find the induced orientation ∂w on $\partial\mathcal{M}$.

   (b) Show that if we take f to be the differential 1-form $-y\,dx + x\,dy$, then Stokes' theorem fails to hold on $\mathcal{M}$. Explain this fact.

4. Let $\mathcal{M} = \{(x, y) \in \mathbb{R}^2 : x^2 + y^2 \geq 1\}$.

(a) If $\mathcal{M}$ has orientation $w = dx \wedge dy$, find ∂w.

(b) If f is the differential 1-form $-y\ e^{-(x^2+y^2)}dx + x\ e^{-(x^2+y^2)}dy$, show that even though $\mathcal{M}$ is not compact, we have

$$\int_{\mathcal{M}} df = \int_{\partial\mathcal{M}} f.$$

5. Supply the details of the proof that

$$E(x) = \begin{cases} e^{-1/(1-x^2)} & \text{if } -1 < x < 1 \\ 0 & \text{otherwise} \end{cases}$$

is C^∞.

6. Let $n(x, y)$ be the outward unit normal vector at every point of the unit circle $x^2 + y^2 = 1$ in $\mathbb{R}^2$. Show that $^*n(x, y)$ points in the counterclockwise direction around the unit circle.

7. Let $\mathcal{M}$ be an oriented K-manifold and $g_0 : U_0 \to \mathcal{M}$ be a coordinate patch such that U_0 is connected. Suppose the orientation of g_0 does not agree with that of $\mathcal{M}$. Define U_1 and $g_1 : U_1 \to \mathcal{M}$ by

$$U_1 = \{(-x_1, x_2, \ldots, x_K) : (x_1, x_2, \ldots, x_K) \in U_0\}$$

and

$$g_1(-x_1, x_2, \ldots, x_K) = g_0(x_1, x_2, \ldots, x_K).$$

Show that g_1 is a coordinate patch such that $g_1(U_1) = g_0(U_0)$ and the orientation of g_1 agrees with that of $\mathcal{M}$.

8. Let $\mathcal{M}$ be the manifold in $\mathbb{R}^3$ defined by $x^2 + y^2 - z^2 = 1$ and $-1 \le z \le 1$.

(a) Let $f(x, y, z) = x^2 + y^2 - z^2 - 1$ and show that ∇f is never zero on $\mathcal{M}$. Define the orientation of $\mathcal{M}$ to be

$$w(p) = \frac{^*\big(\nabla f(p)\big)}{|\nabla f(p)|}.$$

Find an analytic expression for w.

(b) Give a sketch of $\mathcal{M}$ indicating the direction of the induced orientation, ∂w, on $\partial\mathcal{M}$.

9. Recall that S^1 is the unit circle in $\mathbb{R}^2$. If S^1 has the counterclockwise orientation, then without making use of any parametrization, compute

(a) $\int_{S^1} dx$.

(b) $\int_{S^1} x\,dy - y\,dx$.

10. If $\mathcal{M}$ is a compact N-manifold in $\mathbb{R}^N$ with the standard orientation of $\mathbb{R}^N$, show that its N-dimensional volume is given by $\int_{\partial\mathcal{M}} f$ where

$$f(x_1, \dots, x_N) = \frac{1}{N}\sum_{i=1}^{N}(-1)^{i+1}x_i\, dx_1 \wedge \cdots \wedge \overline{dx_i} \wedge \cdots \wedge dx_N.$$

7.6 Applications

The standard theorems of vector analysis, namely Green's theorem, the divergence theorem, and Stokes' theorem in $\mathbb{R}^3$, turn out to be special instances of Stokes' theorem for manifolds.

The first of these we leave as an exercise:

Theorem 7.6.1 (Green's theorem) *If $\mathcal{M}$ is a compact 2-manifold in $\mathbb{R}^2$ with the standard orientation $dx_1 \wedge dx_2$, then for real-valued functions f_1 and f_2 we have*

$$\int_{\partial\mathcal{M}} f_1 dx_1 + f_2 dx_2 = \int_{\mathcal{M}} \left(D_1 f_2 - D_2 f_1\right) dx_1 \wedge dx_2.$$

Note: We know from an earlier discussion that $\partial\mathcal{M}$ must have a "counterclockwise" orientation if it is a closed curve.

Our next theorem requires us to talk about the outward unit normal vector from the boundary of a manifold. We have never defined that idea carefully; however we have used it in heuristic discussions to motivate the definition of induced orientation. We will now reverse the flow of those ideas and use induced orientation to define outward normal vector.

Suppose $\mathcal{M}$ is an N-manifold with boundary in $\mathbb{R}^N$. Recall that the Hodge star operator assigns to a simple K-vector in $\mathbb{R}^N$ an $(N-K)$-vector which is its "orthogonal" complement. If $\mathcal{M}$ has orientation w, then at each point x in $\partial\mathcal{M}$ we want the outward unit normal vector $n(x)$ to be such that ${}^*n(x) = \partial w(x)$. By Corollary 6.4.1 we know that ${}^{**}n(x) = (-1)^{N-1}n(x) ={}^* \partial w(x)$.

Clearly we want the following:

Definition 7.6.1 If $\mathcal{M}$ is an N-manifold with boundary in $\mathbb{R}^N$ and has orientation $dx_1 \wedge dx_2 \wedge \cdots \wedge dx_N$, then at every point x of $\partial\mathcal{M}$, the unit *outward normal vector* is

$$n(x) = (-1)^{N-1}\left({}^*\partial w(x)\right).$$

In the case N=3 we have both ${}^*n = \partial w$ and $n ={}^* \partial w$.

If F is a 1-vector field in $\mathbb{R}^N$ and $F = (F_1, \dots, F_N)$, we define

$$\operatorname{div} F = D_1 F_1 + D_2 F_2 + \cdots + D_N F_N.$$

Theorem 7.6.2 (The divergence theorem) *Let $\mathcal{M}$ be a compact 3-manifold in $\mathbb{R}^3$ with the standard orientation $dx_1 \wedge dx_2 \wedge dx_3$. If n is the outward unit normal vector at every point of $\partial\mathcal{M}$ and F is a 1-vector field on $\mathcal{M}$, then*

$$\int_{\partial\mathcal{M}} F \cdot n = \int_{\mathcal{M}} \operatorname{div} F.$$

Proof. We have

$$\int_{\partial\mathcal{M}} F \cdot n = \int_{\partial\mathcal{M}} {}^*F \cdot {}^*n = \int_{\partial\mathcal{M}} {}^*F \cdot \partial w = \int_{\partial\mathcal{M}} {}^*F = \int_{\mathcal{M}} d({}^*F).$$

Since $d({}^*F) = \operatorname{div} F \, dx_1 \wedge dx_2 \wedge dx_3$, we are done. □

If $F = (F_1, F_2, F_3)$ is a 1-vector field in $\mathbb{R}^3$, we define

$$\operatorname{curl} F = (D_2 F_3 - D_3 F_2)\, dx_1 + (D_1 F_3 - D_3 F_1)\, dx_2 + (D_1 F_2 - D_2 F_1)\, dx_3.$$

Theorem 7.6.3 (The classical Stokes' theorem) *Let $\mathcal{M}$ be a compact 2-manifold in $\mathbb{R}^3$ with orientation w. If F is a 1-vector field on $\mathcal{M}$ and we define a unit normal vector n to $\mathcal{M}$ by $n =^* w$, then*

$$\int_{\partial\mathcal{M}} F \cdot T = \int_{\mathcal{M}} \operatorname{curl} F \cdot n,$$

where T is the unit tangent vector to $\partial\mathcal{M}$ having the direction of the induced orientation.

(**Note:** Of course T is just another name for ∂w. The only reason we use T is to give a statement which looks more like the traditional Stokes' theorem.)

Proof. We have

$$\int_{\partial\mathcal{M}} F \cdot \partial w = \int_{\partial\mathcal{M}} F = \int_{\mathcal{M}} dF = \int_{\mathcal{M}} (dF) \cdot w = \int_{\mathcal{M}} {}^*(dF) \cdot {}^*w.$$

Since ${}^*w = n$ and ${}^*(dF) = \operatorname{curl} F$, we are done. □

Now we consider a particularly simple yet important example of an integral of a differential form. Suppose f is a real-valued function defined over an oriented 1-manifold $\mathcal{M}$ in $\mathbb{R}^N$. Suppose we can realize this 1-manifold as a 1-cell $c : I \to \mathcal{M}$, where the orientation of c agrees with that of $\mathcal{M}$, so that for any 1-form g over $\mathcal{M}$ we have

$$\int_{\mathcal{M}} g = \int_c g.$$

Then it is straightforward to show that

$$\int df = f(c(1)) - f(c(0)).$$

One might rephrase this as

$$\int_{\mathcal{M}} df = f\,(\text{terminal point on } \mathcal{M}) - f\,(\text{initial point on } \mathcal{M}).$$

An interesting point to notice is that the value of the integral does not depend on which curve (or 1-manifold) $\mathcal{M}$ one evaluates along, only on the starting and stopping points. This sort of phenomenon is of great importance in physics where, for example, it turns out that work done in pushing a particle through a gravitational or electric or magnetic field is path-independent. This is generally associated with the fact that the field g is *exact*, that is, that $g = df$ for some real-valued function f. Suppose we have two paths P_1 and P_2 which run from p to q through the vector field g. Suppose further that P_1 and P_2 taken together make up the boundary of some surface $\mathcal{S}$, a surface which is oriented in such a way that the orientation of $\partial\mathcal{S}$ agrees with the orientation of P_1 and runs counter to that of P_2. In effect, $\partial\mathcal{S} = P_1 - P_2$. Then

$$\int_{P_1} g - \int_{P_2} g = \int_{\partial\mathcal{S}} g = \int_{\partial\mathcal{S}} df = \int_{\mathcal{S}} d^2 f = 0.$$

Definition 7.6.2 A differential $(K+1)$-form g is *exact* if there is a K-form f such that $g = df$. A K-form h is *closed* if $dh = 0$.

If a form is exact, it must be closed, because $g = df$ implies $dg = d^2 f = 0$. However the converse does not hold.

Consider, for example, the 1-form

$$h = \frac{-y}{x^2+y^2}\,dx + \frac{x}{x^2+y^2}\,dy$$

defined on $\mathbb{R}^2 - \{0\}$ where x and y have their usual meaning. It is straightforward to calculate that $dh = 0$. However although h is closed, it is not exact. To see this, let us first define $g : \mathbb{R}^2 \to \mathbb{R}^2$ by

$$g(r,\theta) = (r\ \cos(\theta), r\ \sin(\theta)) = (x, y).$$

In effect g changes polar coordinates into the corresponding Cartesian coordinates. It may be calculated that

$$g^*(h) = d\theta.$$

Now suppose we start with a curve $c : I \to \mathbb{R}^2$ in the (r,θ)-plane and carry it over to a curve $g \circ c : I \to \mathbb{R}^2$ in the (x, y)-plane. (It can be shown that every curve in the (x, y)-plane which does not pass through the origin has this form.) If $g \circ c(I)$ never intersects the origin, then integrating h along this curve we obtain

$$\int_{g \circ c} h = \int_c g^*(h) = \int_c d\theta = \theta(c(1)) - \theta(c(0)).$$

That is, the integral of h along the curve amounts to the change of the angular coordinate as one proceeds from the first to the last point of the curve. In particular, closed curves which circle the origin must lead to integrals that are positive or negative integral multiples of 2π depending on how many times the origin is circled and in which direction. But if h were exact, then the integral of h over any closed curve would have to be zero. We conclude that h cannot be exact.

A natural question, then, is this: When is a closed differential form also exact? The answer, strangely enough, depends on the "shape" or "topology" of the domain of the differential form.

Before showing this, we first prove a technical lemma. To see what the lemma is about, imagine f is a differential K-form that changes in a continuously differentiable manner over time. We can think of such an f as consisting of terms of the form $g(x,t)\,dx_{j_1} \wedge \cdots \wedge dx_{j_K}$ where $g(x,t)$ is, for fixed x, a coefficient that varies with time t. The question the lemma addresses is this: How is f at time $t = 0$ related to f at time $t = 1$?

The answer we give to this question involves adding an extra dimension to our manifold, a t-dimension, so in the statement of the lemma, it is convenient to think of f as also including terms of the form $g(x,t)\,dt \wedge dx_{j_1} \wedge \cdots \wedge dx_{j_{K-1}}$. We introduce a map j_t defined in such a way that $j_t^*(f)$ is, in effect, an answer to the question, "What does f look like at time t?" Then the roles of f at times $t = 0$ and $t = 1$ are played by $j_0^*(f)$ and $j_1^*(f)$.

Lemma 7.6.1 *Let U be an open subset of $\mathbb{R}^N$ and define $j_t : U \to U \times [0,1]$ by $j_t(x) = (x,t)$ where $t \in [0,1]$. Let $\mathcal{F}^K(W)$ stand for the set of C^1 differential K-forms on W. For each of $K = 0, 1, 2, \ldots, N$, define a map $H : \mathcal{F}^{K+1}(U \times [0,1]) \to \mathcal{F}^K(U)$ in the following way, where $g(x,t)$ is a C^1, real-valued function:*

(a) Set $H(g(x,t)\,dx_{j_1} \wedge \cdots \wedge dx_{j_{K+1}}) = 0.$

(b) Set $H(g(x,t)\,dt \wedge dx_{j_1} \wedge \cdots \wedge dx_{j_K}) = \left(\int_0^1 g(x,t)\,dt\right) dx_{j_1} \wedge \cdots \wedge dx_{j_K}.$

(c) Now extend H linearly to $\mathcal{F}^{K+1}(U \times [0,1])$.

Then for every C^1 differential K-form f on $U \times [0,1]$ we have

$$dH(f) + H(df) = j_1^*(f) - j_0^*(f).$$

Proof. By the definition of pullback, we must have $j_t^*(dx_i) = dx_i$ for $i = 1, \ldots, N$ and $j_t^*(dt) = 0$. So if f has the form $g(x,t)\,dx_{j_1} \wedge \cdots \wedge dx_{j_K}$, where $g(x,t)$ is a real-valued function, we have $j_t^*(f) = f$. If f has the form $g(x,t)\,dt \wedge dx_{j_1} \wedge \cdots \wedge dx_{j_K}$, then $j_t^*(f) = 0$.

Now suppose $f = g(x,t)\,dx_{j_1}\wedge\cdots\wedge dx_{j_K}$. By definition of H, we have $H(f) = 0$ and hence $dH(f) = 0$. We see that

$$\begin{aligned}
H(df(x,t)) &= H\left(\left(\sum_{j=1}^{N}\frac{\partial g}{\partial x_j}(x,t)dx_j + \frac{\partial g}{\partial t}(x,t)dt\right)\wedge dx_{j_1}\wedge\cdots\wedge dx_{j_K}\right)\\
&= H\left(\frac{\partial g}{\partial t}(x,t)dt\wedge dx_{j_1}\wedge\cdots\wedge dx_{j_K}\right)\\
&= \left(\int_0^1 \frac{\partial g}{\partial t}(x,t)\,dt\right)dx_{j_1}\wedge\cdots\wedge dx_{j_K}\\
&= [g(x,1) - g(x,0)]\,dx_{j_1}\wedge\cdots\wedge dx_{j_K}.
\end{aligned}$$

Then

$$\begin{aligned}
[dH(f) + H(df)](x) &= g(x,1)\,dx_{j_1}\wedge\cdots\wedge dx_{j_K} - g(x,0)\,dx_{j_1}\wedge\cdots\wedge dx_{j_K}\\
&= j_1^*(f) - j_0^*(f).
\end{aligned}$$

Suppose on the other hand that $f(x,t) = g(x,t)\,dt\wedge dx_{j_1}\wedge\cdots\wedge dx_{j_{K-1}}$. We know that $j_1^*(f) = j_0^*(f) = 0$. Now

$$\begin{aligned}
[H(df)](x) &= H\left(\left(\sum_{j=1}^{N}\frac{\partial g}{\partial x_j}(x,t)\,dx_j + \frac{\partial g}{\partial t}(x,t)\,dt\right)\right.\\
&\quad\left.\sum_{j=1}^{N}\quad\wedge dt\wedge dx_{j_1}\wedge\cdots\wedge dx_{j_{K-1}}\right)\\
&= -H\left(\sum_{j=1}^{N}\frac{\partial g}{\partial x_j}(x,t)\,dt\wedge dx_j\wedge dx_{j_1}\wedge\cdots\wedge dx_{j_{K-1}}\right)\\
&= -\left(\sum_{j=1}^{N}\left(\int_0^1\frac{\partial g}{\partial x_j}(x,t)\,dt\right)dx_j\wedge dx_{j_1}\wedge\cdots\wedge dx_{j_{K-1}}\right).
\end{aligned}$$

Also,

$$\begin{aligned}
[dH(f)](x) &= d\left(\left(\int_0^1 g(x,t)\,dt\right)dx_{j_1}\wedge\cdots\wedge dx_{j_{K-1}}\right)\\
&= \sum_{j=1}^{N}\left(\int_0^1\frac{\partial g}{\partial x_j}(x,t)dt\right)dx_j\wedge dx_{j_1}\wedge\cdots\wedge dx_{j_{K-1}}.
\end{aligned}$$

Thus $dH(f) + H(df) = 0 = j_1^*(f) - j_0^*(f)$. □

Note that in the last part of this proof we moved partial differentiation operators from outside an integral to inside. We leave the justification of this step as an exercise.

Definition 7.6.3 We will say a set U in $\mathbb{R}^N$ is *deformable to a point* $p \in U$ provided there is a C^1 map $h: U \times [0, 1] \to U$ such that for all $x \in U$ we have

$$h(x, 0) = x \quad and \quad h(x, 1) = p.$$

Suppose we designate the set $h(U \times \{t\})$ as U_t. Think of t as time. It makes sense to identify U_0 with U and U_1 with the singleton $\{p\}$, and we can imagine the map $t \mapsto U_t$ as describing a continuous evolution in time of the set U into a single point. See Figure 7.6.1.

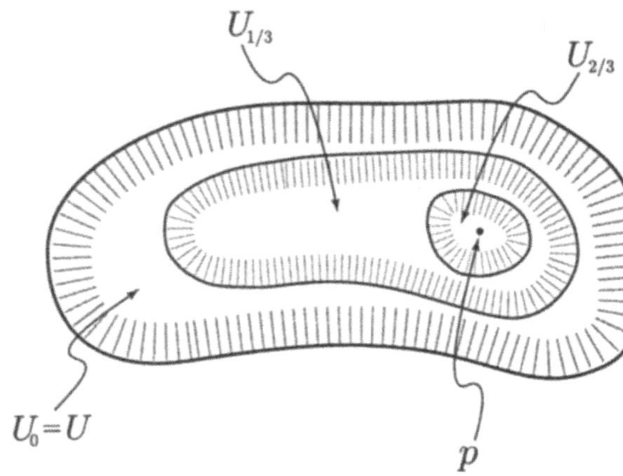

FIGURE 7.6.1.

Example 7.6.1 If we take U to be the square $[-1, 1] \times [-1, 1]$, then it is deformable to $p = (0, 0)$ by the map $h(x, t) = (1 - t)x$. See Figure 7.6.2.

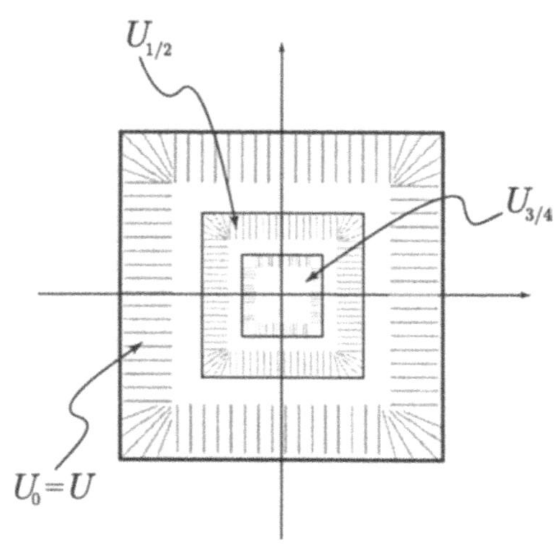

FIGURE 7.6.2.

Example 7.6.2 If we take U to be the annulus in $\mathbb{R}^2$ defined by $1 \le x_1^2 + x_2^2 \le 9$ and $p = (2, 0)$, there is no way U can be deformed to p. This is because U contains closed curves which enclose the origin, and in the continuous deformation of the annulus to p, those closed curves would have to pass through the origin. But this violates the requirement that $U_t \subseteq U$ for all $t \in [0, 1]$. See Figure 7.6.3.

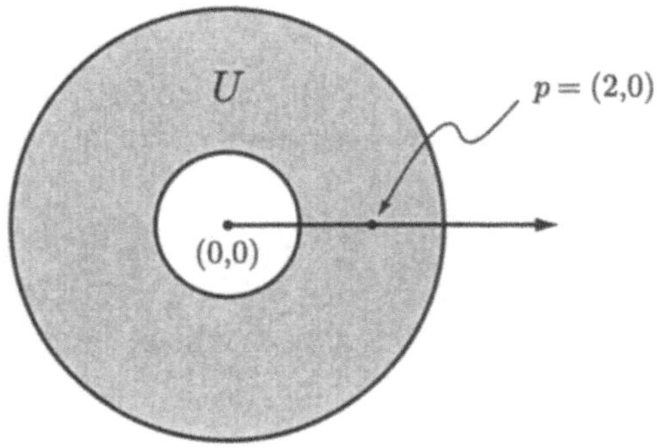

FIGURE 7.6.3.

Theorem 7.6.4 (Poincaré lemma) *If U is an open set in $\mathbb{R}^N$ which can be deformed to a point $p \in U$ and f is a C^1 $(K + 1)$-form on U which satisfies $df = 0$, then there is K-form F on U for which $f = dF$.*

Proof. There is a C^1 map $h: U \times [0, 1] \to U$ such that for all $x \in U$ we have

$$h(x, 0) = x \quad \text{and} \quad h(x, 1) = p.$$

Let H and j_t be as defined in Lemma 7.6.1. We see that $(h \circ j_0)(x) = x$ and $(h \circ j_1)(x) = p$ for all $x \in U$. It follows that if g is a differential form on U, we have

$$j_0^*(h^*(g)) = (h \circ j_0)^*(g) = g$$

and

$$j_1^*(h^*(g)) = (h \circ j_1)^*(g) = 0.$$

By Lemma 7.6.1,

$$dH(h^*(f)) + H(dh^*(f)) = j_1^*(h^*(f)) - j_0^*(h^*(f)) = -f.$$

We also have $H(dh^*(f)) = H(h^*(df)) = 0$. If we set $F = -H(h^*(f))$, we obtain $f = dF$. □

Exercises

1. Prove Green's theorem.

2. Prove that if F is a 1-vector field in $\mathbb{R}^N$, then $d(^*F) = \text{div } F \, dx_1 \wedge \cdots \wedge dx_N$.

3. Generalize the divergence theorem to $\mathbb{R}^N$.

4. Prove that for a 1-vector field F in $\mathbb{R}^3$ we have $*(dF) = \text{curl } F$.

5. Suppose $\mathcal{M}$ is an oriented 1-manifold (curve) in $\mathbb{R}^N$ running from a point p to a point q and f is a real-valued function defined over $\mathcal{M}$. Prove that

$$\int_{\mathcal{M}} df = f(q) - f(p).$$

(Hint: You may assume there is a one-to-one, onto map $c : [0, 1] \to \mathcal{M}$ such that c' is never zero, c' agrees with the orientation of $\mathcal{M}$, and $c(0) = p$ and $c(1) = q$.)

6. If a point mass is situated at 0 in $\mathbb{R}^3$, it produces a gravitational field at every other point x of $\mathbb{R}^3$ which is directed from x to 0 and inversely proportional in magnitude to the square of the distance from x to the origin. Show that this field is exact on $\mathbb{R}^3 - \{0\}$.

7. Show that if

$$h = \frac{-y}{x^2 + y^2}\, dx + \frac{x}{x^2 + y^2}\, dy,$$

on $\mathbb{R}^2 - \{0\}$, then $dh = 0$.

8. Show that if $g : \mathbb{R}^2 \to \mathbb{R}^2$ is defined by

$$g(r, \theta) = (r\ \cos(\theta), r\ \sin(\theta)) = (x, y)$$

and

$$h = \frac{-y}{x^2 + y^2}\, dx + \frac{x}{x^2 + y^2}\, dy,$$

then

$$g^*(h) = d\theta.$$

9. Prove that if f is C^1 on $[a_0, b_0] \times [a_1, b_1] \times \cdots \times [a_N, b_N]$, then

$$D_i \int_{a_0}^{b_0} f(t, x_1, \ldots, x_N)\, dt = \int_{a_0}^{b_0} D_i f(t, x_1, \ldots, x_N)\, dt$$

for $i = 1, 2, \ldots, N$.

10. (Poincaré lemma for 1-forms) Suppose $f = \sum_{i=1}^{N} f_i\, dx_i$ is a 1-form on a subset S of $\mathbb{R}^N$ which is star-shaped with respect to the origin. (To say that S is star-shaped with respect to the origin means that for every $x \in S$, the line segment from 0 to x lies in S.) Define a 0-form g by

$$g(x) = \int_0^1 \sum_{i=1}^{N} f_i(tx)\, x_i\, dt.$$

Show that $f = dg$.

7.7 Manifolds and Differential Forms: an Intrinsic Viewpoint

Our development of manifold theory and differential forms has made essential use of the fact that we consider each K-manifold $\mathcal{M}$ to be lying in some $\mathbb{R}^N$. This embedding of $\mathcal{M}$ in some larger Euclidean space is an *extrinsic* circumstance. If we translate or rotate or deform such a manifold in a continuous, one-to-one fashion, there is an intuitive sense in which it is still essentially "the same" manifold.

The theory of manifolds and differential forms is usually developed in an intrinsic manner, one that is independent of the way in which a manifold $\mathcal{M}$ may be embedded in a larger Euclidean space, one that does not even assume there is such an embedding. But the price that is paid for this approach is, on the one hand, greater abstraction and logical complexity and, on the other hand, a difficulty in seeing the geometric intuition that lies behind the machinery. Our development to this point has been driven by a desire to develop the machinery of differential forms in as concrete a manner as possible and at the same time to develop that all-important intuition.

In this section we indicate how the theory of manifolds and differential forms can be rephrased in the spirit of the more standard, more intrinsic approach. We hope that anyone who has come this far will find it both reasonable and intelligible. Indeed it may even seem simpler than what we have done so far, and if so, we would suggest that this is because most of the proofs are relegated to the exercises, some developments are only indicated or are omitted altogether, and the difficult matter of building intuition has already been attended to.

Let us begin.

Consider for a moment two coordinate patches, $u : U \to \mathcal{M}$ and $v : V \to \mathcal{M}$. We may suppose u assigns a point $p \in \mathcal{M}$ coordinates $(x_1, \ldots, x_K)$ by the rule $u(x_1, \ldots, x_K) = p$, while v assigns the same p coordinates $(y_1, \ldots, y_K)$ by the rule $v(y_1, \ldots, y_K) = p$. (We sometimes call these *local coordinates* because they make sense only for points in the sets $u(U)$ or $v(V)$.) The rule by which we change from one set of coordinates to another, namely $(y_1, \ldots, y_K) = (v^{-1} \circ u)(x_1, \ldots, x_K)$, is an *intrinsic* concept. That is, the rule for coordinate change depends only on how the coordinates $(x_1, \ldots, x_K)$ and $(y_1, \ldots, y_K)$ are assigned to points in M and has absolutely nothing to do with the particular way in which $\mathcal{M}$ is embedded in $\mathbb{R}^N$.

In attempting an intrinsic description of the machinery of analysis on manifolds, we shall rely on the use of local coordinates. We then have two points we must check on: First, we must show that key concepts, such as the differential operator or the integration of differential forms, are independent of our choice of local coordinates. Second, we must see that our new definitions give the same results as our old definitions, that is, that there is an "isomorphism" between the two approaches. However, we shall often only indicate how these tasks can be carried out, leaving the details either as exercises or for another time.

The reader should also be aware that we shall treat the question of differentiability of functions, vector fields, or differential forms in a very cavalier manner. If, for example, we have a real-valued function on a manifold, $f : \mathcal{M} \to \mathbb{R}$, then to say that f is $\mathcal{C}^\infty$ on $\mathcal{M}$ means that for every coordinate patch $u : U \to \mathcal{M}$ which we consider, we assume that $f \circ u : U \to \mathbb{R}$ is $\mathcal{C}^\infty$. Since U is a subset of some $\mathbb{R}^M$, this makes sense. From now on, everything that needs to be differentiable or smooth is automatically assumed to be $\mathcal{C}^\infty$ in the sense indicated here.

The concepts we wish to describe in an intrinsic manner are these:

- Tangent vector to a manifold,
- Push-forward,
- Differential form,
- Pullback,
- Integration of a differential form over a manifold,
- The differential operator.

Our first step in describing any of these concepts is to define differentiation of a real-valued function with respect to local coordinates. Let $\mathcal{M}$ be an M-dimensional manifold, let $f : \mathcal{M} \to \mathbb{R}$, and assume $u : U \to \mathcal{M}$ is a coordinate patch. Let $(x_1, \ldots, x_M)$ be the local coordinate system this coordinate patch assigns to the points of $\mathcal{M}$ lying in $u(U)$. We may regard each x_i as a function from an open subset of $\mathcal{M}$ into $\mathbb{R}$, $x_i : \mathcal{M} \rightarrowtail \mathbb{R}$, namely, for $p \in u(U)$, $x_i(p)$ is the ith coordinate of $u^{-1}(p)$. More symbolically,

$$x_i\ (p) = (u^{-1})_i\ (p),$$

Then for $p \in u(U)$, we define the partial of f with respect to x_i at p by

$$\frac{\partial f}{\partial x_i}\ (p) = D_i\ (f \circ u)\ (x)$$

where $p = u(x)$.

A particularly important instance of this idea occurs when we have two overlapping coordinate patches on $\mathcal{M}$, $u : U \to \mathcal{M}$ and $v : V \to \mathcal{M}$. Let $(x_1, \ldots, x_M)$ be the local coordinate system induced by u and $(y_1, \ldots, y_M)$ be that induced by v. Then

$$\frac{\partial y_i}{\partial x_j}\ (p) = D_j\ (y_i \circ u)\ (x) = D_j\ ((v^{-1})_i \circ u)\ (x) = D_j\ ((v^{-1} \circ u)_i)\ (x)$$

where $p = u(x)$ and $p \in u(U)\ \cap v(V)$.

Theorem 7.7.1 *Suppose f is a real-valued function on the M-dimensional manifold $\mathcal{M}$ and $(x_1, \ldots, x_M)$ and $(y_1, \ldots, y_M)$ are local coordinates on $\mathcal{M}$. Assuming p lies in the domain of both sets of local coordinates, we have the following:*

1. $$\frac{\partial f}{\partial x_i}\ (p) = \sum_{j=1}^{M} \frac{\partial f}{\partial y_j}\ (p)\ \frac{\partial y_j}{\partial x_i}\ (p)$$

2. $\displaystyle\sum_{k=1}^{M} \frac{\partial y_i}{\partial x_k}(p)\,\frac{\partial x_k}{\partial y_j}(p) = \delta_{ij}$ *(Kronecker's delta).*

Proof. We prove only (a) and leave (b) as an exercise. Suppose the local coordinates $(x_1, \ldots, x_M)$ and $(y_1, \ldots, y_M)$ are induced by the coordinate patches $u : U \to \mathcal{M}$ and $v : V \to \mathcal{M}$ respectively and that $p = u(x) = v(y)$. Then

$$\begin{aligned}
\frac{\partial f}{\partial x_i}(p) &= D_i\,(f \circ u)\,(x) \\
&= D_i\,(f \circ v \circ v^{-1} \circ u)\,(x) \\
&= \sum_{j=1}^{M} D_j\,(f \circ v)\,(y)\, D_i\,((v^{-1})_j \circ u)\,(x) \\
&= \sum_{j=1}^{M} \frac{\partial f}{\partial y_j}(p)\,\frac{\partial y_j}{\partial x_i}(p).
\end{aligned}$$

□

Example 7.7.1 Suppose that $\mathcal{M}$ is a 2-manifold with local coordinate system (x, y) defined by the coordinate patch $u : U \to \mathcal{M}$ where $u(x, y) = p$. Consider the function $f : \mathcal{M} \to \mathbb{R}$ that satisfies $f(p) = x^2 + y^2$ on the domain of the local coordinates. Then

$$\frac{\partial f}{\partial x}(p) = D_1(f \circ u)(x, y) = D_1(x^2 + y^2) = 2x$$

and similarly $\frac{\partial f}{\partial y} = 2y$.

The important thing here is that one can, in practice, ignore u. The partial derivatives of f are determined by the form of f in terms of the local coordinates. But it is precisely u that gives information on how $\mathcal{M}$ is embedded in some $\mathbb{R}^N$. The local coordinates carry no such information. One can think of them as being painted on $\mathcal{M}$, like grid lines on a balloon, and $\mathcal{M}$ being moved around $\mathbb{R}^N$ and deformed in various ways. Of course the painted lines of the local coordinates are carried along on $\mathcal{M}$ during these various maneuvers. As long as the formula for f remains the same *in terms of the local coordinates*, such quantities as $\frac{\partial f}{\partial x}$ will remain the same regardless of how $\mathcal{M}$ is embedded. This is the sense in which the partial differentiation discussed here is an intrinsic concept.

Now let us suppose that $\mathcal{M}$ has a second local coordinate system (r, θ) and that at p these local coordinates are related by the equations $x = r\cos\theta$ and $y = r\sin\theta$. Our function f must satisfy $f(p) = r^2$ and we readily calculate $\frac{\partial f}{\partial r}(p) = 2r$. However we can also, without bothering to refer the coordinate patches that define (x, y) and (r, θ), calculate $\frac{\partial x}{\partial r} = \cos\theta$ and $\frac{\partial y}{\partial r} = \sin\theta$ and then use the chain rule as follows:

$$\begin{aligned}\frac{\partial f}{\partial r} &= \frac{\partial f}{\partial x}\frac{\partial x}{\partial r} + \frac{\partial f}{\partial y}\frac{\partial y}{\partial r}\\ &= 2x\cos\theta + 2y\sin\theta\\ &= 2r.\end{aligned}$$

Note that one result of our elaborate machinery is to reduce our calculations to those of elementary calculus.

We now introduce a "new" concept of tangent vector in which a tangent vector is thought of as a kind of directional differentiation operator. Let $u : U \to \mathcal{M}$ and suppose p is a point of $\mathcal{M}$ lying in this coordinate patch. By a *tangent vector to $\mathcal{M}$ at p* we mean a differential operator of the form $\sum_{i=1}^{M} a_i \left.\frac{\partial}{\partial x_i}\right|_p$ where each $a_i \in \mathbb{R}$. The action of this operator on a real-valued function $f : \mathcal{M} \to \mathbb{R}$ is defined by

$$\left(\sum_{i=1}^{M} a_i \left.\frac{\partial}{\partial x_i}\right|_p\right)(f) = \sum_{i=1}^{M} a_i \frac{\partial f}{\partial x_i}(p).$$

Of course by defining tangent vectors in terms of local coordinates, they become objects which are defined intrinsically, not in terms of how $\mathcal{M}$ is embedded in a Euclidean space.

Notice that $\left.\frac{\partial}{\partial x_i}\right|_p$ means differentiation carried out at the point p. (If we were to write $\sum_{i=1}^{M} a_i \frac{\partial}{\partial x_i}$ without the point p, we could interpret our operator as a tangent vector field; the a_i's could then either be constants or real-valued functions whose values changed from point to point of $\mathcal{M}$.)

Our "old" concept of a tangent vector at p was anything of the form $u'(x)\left(\sum_{i=1}^{M} a_i\, e_i\right)$ where $p = u(x)$. We claim these two concepts are, in essence, the same thing in different form. Every "old" tangent vector corresponds to a "new" tangent vector via the one-to-one correspondence

$$u'(x)\left(\sum_{i=1}^{M} a_i e_i\right) \longleftrightarrow \sum_{i=1}^{M} a_i \left.\frac{\partial}{\partial x_i}\right|_p$$

where $p = u(x)$. One important thing we would like to see is that these two concepts both transform in the same way under a change of local coordinates. Suppose $v : V \to \mathcal{M}$ is a second coordinate patch inducing local coordinates $(y_1, \ldots, y_M)$ and that $p \in u(U) \cap v(V)$. Using the fact that

$$\frac{\partial y_i}{\partial x_j}(p) = D_j\left((v^{-1} \circ u)_i\right)(x),$$

it is a straightforward exercise to show that

$$D_j\ u_i\ (x) = \sum_{k=1}^{M} \frac{\partial y_k}{\partial x_j}\ (p)\ D_k\ v_i\ (y)$$

where $p = u(x) = v(y)$. From this it can be shown (another exercise) that

$$u'(x)\ e_i = v'(y)\ \left(\sum_{j=1}^{M} \frac{\partial y_j}{\partial x_i}\ (p)\ e_j \right).$$

If we compare this to differential operators, we find from Theorem 7.7.1 that

$$\frac{\partial}{\partial x_i}\bigg|_p = \sum_{j=1}^{M} \frac{\partial y_j}{\partial x_i}\ (p)\ \frac{\partial}{\partial y_j}\bigg|_p .$$

Thus the "old" and "new" tangent vectors have the same transformation law.

We now indicate some of the important things that can be said about these "new" tangent vectors. We have indicated how to verify one of the properties and the others are left as exercises.

Theorem 7.7.2 *Let $(x_1, \ldots, x_M)$ be a local coordinate system on the manifold $\mathcal{M}$, and let p be a point lying in this local coordinate system. Then the following hold:*

1. *With the operations of addition and scalar multiplication defined by*

$$\left(\sum_{i=1}^{M} a_i\ \frac{\partial}{\partial x_i}\bigg|_p \right) + \left(\sum_{i=1}^{M} b_i\ \frac{\partial}{\partial x_i}\bigg|_p \right) = \sum_{i=1}^{M} (a_i + b_i)\ \frac{\partial}{\partial x_i}\bigg|_p \qquad \textit{and}$$

$$c \left(\sum_{i=1}^{M} a_i\ \frac{\partial}{\partial x_I}\bigg|_p \right) = \sum_{i=1}^{M} c\ a_i\ \frac{\partial}{\partial x_i}\bigg|_p ,$$

where each of c, a_i, b_i is a real number, the set of tangent vectors to $\mathcal{M}$ at p is a vector space.

2. $\left\{ \frac{\partial}{\partial x_1}\Big|_p , \ldots , \frac{\partial}{\partial x_M}\Big|_p \right\}$ *is a basis for the space of tangent vectors to $\mathcal{M}$ at p.*

3. *If p lies in a second local coordinate system $(y_1, \ldots, y_M)$, then*

$$\frac{\partial}{\partial x_i}\bigg|_p = \sum_{j=1}^{M} \frac{\partial y_j}{\partial x_i}\ (p)\ \frac{\partial}{\partial y_j}\bigg|_p .$$

We now consider the notion of a push-forward.

Suppose $g : \mathcal{M} \to \mathcal{N}$ where $\mathcal{M}$ and $\mathcal{N}$ are M- and N-manifolds, respectively. If t is a tangent vector to $\mathcal{M}$ at $p \in \mathcal{M}$, then the "old" push-forward of t is given by

$$g_*(p)t = g'(p)t.$$

Notice that $g'(p)$ makes sense because $\mathcal{M}$ and $\mathcal{N}$ are embedded in Euclidean spaces and $g'(p)$ can be calculated for a map between Euclidean spaces. The vector $g'(p)t$ will be tangent vector to $\mathcal{N}$ at the point $q = g(p)$.

Our "new" push-forward must be describable in terms of differential operators, so first we introduce a real-valued function for operators to operate on. Let $f : \mathcal{N} \to \mathbb{R}$. Notice that f generates a real-valued function on $\mathcal{M}$, namely $f \circ g : \mathcal{M} \to \mathbb{R}$. Then we define our "new" push-forward by

$$g_*(D\big|_p)f = D\big|_p(f \circ g) = D(f \circ g)(p)$$

where D is a differential operator of the type described before and the p beside the operator indicates that the function is to be evaluated at p after the differential operator has been applied.

It is not immediately clear that we have defined a differential operator on f in the sense of our previous discussion. That is, if we are given local coordinates $(x_1, \ldots, x_M)$ on $\mathcal{M}$ and $(y_1, \ldots, y_N)$ on $\mathcal{N}$, p and q lying in the domains of these local coordinates, and $q = g(p)$, then we would like to see that

$$g_*\left(\frac{\partial}{\partial x_i}\bigg|_p\right) = \sum_{j=1}^{N} a_j \frac{\partial}{\partial y_j}\bigg|_q$$

where the a_j's are scalars that possibly depend on p. To see this, suppose that $(x_1, \ldots, x_M)$ and $(y_1, \ldots, y_N)$ are induced on $\mathcal{M}$ and $\mathcal{N}$ by the respective coordinate patches $u : U \to \mathcal{M}$ and $v : V \to \mathcal{N}$. Then one calculates

$$\begin{aligned}
g_*\left(\frac{\partial}{\partial x_i}\bigg|_p\right)(f) &= \frac{\partial}{\partial x_i}(f \circ g)(p) \\
&= D_i(f \circ g \circ u)(x) \\
&= D_i(f \circ v \circ v^{-1} \circ g \circ u)(x) \\
&= \sum_{j=1}^{N} D_j(f \circ v)(y)\, D_i((v^{-1} \circ g \circ u)_j)(x) \\
&= \sum_{j=1}^{N} \frac{\partial f}{\partial y_j}(q)\, \frac{\partial(y_j \circ g)}{\partial x_i}(p)
\end{aligned}$$

where $p = u(x)$ and $q = v(y)$. Thus we see that

$$g_*\left(\frac{\partial}{\partial x_i}\bigg|_p\right) = \sum_{j=1}^{N} \frac{\partial(y_j \circ g)}{\partial x_i}(p)\left(\frac{\partial}{\partial y_j}\bigg|_q\right)$$

as desired. Notice that $\frac{\partial}{\partial y_j}$ and $\frac{\partial(y_j \circ g)}{\partial x_i}$ are intrinsically defined objects and that $\frac{\partial(y_j \circ g)}{\partial x_i}$ depends not on how $\mathcal{M}$ and $\mathcal{N}$ are embedded in Euclidean spaces but rather on how one set of local coordinates transforms into another set of local coordinates. Thus the spirit of the transformation formula is an intrinsic one.

Example 7.7.2 Suppose $\mathcal{M}$ and $\mathcal{N}$ are 2-manifolds where $\mathcal{M}$ has a local coordinate system (ρ, θ) induced by the coordinate patch $u : U \to \mathcal{M}$ and $\mathcal{N}$ has a local coordinate system (x, y) induced by $v : V \to \mathcal{N}$. Suppose further that these local coordinates are related as follows:

$$(x, y) = (v^{-1} \circ g \circ u)(\rho, \theta) = (\rho e^{\theta}, \rho e^{-\theta}).$$

In this last equation we treat x and y as real numbers. Now let p be a point in $\mathcal{M}$ and suppose that $u(\rho, \theta) = p$ and $v(x, y) = q = g(p)$. We know that $\left\{ \frac{\partial}{\partial \rho}\big|_p, \frac{\partial}{\partial \theta}\big|_p \right\}$ is a basis for $T_p\mathcal{M}$ and $\left\{ \frac{\partial}{\partial x}\big|_q, \frac{\partial}{\partial y}\big|_q \right\}$ is a basis for $T_q\mathcal{N}$. In what follows, we treat x and y as functions. We see that

$$\frac{\partial(x \circ g)}{\partial \rho}(p) = D_1(x \circ g \circ u)(\rho, \theta) = D_1(\rho\, e^{\theta}) = e^{\theta}.$$

Similarly, $\frac{\partial(x \circ g)}{\partial \rho}(p) = e^{-\theta}$. We can now compute push-forwards:

$$\begin{aligned} g_*\left(\frac{\partial}{\partial \rho}\bigg|_p \right) &= \frac{\partial(x \circ g)}{\partial \rho}(p)\, \frac{\partial}{\partial x}\bigg|_q + \frac{\partial(y \circ g)}{\partial \rho}(p)\, \frac{\partial}{\partial y}\bigg|_q \\ &= e^{\theta}\, \frac{\partial}{\partial x}\bigg|_q + e^{-\theta}\, \frac{\partial}{\partial y}\bigg|_q . \end{aligned}$$

If we do a little algebra and assume all quantities involved are positive, we can express this a bit differently and slightly more concisely as

$$g_*\left(\frac{\partial}{\partial \rho} \right) = \sqrt{\frac{x}{y}}\, \frac{\partial}{\partial x} + \sqrt{\frac{y}{x}}\, \frac{\partial}{\partial y}.$$

Similarly,

$$g_*\left(\frac{\partial}{\partial \theta} \right) = x\, \frac{\partial}{\partial x} - y\, \frac{\partial}{\partial y}.$$

Returning to a discussion of push-forwards in general, we now recall that this is the "new" push-forward we are dealing with. How does it compare with the "old" push-forward? Keeping in mind the correspondence

$$\frac{\partial}{\partial x_i}\bigg|_p \longleftrightarrow u'(x)e_i \quad \text{and} \quad \frac{\partial}{\partial y_j}\bigg|_q \longleftrightarrow v'(y)e_j \ ,$$

we leave it as an exercise for the reader to show that

$$g_*(p)\ (u'(x)\ e_i) = \sum_{j=1}^{N} \frac{\partial(y_j \circ g)}{\partial x_i}(p)\ v'(y)\ e_j,$$

and thus the two concepts of push-forward give the same results.

A standard modern treatment of differential forms requires us to introduce the idea of a dual space.

Let V be a finite dimensional vector space over $\mathbb{R}$. By the *dual space* of V, V^*, we mean the set of linear maps $\varphi : V \to \mathbb{R}$. We define operations of addition and scalar multiplication on V^* in a pointwise fashion:

$$(\varphi + \psi)(x) = \varphi(x) + \psi(x) ,$$
$$(c\,\varphi)(x) = c\,\varphi(x) ,$$

where φ , $\psi \in V^*$ and $c \in \mathbb{R}$. We leave the establishment of the following properties of V^* as an exercise:

Theorem 7.7.3 *With the above definitions of addition and scalar multiplication, we have the following:*

1. *V^* is a vector space over $\mathbb{R}$,*
2. *If $b_1, \ldots, b_N$ is a basis for V, let $\beta_1, \ldots, \beta_2$ be the unique elements of V^* satisfying $\beta_i(b_j) = \delta_{ij}$ (Kronecker's delta). Then $\{\beta_1, \ldots, \beta_N\}$ is a basis for V^*.*

We are now in a position to define 1-forms on a manifold. Recall that when we considered a differential form f earlier, our interest was ultimately in integrating $f \cdot w$ where w was an orientation of the manifold. Notice that the map $w \mapsto f \cdot w$ is a linear transformation sending vectors (or M-vectors) to real numbers. Building on this idea, we would like to start with tangent vectors to a manifold and consider 1-forms as elements of the dual space of the spaces of tangent vectors. However, this way of expressing things does not capture the full complexity of the situation, for there are many different spaces of tangent vectors, one for every point on the manifold. We must proceed with greater care and subtlety.

If p is a point on the manifold $\mathcal{M}$, let us denote the space of tangent vectors to $\mathcal{M}$ at p by $T_p\mathcal{M}$. By a *1-form* on $\mathcal{M}$, we shall then mean a map $f : \mathcal{M} \to \bigcup_{p \in \mathcal{M}} (T_p\mathcal{M})^*$ having the property that $f(p) \in (T_p\mathcal{M})^*$ for every $p \in \mathcal{M}$. (There is also a sense in which we want f to be C^∞, but we shall not pursue that line of thought.)

One of the simplest and most important examples of this is found by starting with local coordinates $(x_1, \ldots, x_M)$ on $\mathcal{M}$ and defining 1-forms $dx_1, dx_2, \ldots, dx_M$ on the domain of $(x_1, \ldots, x_M)$ by requiring each $dx_i(p)$ to be the unique element of $(T_p\,\mathcal{M})^*$ satisfying

$$dx_i\;(p)\left(\frac{\partial}{\partial x_j}\bigg|_p\right) = \delta_{ij} .$$

We know from Theorem 7.7.3 that $\{dx_1(p), \ldots, dx_M(p)\}$ is a basis for $(T_p\mathcal{M})^*$, therefore any 1-form f defined on the domain of $(x_1, \ldots, x_M)$ can be written in the form

$$f = \sum_{i=1}^{M} f_i\; dx_i$$

where each f_i is a real-valued function. An obvious and important question is this: What is the transformation law for 1-forms when switching between different local coordinate systems? We leave the following as an exercise:

Theorem 7.7.4 *Suppose* $(x_1, \ldots, x_M)$ *and* $(y_1, \ldots, y_M)$ *are overlapping local coordinate systems on the manifold* $\mathcal{M}$*. Then whenever both sets of local coordinates are defined, we have*

$$dy_i = \sum_{j=1}^{M} \frac{\partial y_i}{\partial x_j}\, dx_j \,.$$

We can also ask, what do our "new" 1-forms correspond to in terms of our "old" 1-forms? We know that given a point p on a manifold $\mathcal{M}$ and local coordinates $(x_1, \ldots, x_M)$ induced by a coordinate patch $u : U \to \mathcal{M}$, then the correspondence between "new" and "old" tangent vectors is given by

$$\left.\frac{\partial}{\partial x_i}\right|_p \longleftrightarrow u'(x)\, e_i$$

where $u(x) = p$. Because $dx_i\ (p) \left(\left.\frac{\partial}{\partial x_j}\right|_p \right) = \delta_{ij}$, we would like to set up our correspondence between "new" and "old" 1-forms in such a way that

$$dx_i\ (p) \longleftrightarrow w \in \mathbb{R}^M$$

where

$$w \cdot u'(x)\, e_j = \delta_{ij}.$$

It is easily seen that the correct choice for our correspondence is given by

$$dx_i(p) \longleftrightarrow (u'(x)^{-1})^{\diamond}\, e_i$$

where, we recall, $(u'(x)^{-1})^{\diamond}$ is the dual of $u'(x)^{-1}$.

It is now easy to define the pullback of 1-forms. Suppose we have a map between manifolds, $g : \mathcal{M} \to \mathcal{N}$, and f is a 1-form on $\mathcal{N}$. Then we define the *pullback* of f, $g^*(f)$, to be the unique 1-form on $\mathcal{M}$ satisfying

$$g^*(f)(t) = f(g_*(t)) \tag{7.1}$$

where t ranges over all tangent vectors to $\mathcal{M}$ at which $f(g_*(t))$ can be evaluated. Of course it must be understood that if t is a tangent vector to $\mathcal{M}$ at p, then it must be possible to evaluate f at $q = g(p)$, we must have $f(q) \in (T_q\, \mathcal{N})^*$, and in this case (1) can be more precisely written as

$$(g^*(f(q)))(t) = (f(q))\ (g_*(t)).$$

If we were to identify the process of evaluating a dual vector at a vector, $f(t)$, with that of taking a dot product of two vectors, $f \cdot t$, then (1) becomes

$$g^*(f)\ \cdot t = f\ \cdot\ (g_*(t)).$$

Since this is essentially the way we defined our "old" pullback, there is no question that our "new" pullback corresponds to the "old" pullback.

We leave establishment of the following properties of the pullback as an exercise:

Theorem 7.7.5 *Let $g : \mathcal{M} \to \mathcal{N}$ where $\mathcal{M}$ and $\mathcal{N}$ are manifolds. Then the following hold:*

1. *g^* is linear on 1-forms.*
2. *If $(x_1, \ldots, x_M)$ and $(y_1, \ldots, y_N)$ are local coordinates on $\mathcal{M}$ and $\mathcal{N}$ respectively, then*

$$g^*(dy_i) = \sum_{j=1}^{M} \frac{\partial(y_i \circ g)}{\partial x_j}\, dx_j$$

whenever the indicated operations can be carried out.

Example 7.7.3 Suppose $\mathcal{M}$ and $\mathcal{N}$ are 2-manifolds with local coordinate systems (x_1, x_2) and (y_1, y_2), respectively. Suppose further that g is a map, $g : \mathcal{M} \to \mathcal{N}$, with the property that if p has coordinates (x_1, x_2) and $g(p)$ has coordinates (y_1, y_2), then $y_1 = x_1^2 - x_2^2$ and $y_2 = 2x_1x_2$. One readily calculates, as in Example 7.7.2, that $\frac{\partial(y_1 \circ g)}{\partial x_1} = 2x_1$ and $\frac{\partial(y_1 \circ g)}{\partial x_2} = -2x_2$. Then

$$g^*(dy_1) = \frac{\partial(y_1 \circ g)}{\partial x_1}\, dx_1 + \frac{\partial(y_1 \circ g)}{\partial x_2}\, dx_2 = 2x_1\, dx_1 - 2x_2\, dx_2.$$

Similarly one shows that

$$g^*(dy_2) = 2x_2\, dx_1 + 2x_1\, dx_2.$$

Before proceeding further, we must revisit the idea of the wedge product.

For any finite dimensional vector space V over $\mathbb{R}$, one may construct a wedge product and spaces of k-vectors, $\wedge^K V$. One may do this by imitating the construction of Chapter 6 or by using the more general methods of multilinear algebra. The formal, algebraic properties of such a construction will just be those we have seen before. For example, if $\{v_1, \ldots, v_N\}$ is a basis for V, then a basis for $\wedge^K V$ is given by the set of $v_{i_1} \wedge \cdots \wedge v_{i_K}$ such that $i_1 < i_2 < \cdots < i_K$, assuming $1 \le K \le N$. We do not attempt such a construction but simply take it for granted.

Because of this, if $(x_1, \ldots, x_M)$ is a local coordinate system on the manifold $\mathcal{M}$ and p lies in the domain of $(x_1, \ldots, x_M)$, then we accept without discussion that

$$\left\{ \frac{\partial}{\partial x_{i_j}}\bigg|_p \wedge \cdots \wedge \frac{\partial}{\partial x_{i_K}}\bigg|_p : i_1 < i_2 < \cdots < i_K \right\}$$

is a basis for $\bigwedge^K (T_p\mathcal{M})$ and that every K-form defined on the domain of $(x_1, \ldots, x_M)$ can be written in the form

$$\sum_{i_1 < \cdots < i_K} f_{i_1 \ldots i_K}\, dx_{i_1} \wedge \cdots \wedge dx_{i_K}$$

where each $f_{i_1 \dots i_K}$ is a real-valued function and $1 \le K \le M$. We note also that definition of the "new" push-forward and pullback can be extended to wedge products in such a fashion that

$$g_*(t_1 \wedge \cdots \wedge t_K) = g_*(t_1) \wedge \cdots \wedge g_*(t_K)$$

and

$$g^*(f_1 \wedge \cdots \wedge f_K) = g^*(f_1) \wedge \cdots \wedge g^*(f_K)$$

where the t_i's are tangent vectors and the f_i's are 1-forms.

One of the most important things one can do with differential forms is integrate them over manifolds. In describing how this is done, we ignore the problem of describing an orientation of a manifold in terms of the "new" concepts and consider only how one might integrate over the domain of a set of local coordinates $(x_1, \dots, x_M)$ on $\mathcal{M}$. Let f be an M-form defined over this region. We can write

$$f = g\, dx_1 \wedge \cdots \wedge dx_M$$

where we are thinking of each x_i as a real-valued function, $x_i : \mathcal{M} \rightarrowtail \mathbb{R}$, and $g = h(x_1, \dots, x_M)$ is the composition of a real-valued function h and the one-to-one vector-valued function $(x_1, \dots, x_M)$. If R is the domain of $(x_1, \dots, x_M)$, we then define

$$\int_R f = \pm \int_U h(t_1, \dots, t_M) dt_1 \dots dt_M$$

where the last integral is understood to be a standard integral over the M-dimensional set U and $t_1, \dots, t_M$ are now variables rather than functions. We write $+$ or $-$ in front of the integral depending on whether we think of the local coordinates $(x_1, \dots, x_M)$ as "agreeing" with the orientation of $\mathcal{M}$ or not, though this last remark is rendered less than clear by the fact that we have not defined orientation in terms of "new" concepts.

Example 7.7.4 Let D be the unit disk in $\mathbb{R}^2$ defined by $|t| \le 1$ where $t \in \mathbb{R}^2$. We take $\mathcal{M}$ to be a half-sphere sitting on D,

$$\mathcal{M} = \{(t_1, t_2, z) \in \mathbb{R}^3 : z = \sqrt{1 - t_1^2 - t_2^2} \text{ where } (t_1, t_2) \in D\}$$

and (x_1, x_2) to be the local coordinate system defined by the coordinate patch $u(t_1, t_2) = (t_1, t_2, \sqrt{1 - t_1^2 - t_2^2}\,)$; thus $x_i(t_1, t_2, z) = t_i$ for $i = 1, 2$. Let f be the 2-form on $\mathcal{M}$ defined by

$$f(p) = z\, dx_1 \wedge dx_2$$

where $p = (t_1, t_2, z)$. Then

$$\int_{\mathcal{M}} f = \pm \int_D \sqrt{1 - t_1^2 - t_2^2}\, dt_1\, dt_2 .$$

We choose $+$ if we assign $\mathcal{M}$ the orientation agreeing with $u_*(t)(e_1 \wedge e_2)$ and otherwise we take $-$.

It can be shown that our "new" concept of an integral of a differential form over a manifold agrees with the "old" one. We illustrate this fact in the following example:

Example 7.7.5 Let $\mathcal{M}$ be the 2-manifold in $\mathbb{R}^3$ given by

$$\mathcal{M} = \{(t_1, t_2, z) \in \mathbb{R}^3 : z = \varphi(t_1, t_2)\}$$

where $\varphi : \mathbb{R}^2 \to \mathbb{R}$. Take (x_1, x_2) to be the local coordinate system defined by the coordinate patch $u(t_1, t_2) = (t_1, t_2, \varphi(t_1, t_2))$ and f to be the 2-form

$$f = g\, dx_1 \wedge dx_2$$

where g is a real-valued function on $\mathcal{M}$. If we take the orientation of $\mathcal{M}$ to be the one that agrees with $u_*(t)(e_1 \wedge e_2)$, where $t = (t_1, t_2) \in \mathbb{R}^2$, then our "new" definition of an integral of a differential form gives us

$$\int_{\mathcal{M}} f = \int_{\mathbb{R}^2} g\,(t_1, t_2, \varphi(t_1, t_2))\, dt_1\, dt_2.$$

To compare this with the "old" definition of an integral of a differential form, first notice that the orientation of $\mathcal{M}$ is

$$w(p) = \frac{1}{D(u'(t))}\, u_*(t)(e_1 \wedge e_2)$$

where $p = (t, z) = (t_1, t_2, z) \in \mathcal{M}$. Recalling the correspondences between "old" and "new" tangent vectors and forms, we see that the "old" definition of an integral of a differential form gives us

$$\begin{aligned}
\int_{\mathcal{M}} f &= \int_{\mathbb{R}^2} (f \circ u) \cdot (w \circ u) D(u') \\
&= \int_{\mathbb{R}^2} (g \circ u)\{\wedge^2 (u'(t)^{-1})^{\diamond}(e_1 \wedge e_2)\} \cdot \frac{1}{D(u')}\{\wedge^2 u'(t)(e_1 \wedge e_2)\}\, D(u') \\
&= \int_{\mathbb{R}^2} g \circ u \\
&= \int_{\mathbb{R}^2} g(t_1, t_2, \varphi(t_1, t_2)) dt_1\, dt_2\, .
\end{aligned}$$

We turn finally to the concept of the differential operator. We need consider this only for 0-form since the extension to higher order forms follows readily from the formula

$$df = \sum_{i_1 < \cdots < i_k} df_{i_1 \ldots i_k} \wedge dx_{i_1} \wedge \cdots \wedge dx_{i_k}$$

where

$$f = \sum_{i_1 < \cdots < i_k} f_{i_1 \ldots i_k}\, dx_{i_1} \wedge \cdots \wedge dx_{i_k}$$

and each $f_{i_1 \ldots i_k}$ is a real-valued function.

Let f be a real-valued function defined on the manifold $\mathcal{M}$. We define df, the *differential* of f, to be the 1-form whose value over the domain of the local coordinate system $(x_1, \ldots, x_M)$ is given by

$$df = \sum_{i=1}^{M} \frac{\partial f}{\partial x_i}\, dx_i .$$

We need to see that this definition is independent of our choice of local coordinates. Let $(y_1, \ldots, y_M)$ be a second local coordinate system whose domain overlaps that of $(x_1, \ldots, x_M)$. Then on that overlap we have

$$\begin{aligned} \sum_{i=1}^{M} \frac{\partial f}{\partial y_i}\, dy_i &= \sum_{i=1}^{M} \frac{\partial f}{\partial y_i} \left(\sum_{j=1}^{M} \frac{\partial y_i}{\partial x_j}\, dx_j \right) \\ &= \sum_{j=1}^{M} \left(\sum_{i=1}^{M} \frac{\partial f}{\partial y_i} \frac{\partial y_i}{\partial x_j} \right) dx_j \\ &= \sum_{j=1}^{M} \frac{\partial f}{\partial x_j}\, dx_j \, . \end{aligned}$$

Thus the differential of f is well defined.

An important instance of a differential occurs when considering overlapping local coordinates $(x_1, \ldots, x_M)$ and $(y_1, \ldots, y_M)$. We may either consider dy_i as the dual vector satisfying

$$dy_i \left(\frac{\partial}{\partial y_j} \right) = \delta_{ij}$$

or, applying the differential operator d to the real-valued function y_i, as the 1-form

$$dy_i = \sum_{j=1}^{M} \frac{\partial y_i}{\partial x_j}\, dx_j .$$

This last equation is, of course, the transformation law for 1-forms that was derived before the introduction of the differential operator. We see from this that it is appropriate to refer to K-forms on a manifold as *differential* forms.

The sense in which our "new" definition of d corresponds to our "old" definition is touched upon in the exercises.

Exercises

1. Prove $\sum_{k=1}^{M} \frac{\partial y_i}{\partial x_k} \frac{\partial x_k}{\partial y_j} = \delta_{ij}$ (Kronecker's delta) where $(x_1, \ldots, x_M)$ and $(y_1, \ldots, y_M)$ are overlapping local coordinates on a manifold $\mathcal{M}$.

2. Suppose that $(x_1, \ldots, x_M)$ and $(y_1, \ldots, y_M)$ are local coordinates on the manifold $\mathcal{M}$ induced by the coordinate patches $u : U \to \mathcal{M}$ and $v : V \to \mathcal{M}$

respectively. If $p \in u(U) \cap v(V)$ and $p = u(x) = v(y)$, show that

$$u'(x)e_i = \sum_{j=1}^{M} \frac{\partial y_j}{\partial x_i}(p)\, v'(y)e_j.$$

3. Suppose that $g : \mathcal{M} \to \mathcal{N}$ where $\mathcal{M}$ and $\mathcal{N}$ are M- and N-manifolds respectively. Suppose further that $u : U \to \mathcal{M}$ and $v : V \to \mathcal{N}$ are coordinate patches, u and v induce local coordinates $(x_1, \ldots, x_M)$ and $(y_1, \ldots, y_M)$ respectively, and $p = u(x) \in u(U)$ and $q = g(p) = v(y) \in v(V)$. Show that

$$g_*(p)(u'(x)e_i) = \sum_{j=1}^{M} \frac{\partial(y_j \circ g)}{\partial x_i}(p)\, v'(y)e_j\ .$$

4. Prove the first and second parts of Theorem 7.7.2.

5. Prove Theorem 7.7.3.

6. Prove Theorem 7.7.4.

7. Prove Theorem 7.7.5

8. Suppose that a local coordinate system $(x_1, \ldots, x_M)$ is induced on the manifold $\mathcal{M}$ by the coordinate patch $u : U \to \mathcal{M}$ and that $\mathcal{M}$ has an orientation in agreement with that induced by u. Let f be an M-form defined on the domain of $(x_1, \ldots, x_M)$, and show that the "old" and "new" concepts of integration of a differential form over a manifold yield the same value for $\int_{u(U)} f$. (Hint: Look at Example 7.7.5.)

9. Suppose that f is a real-valued function on the manifold $\mathcal{M}$ and that $(x_1, \ldots x_M)$ is a local coordinate system on $\mathcal{M}$ that is induced by the coordinate patch $u : U \to \mathcal{M}$. Show that there is correspondence between the "old" and "new" definitions of d in that we have

$$df\left(\frac{\partial}{\partial x_i}\right) = df \cdot u_*(e_i)$$

where the df on the left is understood in the "new" sense while that on the right is understood in the "old" sense.

References

The following books are suitable for those who wish to supplement their understanding of the material in this book or deepen their studies.

[1] Bishop, R. L. and Goldberg, S. I., *Tensor Analysis on Manifolds*, Dover, 1980.

[2] Buck, R. C., *Advanced Calculus*, 3rd ed., McGraw-Hill, 1978.

[3] Douglass, S. A., *Introduction to Mathematical Analysis*, Addison-Wesley, 1996.

[4] Edwards, H. M., *Advanced Calculus: A Differential Forms Approach*, Birkhäuser, 1994.

[5] Flanders, H., *Differential Forms with Applications to the Physical Sciences*, Academic Press, 1963.

[6] Guillemin, V. and Pollack, A., *Differential Topology*, Prentice-Hall, 1974.

[7] Loomis, L. and Sternberg, S., *Advanced Calculus*, Jones and Bartlett, 1989.

[8] Mikusiński, J. and Mikusiński, P., *An Introduction to Analysis: From Number to Integral*, John Wiley & Sons, 1993.

[9] Munkres, J. R., *Analysis on Manifolds*, Addison-Wesley, 1991.

[10] Schreiber, M., *Differential Forms: A Heuristic Introduction*, Springer-Verlag, 1977.

[11] Spivak, M., *Calculus on Manifolds*, Addison-Wesley, 1965.

[12] Taylor, A. E. and Mann, W. R., *Advanced Calculus*, 3rd ed., John Wiley & Sons, 1983.

Index

GPSR Compliance

The European Union's (EU) General Product Safety Regulation (GPSR) is a set of rules that requires consumer products to be safe and our obligations to ensure this.

If you have any concerns about our products, you can contact us on ProductSafety@springernature.com

In case Publisher is established outside the EU, the EU authorized representative is:

Springer Nature Customer Service Center GmbH
Europaplatz 3
69115 Heidelberg, Germany

Batch number: 10577880

Printed by Printforce, the Netherlands